El Niño, Catastrophism,

and Culture Change

in Ancient America

El Niño, Catastrophism, and Culture Change in Ancient America

EDITORS

Daniel H. Sandweiss

Jeffrey Quilter

PUBLISHED BY Dumbarton Oaks Research Library and Collection

DISTRIBUTED BY Harvard University Press, 2008

General editor: Joanne Pillsbury
Managing editor: Grace Morsberger
Art director: Denise Arnot

Library of Congress Cataloging-in-Publication Data

El Niño, catastrophism, and culture change in ancient America : a symposium
at Dumbarton Oaks, 12th-13th October 2002 / Daniel H. Sandweiss and
Jeffrey Quilter, editors, Joanne Pillsbury, general editor.
       p. cm.
    ISBN 978-0-88402-353-1
    1. El Niño Current—Environmental aspects—Latin
America—History—Congresses. 2. El Niño Current—
Social aspects—Latin America—History—Congresses.
3. Climatic changes—Environmental aspects—Latin
America—History—Congresses. 4. Climatic changes—
Social aspects—Latin America—History—Congresses.
5. Climate and civilization—Latin America—Congresses.
I. Sandweiss, Daniel H. II. Quilter, Jeffrey, 1949-
III. Pillsbury, Joanne. IV. Dumbarton Oaks.
    GC296.8.E4E3955 2006
    304.2'5098—dc22   2007039441

Printed in the United States of America

TABLE OF CONTENTS

Central America and Mesoamerica

Over three years later, the devastation left by Hurricane Katrina still reminds us of the impact of severe meteorological events on populations. The hurricane was one of the deadliest on record, and the true scope of the destruction is still being calculated. New Orleans and other areas of the Gulf Coast of the United States were profoundly affected, and the economic losses as a consequence of this event made it one of the costliest disasters in U.S. history. The impact upon future settlement and economic patterns in this region are still unknown, but Katrina provides us with an all-too-real glimpse of the power of these events to shape history.

This volume is concerned with questions of climate change, resulting catastrophes, and the cultural responses to them. The primary focus is on the El Niño–Southern Oscillation phenomenon, a fluctuation in the normal ocean temperatures and related atmospheric conditions that can trigger unusual weather patterns. Traditionally known as El Niño (the Christ child) for the time of year when the first signs of the onset of an event are noted on the west coast of South America, this global phenomenon can result in events that range from mild to catastrophic in their outcome. Geological events such as volcanic eruptions and other phenomena with potentially devastating consequences are also taken into consideration in this examination of catastrophes and culture change.

Recent advances in geoarchaeology and paleoclimatology have provided researchers with new data for understanding the long-term environmental history of the ancient Americas. But what is the effect of climate change on cultures? Are the correlations between geophysical indicators of extreme events and cultural shifts evident in the archaeological record indicative of a causal relationship? What are the social responses to such events? How do we tease out the variables in the complex interactions between humans and their environment?

Dumbarton Oaks has historically been a center for the humanities, and as such, the natural sciences have not been a central focus for research activities here. Dumbarton Oaks has, however, long been a place where scholars from different disciplines have gathered to discuss a common problem. Indeed, one of the strengths of the institution has been the interest in gathering scholars who do not normally have a chance to meet in the specialized conferences of their specific disciplines. In this sense, the symposium on which this volume is based was very much in keeping with this spirit of multidisciplinary inquiry.

The papers in this volume were written by scholars from an array of fields, including cultural anthropology, geology, archaeology, and epidemiology, and each author contributed a distinct perspective to the discussion. As with any volume embracing such a diversity of approaches, the contributions are varied, ranging from a distillation of complex data on understanding the paleoclimatic records through ice cores to an extraordinary first-person narrative of living through a major disaster. Through these diverse approaches, we have reached a new level of understanding of the principal questions surrounding climate, catastrophe, and culture.

I would like to thank Jeffrey Quilter and Daniel Sandweiss for their work in organizing the symposium and the present volume. Thanks are also due to Cecilia Montalvo for her help with the practical matters of the meeting, and to Jai Alterman and Emily Gulick for their assistance in preparing the volume for publication. I am grateful to Lynne Shaner, the director of publications, and her staff, especially our managing editor, Grace Morsberger, and Hilary Parkinson and Denise Arnot, for their fine work on this volume.

Joanne Pillsbury
Dumbarton Oaks

INTRODUCTION

CLIMATE, CATASTROPHE, AND CULTURE
IN THE ANCIENT AMERICAS

Daniel H. Sandweiss
Jeffrey Quilter

Natural systems are neither more nor less stable than human societies. Catastrophes occur and effect changes in both realms. As debates over the environmental and social implications of global warming, abrupt climatic changes, and related topics continue, understanding the relationships between change in human and natural systems in the past takes on ever greater significance in understanding the future. We developed the meeting on which this book is based primarily because we believed that a critical mass of information had been acquired, making a Dumbarton Oaks symposium appropriate and timely for a scholarly audience. But we also felt that the synthetic nature of the papers might reach a larger audience, beyond archaeologists and environmental scientists, who might find valuable lessons to be learned in the resulting publication.

The Pre-Columbian Studies Program at Dumbarton Oaks serves as a medium through which humanistic scholars may interact with more scientifically oriented colleagues. Commonly, Pre-Columbian symposia focus on a temporal or geographical topic and bring together different scholars with their various approaches. Meetings on the Inca, Classic-Period Veracruz, and the Moche of Peru are all examples of this genre. Other symposia are thematic, such as previous meetings on mortuary practices, the sea in ancient America, or human sacrifice. The symposium on which this volume is based, however, was unique in being the most scientifically based meeting held in this format. As co-organizers of the meeting and now coeditors of this volume, we believed that we had a unique opportunity to identify scholars engaged in "hard" science who would be able to present current theories on climate change and its influence and interactions with culture in ways that would make their data and interpretations more accessible to nonspecialists (Mayewski, Maasch). We also included authors who employ more traditional social science and humanistic approaches to these issues (Roscoe, Kiracofe and Marr, Sheets, Wilkerson). Other papers are written by archaeologist-paleoclimatologist teams (Richardson and Sandweiss, Billman and Huckleberry, Moseley and Keefer, Yaeger and Hodell). Most of these teams have worked together for years, but we take some credit and great pride in bringing together a Mayanist, Jason Yaeger, and a paleoclimatologist working on the Yucatan Peninsula, David Hodell; although these scholars had not previously worked together, they have crafted a single paper that is a model for the nuanced integration of archaeological and paleoclimatic data.

Earth and weather scientists are increasingly able to discuss climatic and related events in the past with greater and greater accuracy. We therefore believe that all students of the past will gain from greater knowledge on the current status of such data and the information derived from it, regardless of an individual scholar's views on the degree to which natural phenomena influence culture. We hope that the chapters in this volume will help to provide this information as well as offer substantive information and food for thought on how climate, catastrophe, and cultures interacted in the past.

The chapters in this volume offer diverse perspectives on climate, catastrophe, and culture in the ancient Americas while at the same time presenting only a small sample of a great number of studies recently accomplished or currently under way. Indeed, when we first began to develop the list of speakers we soon found that there were many potential contributors who could offer substantial and insightful contributions to the meeting. After some discussion and difficult choices, we decided to focus the meeting on the role of the El Niño phenomenon in producing catastrophic events, while adding selected examples of other phenomena to enrich and diversify the volume.

Although neither this collection of articles nor those in any other compendium could fully cover current knowledge on climate, catastrophe, and culture, the discussion of the original papers at the Dumbarton Oaks conference in 2002 and the resulting chapters do stimulate several observations about the state of research on human-environment-climate relations in the Pre-Columbian past and the study of these issues.

First, we were impressed by the great successes that can be achieved by true interdisciplinary research. Granted, the term *interdisciplinary* has become a catchphrase of academia over the last decade. Unlike many fads, however, this one has roots in broad scholarship and has attracted significant funding. For instance, the National Science Foundation has increasingly emphasized and promoted "interdisciplinary programs, programs that are supported by multiple Directorates at NSF, and programs jointly supported by NSF and other Federal agencies" <www.nsf.gov/pubs/2003/nsf03574/nsf03574.txt>.

Interdisciplinary work can be professionally risky for, to be fully effective, scholars must be willing to venture into fields beyond their expertise. In this light, we designed this symposium to demonstrate that students of El Niño and other catastrophes in Pre-Columbian America do engage in such interdisciplinary efforts, to great effect. Among the authors we find not only archaeologists with a variety of topical and geographic specialties, but also a social anthropologist, an architectural historian, an epidemiologist, and geologists working on ice-core records, climatology, earthquake hazards, fluvial geomorphology, and paleolimnology. The results of the kinds of interactions exemplified by this group move us closer to understanding the effects of catastrophe on human societies and point us toward future research programs.

A second observation we have drawn from the conference and the editing of this volume is the difficulty of proceeding from collation to correlation to causation. Collation involves showing that events occurred simultaneously. Correlation, in formal terms, means demonstrating that events of interest co-vary in a statistically significant way. Causation means showing that outcomes are the necessary result of specific conditions and processes—that differences have occurred that make a difference, as social anthropologists sometimes put it. Chapters in this volume point out some of the difficulties as well as partial solutions to this problem.

Working with our natural-science colleagues, we are increasingly able to achieve collation: to demonstrate that particular climatic or environmental events are coincident in time with cultural change. Paul Mayewski gives us the tools to use polar ice-core data to reconstruct certain aspects of global climate that may have impacted past societies at particular times and places. For much of the time humans have been present in the Americas, this record has decadal or better resolution. Research by Mayewski and others has demonstrated that some of the most significant changes in global climate occurred over incredibly short spans, perhaps a few years to a few decades. An example would be the radical warming at the end of the Younger Dryas about twelve thousand years ago (Taylor et al. 1997). Collating the ice-core record with the archaeological record thus requires not only an understanding of climate dynamics but also considerable refinement in the dating of archaeological phenomena. Radiocarbon dates must be calibrated to calendar years, but even then, achieving decadal or better resolution in most New World prehistoric sites will remain a dream for years to come. As Jason Yaeger and David Hodell point out, chronologies built on radiocarbon dates, ceramics, or calendrical texts (the three principal dating methods for the ancient Americas) all present challenges to effective collation that are further complicated by the time-averaged nature of archaeological deposits.

Kirk Maasch's discussion of El Niño dynamics approaches the problem from the opposite direction: by showing how El Niño plays out in different regions of the Americas, he provides criteria for seeking El Niño's signal in the archaeological and coincident natural records of each region. At the same time, he points out another problem in collating records: not all El Niño events behave the same way or leave the usual signal in a given locale, and sometimes non-Niño processes can deposit a Niño-like signal in the absence of the phenomenon, a false positive as it were. To ensure proper interpretation of the climate record, we need to employ a multiproxy approach, to triangulate possible paleo-events with as many independent data sets as possible. To do so requires collaborating with many different specialists.

Several of the chapters offer case studies in collation as well as clues to causation. James Richardson and Daniel Sandweiss's summary of research on Mid-Holocene coastal Peru points to temporal coincidence between changes in El Niño frequency and the construction of monumental religious architecture. Coastal cooling in the northern half of Peru and the onset of El Niño with long recurrence intervals took place at 5,800 cal yr BP, shortly before ancient Peruvians began building temple mounds, while an increase in El Niño frequency at about 3,000 cal yr BP coincides with the abandonment of coastal large-scale constructions for several centuries. Although these hypotheses of climatic change came originally from the remains of biological organisms in archaeological sites, both are now well supported by multiple independent data sets as well as some climate-modeling exercises. The beginning and end of the phase of early monumental construction on the Peruvian coast are now fairly well documented, while the implications of bioindicators have been reconfirmed several times.

Can we go beyond collation in this case? The intriguing temporal coincidence was license enough for Richardson and Sandweiss and for Paul Roscoe to consider interpretive scenarios involving environmental, climatic, and social factors. Because these scenarios are consistent with what we know about conditions, events, and human behavior from both the ethnographical and archaeological record, they can guide future research even though we recognize them as stories rather than histories. If we were to wait until we had full agreement on every aspect of the paleoclimatic and cultural records before venturing an interpretation, we would have a long wait and very dull meetings.

Given that we cannot interview ancient informants about their perceived motivations nor observe their actual behavior directly, in these as in other attempts to suggest prehistoric causation, there are caveats. In the northern highlands of Peru, where direct effects of El Niño are muted, temple mounds have a similar start date but different form. In addition to climate, other factors changed throughout the central Andes after 5,800 cal yr BP: populations seem to have grown, many more domesticated plants came into use, and interregional interaction may have increased. The degree to which climatic change influenced these events is uncertain although, in general, catastrophic events do not appear to have played significant roles. Should we then throw out the Niño with the bath water? We suggest that the answer is no, so long as scenarios such as those proposed by Richardson and Sandweiss and Roscoe are used to orient our data collection and future interpretation rather than as explanation.

Yaeger and Hodell also point to variability in the record of cultural and climatic change in their study of the Terminal Classic Period on the Yucatan Peninsula. Though droughts occurred, each region had a unique history as well as a unique version of the events often glossed as the Maya Collapse. Yaeger and Hodell outline the difficulties in achieving sufficient temporal collation for explanatory purposes. They conclude that climatic change was undoubtedly an important factor in many of the social transformations that occurred during this time but that neither social nor climatic change was uniform or unicausal. They also call for greater consideration of the possible benefits of climatic change for some groups under some circumstances; Roscoe, Payson Sheets, and Brian Billman and Gary Huckleberry elaborate on the same point in different contexts.

The chapters by Michael Moseley and David Keefer and by Sheets are case studies, each with a significant advantage—they deal with point catastrophes that can be tied directly to specific moments in the archaeological record. For El Niño studies in particular, this is an area where we can do more, especially working in an interdisciplinary mode. As seen in the Moseley and Keefer study, in the paper by Billman and Huckleberry, and in work by geologists Lisa Wells (e.g., 1987, 1988), Michel Fontugne et al. (1999), and others, El Niño–generated floods sometimes deposit a recognizable sedimentary signature in or on overlapping archaeological strata. On the Peruvian desert coast, in later sites with mud-plastered floors or walls, El Niño–derived rainfall can lead to erosion tracks and even footprints (e.g., Sandweiss 1995). At the late Pre-Hispanic site of Túcume, in the Lambayeque valley on Peru's north coast, a small entry temple was reconstructed five times through the course of about five centuries (Narváez 1995). Each reconstruction followed a rainfall event that left archaeological indicators.

Harold Rollins and his colleagues have demonstrated particular alterations in the macro- and microstructure and in the stable isotope geochemistry of mollusk shells that survived El Niño (e.g., Rollins et al. 1987), and Andrus et al. (2002) have recently done the same for fish otoliths. Found in archaeological contexts, these bioindicators can demonstrate the presence of El Niño at specific moments in the past. Recently, Andrus, Hodgins, and Sandweiss have begun to work with radiocarbon in mollusk valves as an indicator not only of the presence but also potentially of the intensity of El Niño events (Andrus et al. 2005). Others such as Elera and colleagues (1992) have identified El Niño events from short-term changes in faunal assemblages from midden deposits. Tree rings may eventually provide similar indicators of specific events. In short, we have a number of known and potential approaches to placing El Niño in direct association with archaeological deposits, as long as we know what to look for. With more cases we can move more quickly from collation to correlation.

Although the Moseley and Keefer study of the Miraflores event is a case of collation, it is hard to deny likely correlation with, and even a causal link to the downfall of the Chiribaya culture. S. Jeffrey K. Wilkerson's paper brings home the destructive power of catastrophic floods, their potential to cause cultural change, and some of the processes by which this change can ensue. Some time ago, Sheets discussed the possible influence of population movements from the Maya highlands to the Peten in the Classic-Period florescence as a result of volcanic activities. In the Osmore case, detailed by Moseley and Keefer for southern Peru, style change is dramatically in evidence in the apparent total elimination of the Chiribaya style and its replacement with another style, apparently of highland origin. Thus we are sensitized to the fact that an art style can evolve not only in a battleship-curve mode, quietly fading in popularity, but can also be dramatically changed or eliminated due to catastrophe. It would be worthwhile to look for other examples in the archaeological record.

Sheets's analysis of volcanic impacts on Central and Mesoamerican societies also illustrates the utility of direct physical association between the indicators of catastrophic events and the archaeological record. He has amassed enough cases that we might be able to consider correlation between natural processes and cultural change—except that in some of the cases, there was no archaeologically apparent cultural change.

The lack of observable cultural change after catastrophe in the archaeological record leads us to a third observation on the relationships between sociopolitical organization and responses to catastrophic events as well as to slower-paced processes of environmental change. Sheets's concept of "scaled vulnerability" is critically important and well illustrated by his case studies. In the book *Floods, Famines and Empires: El Niño and the Fate of Civilizations*, Brian Fagan (1999) has made a similar point concerning climatic variation at the El Niño scale: small, mobile, egalitarian groups of foragers have more options than do large, sedentary societies when faced with sudden (or progressive) downturns in resource availability. In other words, social organization, economy, and population density can act as both constraints on and enablers of human responses to environmental catastrophes—as Sheets, Roscoe, Yaeger and Hodell, and Billman and Huckleberry argue in their articles here and elsewhere. Of course, employing this principle requires controlling all three variables effectively, which is often easier to attempt than to attain.

Much is opaque in the archaeological record; put another way, despite the optimism forty years ago, there seems to be a lot of human behavior that is not encoded in the archaeological record. For instance, more than twenty years ago, one of us (Sandweiss) found that an eighty-year trajectory of landscape alteration in coastal Honduras had operated in tandem with the local land-tenure system to alter social structure. The resulting changes would have left little if any material trace. Without prior knowledge of the traditional land-tenure rules, these changes could not be interpreted even were they recognized. And it took several months interviewing live informants before he could even ask the right questions!

The complex ways in which people react to changes in their environment mean that not all climatic change and not all catastrophes are bad for everyone. Sheets makes this point in his discussion of the variable response to volcanism, as do Billman and Huckleberry in their discussion of the agricultural impact of El Niño floods in the different valleys of the Peruvian north coast. Drawing on the work of Sandweiss and Richardson (1999; Sandweiss 1996), Roscoe stresses the latitudinal gradient in El Niño impacts along the Peruvian coast, noting that for each event there is usually a point north of which the constraints imposed by El Niño outweighed the opportunities the event offered for political entrepreneurs, while to the south the reverse obtained. Similar to Yaeger and Hodell, these authors have pointed to the opportunities that "catastrophes" may provide for political entrepreneurs.

Knowing the right questions is critical to pursuing these issues. All of the chapters in this book raise specific as well as general questions about prehistoric human-environment interaction and point to ways to answer them. James Kiracofe and John Marr's paleoepidemiological hypothesis is a case in point. They resurrect an issue long thought resolved and implicate El Niño along with imperial policy in nothing less than the downfall of the

Inca empire, potentially mitigating the "Black Legend" of Spanish destruction in the New World. That El Niño brings disease to the coast of Peru and Ecuador, along with flooding, plagues of insects, snakes, crop rot, and social and economic disruption is well known by now. Those dealing with the effects of the 1982–83 or 1997–98 events are certainly aware of these facts, as are regional historians concerned with climate. Until now, however, no one thought to pursue the prehistoric impact of El Niño on epidemic disease, and there have been only a few studies on the effects of epidemics on empire.

The Kiracofe and Marr chapter also points toward another source of useful information in understanding El Niño and the fate of Andean and other civilizations: the ethnohistoric record. The instrument record for El Niño covers little more than a century, so historic and ethnohistoric sources are critical in gaining a longer-term perspective on events and social consequences to use as analogies in interpreting archaeological evidence. By 1978, when William Quinn and his colleagues used Peruvian and Ecuadorian historical data to make the first modern chronology and rating of El Niño events that have occurred since the Spanish Conquest, archaeologists had already evinced considerable interest in the potential explanatory power of El Niño (e.g., Parsons 1970). Quinn et al.'s (1978) concern was principally climatological, and some of their specific attributions have been challenged. With modifications, however, the revised Quinn et al. (1987) chronology remains the standard against which other proxy records for the last 450 years are judged, for instance, the Nile River discharge record (Eltahir and Wang 1999; Quinn et al. 1992). For the human response, ethnohistoric sources have been tapped but by no means exhausted. A case in point is the first major El Niño event of the Post-Columbian period, which devastated northern Peru in 1578. Huertas Vallejos (1987, 2001; see also Copson and Sandweiss 1999) has published and commented on the extant portions of a Spanish colonial administrative document detailing the effects of this sixteenth-century event in northern Peru.

A related approach to studying the human reaction to El Niño or similar events is conducting research in ethnoarchaeology: to look at what happened to particular sites during recent, documented events and try to recover material correlates of the environmental processes and linked human behaviors. There are few such studies for El Niño in the Andes. For the project that resulted in his co-authored article on subsistence strategies in Late Pre-Ceramic Peru (Quilter and Stocker 1983), Terry Stocker interviewed a number of local fisherman in the central coast village of San Bartolo regarding their observations prior to El Niño events. Informant interviews by Sandweiss and historical records indicate that in the Saña Valley on Peru's north coast, the coastal village of Lagunitas had three different locations during the twentieth century, with the move from one to another precipitated by El Niño flooding. There are surely many more such sources to be tapped to enrich our studies of how earlier inhabitants of the Americas responded to castastrophes. In the present volume, the careful observations made by Wilkerson during massive flooding in Veracruz and in its aftermath are a case study of ethnoarchaeology that, while done under duress, will have many implications for future research in the region and beyond.

Many of the points we raised above are true for any scholarly endeavor. Finding new questions to ask leads to new answers and also to new questions. In raising new questions, we must be sensitive to the fact that analogy is not explanation but offers us interesting ways to think about our subject matter. Interdisciplinary studies almost always are advantageous in examining complex phenomena, but they are difficult to carry out for a variety of reasons. These issues are true for every field, whether it be searching for water on Mars or determining who wrote Shakespeare's plays. The study of past human behavior carries its own special burdens for investigators, however.

The impulse to explain human variation as the result of geographical conditions or natural processes has been with us since at least the time of Hippocrates' (460–377 BC) treatise *On Airs, Waters, and Places*. At the other end of the spectrum of causality, free will has been cited to explain why humans do what they do since the Garden of Eden. More recently, the New Archaeology, with its emphasis on culture as a form of adaptation, trended toward mechanistic, deterministic explanatory modes while Post-Processual archaeology has veered toward the other extreme, seeing human agency as trumping environmental determinism. Granted, individual arguments have been more subtle than either of these two generalized statements, but these do seem fair enough to characterize trends.

In organizing the symposium and editing this volume, we have been sensitive to the extremes of the bell curve of explanatory modes. Archaeology has great resources in tracking change but the nature of its data often makes causation, especially the working of human agency, very hard to discern. We realize that human decisions often conflict with what might appear to be "rational" choices, using the criteria of Western science. For example, Fagan's (1999) argument that small, egalitarian bands have more options than do larger societies with greater investments in immovable, large-scale infrastructure, is only true in a generalized way. In fact, small societies may have their own constraints imposed from within, based on an internal logic that is hard to assess from the outside and at a distance. The ultimate issue is how flexibly a society adapts to new and changed conditions, not its size or a generalized notion of "egalitarianism," per se.

Catastrophism is a "big picture" in more ways than one, whether we are referring to a theory of culture change or a specific event that affected human lives. Such events are, by definition, large-scale, short-term, and, quite often, dramatic in their effect on the landscape and the people who dwell in it. Even in the face of an erupting Vesuvius, however, some people have more options to escape disaster or, sometimes, even turn a profit out of it, than do others. Thus, sociocultural factors always are at play in the face of "natural" events. The anthropology (and archaeology) of disaster is a growing field (e.g., Hoffman and Oliver-Smith 2002; Bawden and Reycraft 2000).

In studying reactions to catastrophes, we are focusing on case studies of how humans, in general, react to sudden events. There are probably as many ways to react to events as the events themselves. We may posit, however, that there are three major approaches to catastrophes (see also Sheets, this volume, on Burton et al.'s [1978] stages of Loss Absorption, Loss Reduction, and Radical Action). The first is flight—abandonment of the area affected by the event in question. The second is adaptation, in which changed conditions are recognized and people reconfigure their strategies for survival and prosperity in this new context. A third approach might be termed restoration, in which attempts are made at regaining material and sociocultural losses in order to reestablish, as much as possible, pre-disaster conditions. How well archaeologists and other scientists are able to identify such strategies is hard to assess. What we might term neo-catastrophism is in its infancy. Such a field recognizes the complex interplay between culture and environment as continuous. It is multidisciplinary incorporating both contemporary events and the record of ancient ones.

From the nineteenth through the twentieth centuries, catastrophism, in one variation or another, has been pitted against evolutionism in its gradualist mode. Geologists and evolutionary biologists, by and large, have rejected this binary opposition and recognized that past change includes periods of both drastic, rapid change and more slowly evolving processes. In our own lives we have witnessed slow cultural processes, such as the economic and political growth of China. We also have seen relatively slow developments capped by very rapid change, such as the slow decline and then quick collapse of communism in the former Soviet Union and Eastern Europe. We also have witnessed what appear to us as slow environmental degradation (which actually is quite rapid in geological terms) and dramatically short changes brought on by environmental disasters that occurred on huge scales, such as the December 26, 2004 Indian Ocean tsunami.

Several authors demonstrate that El Niño patterns are far more irregular than previously thought. This point is as sobering as the message that other environmental changes are equally erratic, at least from our small-scale human perspective and relatively short chronological vita. A number of inferences may be drawn from this volume which may be relevant to national and international environmental policies and other matters; we leave it to individual readers to draw such conclusions as they see fit. For academic purposes this volume teaches that neither a simplistic return to castastrophism nor a complete denial of the role of natural forces in the affairs of humans can explain large-scale changes in the past. Human societies are rarely the passive victims of natural forces, but they also are not completely immune to them. To explain many small- and large-scale changes in the past, we must look to both nature and nurture and their interactions.

The scope of this book and the conference are limited in their ability to raise the questions and explore the issues presented above. Many other case studies, theoretical issues, and methodological considerations could not be included due to space and time limitations. We and all the participants do hope and believe, however, that by raising these issues we will contribute to discussions of human-environmental interactions in our own fields and in the greater realm of scholarly discourse.

References cited

Andrus, C. Fred T., Douglas E. Crowe, Daniel H. Sandweiss, Elizabeth J. Reitz, and Christopher S. Romanek
2002 Otolith ∂¹⁸O Record of Mid-Holocene Sea Surface Temperatures in Peru. *Science* 295: 1508–1511.

Andrus, C. Fred T., Gregory W. L. Hodgins, Daniel H. Sandweiss, and Douglas E. Crowe
2005 Molluscan Radiocarbon as a Proxy for El Niño-Related Upwelling Variation in Peru. In Special Paper 395 (German Mora and Donna Surge, eds.): 13–20. Geological Society of America, Boulder, CO.

Bawden, Garth, and Richard M. Reycraft (eds.)
2000 *Environmental Disaster and the Archaeology of Human Response.* Anthropology Papers 7. Maxwell Museum of Anthropology, Albuquerque.

Copson, Wendy, and Daniel H. Sandweiss
1999 Native and Spanish Perspectives on the 1578 El Niño. In *The Entangled Past: Integrating History and Archaeology* (M. Boyd, J. C. Erwin, and M. Hendrickson, eds.): 208–220. Proceedings of the 1997 Chacmool Archaeological Conference. The Archaeological Association of the University of Calgary, Calgary.

Elera, Carlos, José Pinilla, and Victor Vásquez
1992 Bioindicadores zoológicos de eventos ENSO para el Formativo Medio y Tardío de Puemape—Perú. *Pachacamac* 1 (1): 9–19.

Eltahir, Elfatih A. B., and Guiling Wang
1999 Nilometers, El Niño and Climate Variability. *Geophysical Research Letters* 26: 489–492.

Fagan, Brian C.
1999 *Floods, Famines and Emperors: El Niño and the Fate of Civilizations.* Basic Books, New York.

Fontugne, Michel, Pierre Usselmann, Danièle Lavallée, Michèle Julien, and Christine Hatté
1999 El Niño Variability in the Coastal Desert of Southern Peru during the Mid-Holocene. *Quaternary Research* 52: 171–179.

Hoffman, Susanna M., and Anthony Oliver-Smith (eds.)
2002 *Catastrophe and Culture: The Anthropology of Disaster.* SAR Press, Santa Fe.

Huertas Vallejos, Lorenzo
2001 *Diluvios andinos a través de las fuentes documentales.* Pontificia Universidad Católica del Perú, Fondo Editorial, Lima.

Huertas Vallejos, Lorenzo (ed.)
1987 [1578] *Ecología e Historía: Probanzas de indios y españoles referentes a las catastróficas lluvías de 1578, en los corregimientos de Trujillo y Saña.* CES Solidaridad, Chiclayo, Peru.

Narváez, Alfredo
1995 The Pyramids of Túcume: The Monumental Sector. In *Pyramids of Túcume: The Quest for Peru's Forgotten City* (Thor Heyerdahl, Daniel H. Sandweiss, and Alfredo Narváez): 79–130. Thames and Hudson, London.

Parsons, Mary H.
1970 Preceramic Subsistence on the Peruvian Coast. *American Antiquity* 35: 292–304.

Quilter, Jeffrey, and Terry Stocker
1983 Subsistence Economies and the Origins of Andean Complex Societies. *American Anthropologist* 85: 545–562.

Quinn, William H.
1992 A Study of Southern Oscillation-Related Climatic Activity for AD 622–1900 Incorporating Nile River Flood Data. In *El Niño: Historical and Paleoclimatic Aspects of the Southern Oscillation* (Henry F. Diaz and Vera Markgraf, eds.): 119–149. Cambridge University Press, Cambridge.

Quinn, William H., David O. Zopf, Kent S. Short, and Richard T. W. Yang Kuo
1978 Historical Trends and Statistics of the Southern Oscillation, El Niño, and Indonesian Droughts. *Fisheries Bulletin* 76: 663–678.

Quinn, William H., Victor T. Neal, and Santiago E. Antunez de Mayolo
1987 El Niño Occurrences over the Past Four and a Half Centuries. *Journal of Geophysical Research* 92 (C13): 14,449–14,461.

Rollins, Harold B., Daniel H. Sandweiss, Uwe Brand, and Judith C. Rollins
1987 Growth Increment and Stable Isotope Analysis of Marine Bivalves: Implications for the Geoarchaeological Record of El Niño. *Geoarchaeology* 2: 181–187.

Sandweiss, Daniel H.
1995 Life in Ancient Túcume. In *Pyramids of Túcume: The Quest for Peru's Forgotten City* (Thor Heyerdahl, Daniel H. Sandweiss, and Alfredo Narváez): 142–168. Thames and Hudson, London.

1996 The Development of Fishing Specialization on the Central Andean Coast. In *Prehistoric Hunter-Gatherer Fishing Strategies* (Mark Plew, ed.): 41–63. Boise State University, Boise.

Sandweiss, Daniel H., and James B. Richardson III
1999 Las fundaciones precerámicas de la etapa Formativa en la costa peruana. In *El Formativo Sudamericano: Una revaluación* (Paulina Ledergerber-Crespo, ed.): 179–188. Abya-Yala, Quito, Ecuador.

Taylor, Kendrick C., Paul A. Mayewski, Richard B. Alley, Ed J. Brook, Anthony J. Gow, P. M. Grootes, Debra A. Meese, Eric S. Saltzman, Jeffrey P. Severinghaus, Mark S. Twickler, James W. C. White, Sallie Whitlow, and Gregory A. Zielinski
1997 The Holocene-Younger Dryas Transition Recorded at Summit, Greenland. *Science* 278: 825–827.

Wells, Lisa E.
1987 An Alluvial Record of El Niño Events from Northern Coastal Peru. *Journal of Geophysical Research* 92 (C13): 14,463–14,470.

1988 Holocene History of the El Niño Phenomenon as Recorded in Flood Sediments of Northern Coastal Peru. *Geology* 18: 1134–1137.

ANCIENT AMERICAN CLIMATES

PALEOCLIMATE FROM ICE CORES: A FRAMEWORK FOR ARCHAEOLOGICAL INTERPRETATIONS

Paul Andrew Mayewski

U nderstanding the Earth system and, in particular, its climate, remains one of the major intellectual challenges faced by science. The processes influencing climate, the mechanisms through which they act, and the responses they generate are, in general, as complex and poorly understood as they are important. Because observational records of climatic processes span only the most recent years of Earth's history and, in many instances, are known to be markedly affected by anthropogenic influences, records of past climates are exceedingly important to the development of scientific understanding for local, regional, and global climatic systems.

Records from a variety of sources (e.g., instrumental observations, historical documents, deep-sea and continental sediments, tree rings, and ice cores) can provide the basic boundary conditions (e.g., sea-surface temperature, precipitation, and atmospheric circulation patterns) necessary for robust environmental reconstructions. Along with other high-resolution paleorecords, ice-core records provide detailed descriptions of climatic change that are extremely valuable for comparison with modern observations. Further, they document not only a wide range of environmental parameters that are both measures of and responses to climatic change (e.g., atmospheric chemistry and circulation, temperature, precipitation) but also many of the causes of climatic change (e.g., solar variability, volcanic activity, greenhouse gases). Because of their potential resolution (subannual), length (several glacial cycles), and in some cases precise dating (subannual), they also provide a framework for relating other records of past climate and for investigating associations among change in climate, change in ecosystems, and human activity.

This paper is intended to provide some of the basic information about utilizing the ice-core paleoclimatic record to investigate links between events in the archaeological record and changes in climate. Examples of several now-classic ice-core results are presented along with a brief primer for interpreting ice-core records. The immense scope in terms of resolution, length of record, and types of measurements available from ice-core records requires that those seeking to apply ice-core data to related disciplines (e.g., archaeology, atmospheric chemistry, history) be sufficiently conversant both with this scope and with sample interpretations to fully utilize this robust resource.

Considerable attention has focused on the deep ice cores recovered from central Greenland and several sites in Antarctica. Here the emphasis will be on the GISP2 (Greenland Ice Sheet Project 2) deep ice core from central Greenland. The annual layer dating, resolution, continuity, length of record, preservation, calibration, and range of variables measured make this ice core a classic record from which to define climatic change over the North Atlantic and regions atmospherically teleconnected to the North Atlantic.

On 1 July 1993, the Greenland Ice Sheet Project 2 successfully completed drilling to the base of the Greenland Ice Sheet (72.6° N; 38.5° W; 3200 m.a.s.l.) after five summer seasons of occupation at the site and two previous seasons of reconnaissance activity. In so doing, GISP2 recovered the deepest ice-core record available from the Northern Hemisphere (3053.44 m) and, along with its European companion project, the Greenland Ice Core Project (GRIP), developed the longest paleoenvironmental record, >250,000 years, available from the Northern Hemisphere.

Based on preliminary comparisons, at least the upper 90% (~2800 m) of the two cores display extremely similar, if not absolutely equivalent records. This agreement between the two cores separated by 30 km (~10 ice thicknesses) provides strong support for a climatic origin for even the minor features of the records and implies that investigations of subtle environmental signals (e.g., rapid climatic change events with onset and termination as fast as 1–2 years) can be rigorously pursued.

The GISP2 record provided a new standard for ice-core research and climatic-change interpretations because it was (1) dated by annual layer counting using multivariate techniques[1]; (2) sampled continuously and at high resolution (e.g., biannual and finer [selected sampling at 10 samples/year]) through the Holocene and >3–50 years/sample through the Pre-Holocene for major anions, major cations, insoluble particles, electrical conductivity, methanesulfonate (MS), and stable isotopes; (3) sampled using an integrated sampling plan for all parameters thus providing a multivariate view of past polar atmospheres; (4) rigorously verified in the field for several parameters by selected on-site analysis and the use of processing blanks, container blanks, and replicates followed by complete analysis in home laboratories to verify that no changes had occurred due to differences in environmental conditions

---

1 The depth/age relationship for the GISP2 core was developed from a variety of core parameters, including annual layer counting of visual stratigraphy, electrical conductivity, laser-light scattering of dust, stable isotopes, major anions and cations, insoluble particles, ^{210}Pb, total β activity, and ^{14}C from occluded CO_2 (e.g., Wilson and Donahue 1990; Dibb 1992; Taylor et al. 1992; Alley et al. 1993; Meese et al. 1994a, 1994b) plus ice-dynamics modeling (Schott et al. 1992). Current estimated age error is 2% for 0–11.64 kyr BP, 5% for 11.64–17.38 kyr BP years ago, and 10% for 17.38–40.5 kyr BP years ago (Alley et al. 1993). The age scale back to 40.5 kyr BP comes from a variety of techniques, notably that of Meese et al. (1994a). Below 40 kyr BP the chronology is based on the Vostok chronology of Sowers et al. (1993) using the $\delta^{18}O$ of O_2 of ice (Bender et al. 1994). This approach to correlation invokes the fact that the $\delta^{18}O$ of atmospheric O_2 varies with time but is constant at any one time throughout the atmosphere.

(between field and home) or by contamination during transport; (5) sampled to document >95% of all of the major ions that comprise the soluble fraction of the atmosphere; (6) directly compared to several other ice-core parameters; (7) integrally associated with a companion program to monitor modern atmospheric conditions; (8) of sufficient quality (e.g., sampling resolution, continuous sampling, multispecies) to warrant extensive multivariate statistical and mathematical analysis; and (9) directly calibrated with modern instrumental records to provide proxies of past climatic change (e.g., temperature, sea-level pressure, precipitation).

ICE-CORE PARAMETER PRIMER

To facilitate the use of the GISP2 record as an archaeological tool, a brief primer follows to describe the paleoclimatic reconstructions available from ice-core records. Much of this information comes directly from Mayewski and White (2002).

Stable Isotope Proxy for Temperature

The $\delta^{18}O$ of ice has classically provided the basic stratigraphy and paleoclimatology for ice cores (Mayewski and Bender 1995). Ice originates by evaporation of seawater, and as the air mass travels away from the site of evaporation toward higher latitudes, it cools and is able to hold less water. Water is lost from the air mass as precipitation (rain or snow). During evaporation and subsequent precipitation, the heavy isotope of O (^{18}O, 0.2% natural abundance) is depleted in water vapor relative to the light, major isotope (^{16}O, 99.8% natural abundance). Along the flow path of the air mass, residual water vapor becomes progressively more depleted in ^{18}O as temperature becomes colder and more and more of the original H_2O content is lost as precipitation. Independent calibrations of the oxygen-isotope-temperature relationship have been developed through the analysis of GISP2 borehole temperature, allowing conversion of isotope-derived surface-temperature histories to temperature-depth profiles (Cuffey et al. 1992). Thus, variations with depth in the $\delta^{18}O$ of ice in a core reflect past variations of temperature with time at a study site. Grootes et al. (1993) measured the $\delta^{18}O$ of ice in the GISP2 core and compared their record with the previously published record for GRIP (Dansgaard et al. 1993). Down to a depth of 2790 m in GISP2 (corresponding to an age of about 110 kyr B.P.), the GISP2 and GRIP records are nearly identical in general shape and in most details.

Glaciochemistry and Chemical Species Source Strength

The chemical composition of polar ice is made up almost exclusively of the following: chloride (Cl^-), nitrate (NO_3^-), sulfate (SO_4^+), calcium (Ca^{++}), magnesium (Mg^{++}), sodium (Na^+), potassium (K^{++}), ammonium (NH_4^+), and methanesulfonate (MS) (Legrand and Mayewski 1997). In fact these chemical species make up more than 95% of all of the chemistry dissolved in the atmosphere. Changes in the concentration of chemical species in ice are related to changes in source emissions (figs. 1 and 2) and/or changes in

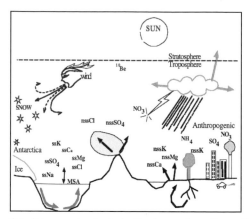

FIG. 1 Chemical species that come from the ocean are transported either as sea salt (ss) (typically as ssNa, ssCl, ssK, ssMg and ssSO$_4$ [where x stands for an excess of sea salt SO$_4$] or as biological emissions from the ocean surface MS or exSO$_4$). Ocean chemicals are incorporated into the atmosphere as wind blows over the ocean surface, incorporating sea spray, and as organisms emit gases from the ocean's surface. Volcanoes emit gases that are often rich in exSO$_4$ and non–sea-salt (nss) Cl, allowing volcanic-event histories to be developed from ice cores. When winds blow over the land, they incorporate dusts in the form of nssK, nssCa, and nssMg. Stronger winds and/or arid regions allow more dusts to be incorporated into the atmosphere and subsequently to be trapped in ice cores. Biological emissions from terrestrial plants, soils, and animals can be traced by viewing levels of NH$_4$, NO$_3$, and nssK in ice cores. By-products of human activities can be seen in ice cores in a wide variety of chemicals including exSO$_4$ from fossil-fuel burning and NO3 from automobile emissions, plus trace metals such as lead, cadmium, and mercury (not listed on figure) (after Mayewski and White 2002, 89).

the transport pathways of the chemical species to an ice-core site (described below and in fig. 3). Therefore, snow and ice from a place like Greenland may contain dust from, for example, Asia or North America. They may also contain volcanic sulfate from, for example, Indonesia or Alaska; Na and Cl transported in sea-salt droplets from the mid-Atlantic; NH$_4$ from terrestrial biota in, for example, arctic Canada, and biological-source sulfate emitted by marine organisms.

Concentrations of these chemical species are measured in ppb (parts per billion) in GISP2 ice-core samples. The very low levels seen in central Greenland ice compared to measurement of these chemical species closer to their sources in ppm (parts per million) or ppt (parts per thousand) and higher is a result of the distance traveled (the farther the distance, the more likely the chemistry will be deposited en route

FIG. 2 (right) The past 500 years of SO$_4$, NO$_3$, nssK, and NH$_4$ developed from the GISP2 ice core. Note dramatic increase in the two major components of acid rain (SO$_4$ and NO$_3$) during the twentieth century. Levels of change in SO$_4$ are closely tied to industrial activity in North America and Europe: (1) beginning of the Industrial Revolution; (2) the Great Depression; (3) World War II; (4) period of most intense burning of sulfur-rich, "dirty" coal; and (5) beginning period of the Clean Air Act. The Clean Air Act did not have a dramatic effect on NO$_3$ levels. Most of the short-term (up to one- to two-year) increases in SO$_4$ are the product of volcanic activity such as the Tambora eruption of 1815 and the Laki eruption of 1783. Data taken from Mayewski et al. (1986, 1990a).

The 1930s dust bowl is clearly seen in the record of nssK dust, in addition to the frequency of dust-bowl–like events in the past. Within the context of the last 400 years the 1930s dust bowl is an extremely dramatic event. The increased levels were no doubt a combination of low precipitation exacerbated by poor farming practices—a combination of natural climate and human activity.

Biomass burning (forest fires) developed from the NH$_4$ record. Note in particular the forest fire produced by the 1908 Tunguska meteorite impact in Siberia. Other periods of forest-fire activity are seen in this record. The GISP2 forest-fire record represents a mix of forest fires from throughout the high latitudes of the Northern Hemisphere.

ICE CORES AND THE ENVIRONMENT

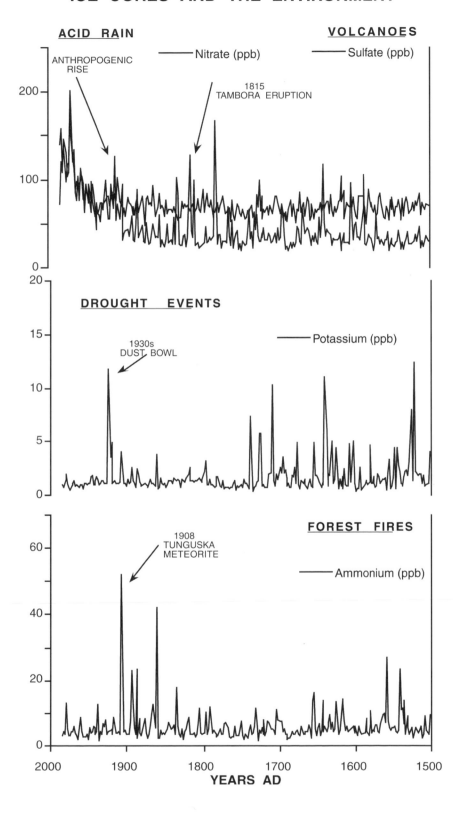

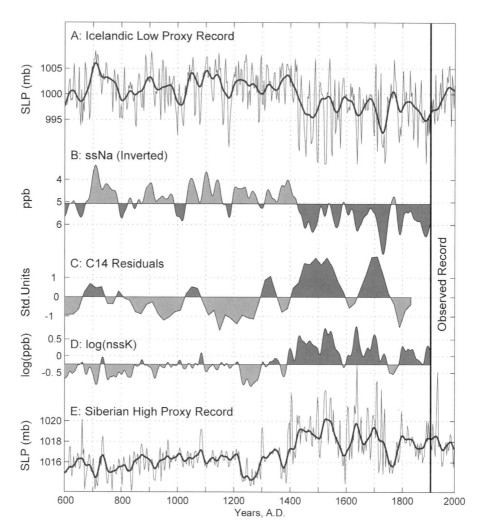

FIG. 3 The observed record of sea-level pressure (SLP) for the Northern Hemisphere extends back only to AD 1899, but ice-core records can be used to extend SLP back significantly longer. In (A) on this figure, a proxy record for a 3-year average of the Icelandic Low SLP extending back to AD 600 is presented based on a 3-year smoothed GISP2 ssNa series (B) following techniques developed in Meeker and Mayewski (2002). Similarly, an approximation for the Siberian High is presented in (E) based on GISP2 nssK data (D). The ssNa and nssK series are smoothed to 20 years, revealing a sharp break in atmospheric circulation regime over the North Atlantic and Eurasia as of AD 1400 (onset of the Little Ice Age). The change is characterized by intensified circulation (a deeper Icelandic Low and stronger Siberian High) as of AD 1400. In (C) of this figure, an approximation for solar variability is provided based on the $\delta^{14}C$ series (Stuiver and Braziunas 1993). High levels of $\delta^{14}C$ correlate closely with periods of low sunspot (low solar output), suggestive of solar forcing of climatic change (after Meeker and Mayewski 2002, 140).

by rain or snow or will just dry-fall out of the atmosphere). Concentrations also increase if the strength at the source increases or if the transport energy increases. As an example, if deserts increase in area, then the source of nssCa, nssMg, and nssK (nss = non–sea salt) increases, so that levels measured in central Greenland increase. If wind speeds over the desert also increase, then these chemicals travel more quickly to central Greenland and less is lost en route, yielding higher concentrations upon deposition in central Greenland.

During glacial periods there is less vegetation to hold soils in place, and stronger winds yield increased levels of continental- and marine-source chemicals. Even marine-biological-source MS increases because the oceans are turned over by higher surface winds, bringing more nutrients from depth to the surface. Terrestrial source NH_4 usually decreases, however, because vegetation on land is generally reduced under colder conditions.

Some chemical species have multiple sources that can be differentiated by chemical or statistical associations. For example, sulfate present in snow can be linked to primary-marine sea salt (as Na_2SO_4) or continental dust (as $CaSO_4$). It can also arise because of the presence of H_2SO_4 produced by atmospheric oxidation of SO_2 introduced directly into the atmosphere during volcanic eruptions, through anthropogenic activity or from various sulfur compounds emitted from the biosphere (fig. 2).

Glaciochemical Proxy for Atmospheric Circulation

Based on an understanding of chemical-species input timing, source, and relationship to other environmental parameters (e.g., temperature via stable isotopes, moisture flux via accumulation rate), we have developed a methodology for interpreting changes in atmospheric circulation patterns (e.g., Meeker and Mayewski 2002: fig. 3) through investigation of changes in multivariate glaciochemical time series. Two primary factors support this interpretation: (1) chemical concentrations in GISP2 snow and ice have been shown to parallel changes in atmospheric flux (Mayewski et al. 1990b) and (2) the chemical composition of an individual air mass over a site is determined both by conditions in the source area(s) over which the transporting air mass passed and the transport process itself. In the simplest case, marine versus continental air masses can be differentiated by comparing sea salt (e.g., NaCl) versus continental dust (e.g., $CaSO_4$) concentrations. More complex atmospheric circulation patterns can be differentiated by including other tracers that record, for example, continental biogenic inputs (e.g., NH_4) and by multivariate discrimination employing a variety of parameters that allow more unique air-mass fingerprinting (e.g., chemical species, particle-size ranges, stable isotopes, accumulation rate).

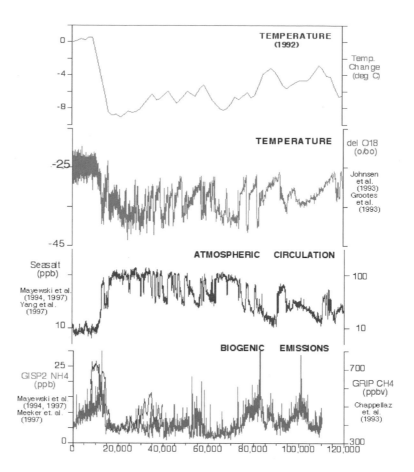

FIG. 4 As recently as 1992, scientists viewed climatic change as slowly evolving (upper plate). For example, temperature change over a glacial/interglacial cycle (~100,000 years) gradually cooled to a minimum (approximately -8°C global mean relative to present), then experienced a relatively fast (2000–4000 years) transition to modern conditions. During our current interglacial (the last ~11,500 years, the Holocene), temperature was assumed to have changed very little. From the GISP2 and GRIP records it is clear that natural climatic variability operates very rapidly and change occurs frequently. Rapid climatic-change events with temperature shifts as much as −30°C affected central Greenland between 11,500 and 110,000 years ago (upper middle plate, data from Johnsen et al. [1992] and Grootes et al. [1993]). These changes in temperature were accompanied by massive shifts in sea-salt concentration, indicating dramatic increases in windiness over the oceans (lower middle portion of figure, data from Mayewski et al. 1994, 1997; Yang et al. 1997). Synchronous changes in dust levels (not on figure) signaled windiness over land (Mayewski et al. 1994, 1997; Yang et al. 1997). The changes in atmospheric circulation over ocean and land have been far more subdued (note logarithmic scale on figure) during the Holocene, but they still represent large enough changes to have influenced human activity. The Greenland ice cores also reveal dramatic shifts in terrestrial biomass seen through the measurement of NH_4 and CH_4 in ice cores (lower part of figure, data from Mayewski et al. 1994, 1997; Chappellaz et al. 1993). Increased levels of ammonium correlate to increased biomass. These variations appear to follow the pattern of Earth's orbital cycles, notably the 23,000-year precession cycle (Meeker et al. 1997), demonstrating how closely plants are controlled by changes in incoming solar radiation.

THE LAST 100,000 YEARS OF CLIMATE HISTORY
AND RAPID CLIMATIC-CHANGE EVENTS

One of the most dramatic recent contributions to our understanding of paleoclimate during the last glacial cycle has come in the millennial-scale range of climatic variability (fig. 4). Unprecedented swings in earth's climate—rapid climatic change events—are recorded in ice cores from central Greenland instigating new, higher-resolution investigations of land and marine paleoclimatic records. The rapid events recorded in the upper 110,000 years of the two central Greenland ice cores are unequivocally climatic events. They represent large climate deviations (massive reorganizations of the ocean-atmosphere system) that occur over decades or less and during which ice-age temperatures in central Greenland may have been as much as 20°C colder than today (Cuffey et al. 1995). These events had their greatest magnitude during the glacial portion of the record, prior to ~14,500 years ago.

Examination of one of the rapid climatic-change events, the Younger Dryas (a near return to glacial conditions during the last deglaciation, previously identified in a variety of paleoclimatic records) reveals the characteristics of such events (fig. 5). Annually resolved sampling over early and late stages of the Younger Dryas indicates that this ~1300-year duration event began and ended in <5–20 years (Alley et al. 1993; Mayewski et al. 1993; Taylor et al. 1993) and perhaps in less than 2 years (Taylor et al. 1997).

FIG. 5 During the Younger Dryas, temperature over central Greenland lowered by 8–10°C, accompanied by twofold and greater changes in snow accumulation (Alley et al. 1993), order-of-magnitude changes in the amount of wind-blown dust and sea salt in the atmosphere (Mayewski et al. 1993; Zielinski and Mershon 1997), and large changes in methane concentration (Brook et al. 1996), synonymous with cold, dry, and dusty conditions (after Mayewski and White 2002, 92).

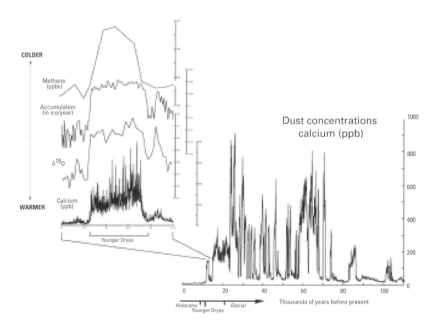

The identification of rapid climatic change event–style variations in the GRIP CH_4 record (Chappellaz et al. 1993; Brook et al. 1996) prompted considerable interest in the identification of such events in other regions because the source areas for CH_4 during the last glaciation may have been in the middle to lower latitudes.

THE HOLOCENE

Paleoclimatic records from the GISP2 ice core demonstrate that Holocene climate is characterized by subdecadal- to millennial-scale variability and is significantly more complex than glacial-age climate (e.g., Barlow et al. 1993; Meese et al. 1994b; O'Brien et al. 1995; Stuiver et al. 1997; White et al. 1997). Time series for the major ions dissolved in the atmosphere, used as tracers for major atmospheric circulation systems (Mayewski et al. 1994, 1997), reveal a strong association between expansions of Northern Hemisphere polar atmospheric circulation systems (zonal and meridional components), and a variety of discontinuous paleoclimatic records (Denton and Karlen 1973; Harvey 1980) record worldwide coolings (fig. 6).

The Holocene rapid-climatic-change events in figure 6 have a quasi-periodicity of ~2600 years similar to previously defined ~2500 year variations in $\delta^{14}C$ residual series from tree rings, suggesting some influence of solar variability on climate (e.g., Denton and Karlen 1973; Damon et al. 1989; Damon and Jirikowic 1992; O'Brien et al. 1995). Preliminary correlations between the Holocene portion of the GISP2 glaciochemical record and Holocene marine-sediment records from the Santa Barbara Basin (Kennett and Mayewski, forthcoming) suggest intriguing similarities of the same nature as those demonstrated between these marine sediments and the GISP2 stable-isotope record covering the last glacial period (Behl and Kennett 1996). Other interesting associations have also been made with the GISP2 Holocene sequence. Namely, ~1450-year recurring events noted in the glacial-age portion of the GISP2 glaciochemical record (Mayewski et al. 1997) have now been described in North Atlantic sediments and correlated with the Holocene glaciochemical record (Bond et al. 1993). In addition, the 8200-year event in the GISP2 Holocene stable-isotope and glaciochemical record has been correlated with several sites in the Northern Hemisphere (Alley et al. 1997) and with climatic events in Lake Victoria sediments and an ice core from East Antarctica (Stager and Mayewski 1997). A similarly timed event is also described in African lake-level records (Street-Perrott and Perrott 1990). Finally, changes in paleoproductivity, used as a proxy for sea-ice extent, from the Antarctic Peninsula and a GISP2 glaciochemical proxy for North Atlantic storminess, display similar centennial-scale variability over the last 4500 years (Domack and Mayewski 1999).

During major Early- to Mid-Holocene coolings recorded in the GISP2 paleo-atmospheric circulation series (fig. 6), the climate-response system operated similarly to Pre-Holocene cooling events (fig. 5)—namely, cooler temperatures (more negative stable isotopes), reduced methane, reduced accumulation rate, and intensification of polar atmospheric circulation

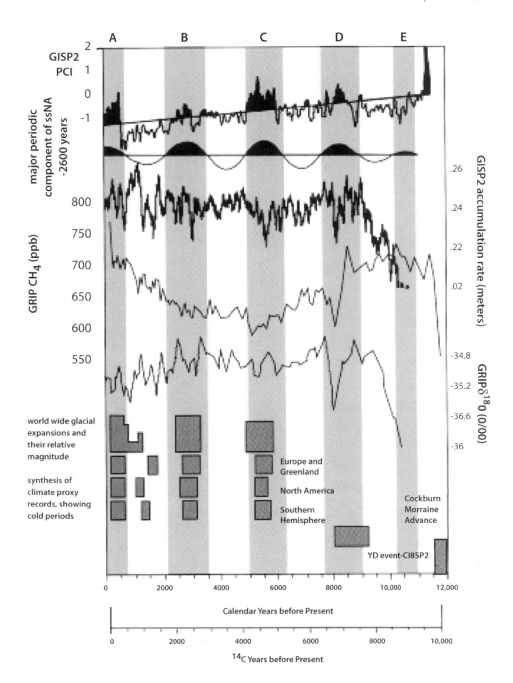

FIG. 6 GISP2 annually dated Holocene PCI (Polar Circulation Index, a proxy for Northern Hemisphere polar-cell intensity) smoothed with a robust spline (equivalent to a 100-year smooth) and a quasi-2600–year periodicity (O'Brien et al. 1995); GISP2 accumulation rate (Meese et al. 1994b); GRIP CH_4 (Blunier et al. 1995); GRIP $\delta^{18}O$ record (Dansgaard et al. 1993); worldwide glacier expansions and their relative magnitude (Denton and Karlen 1973); synthesis of various climate proxy records from Europe, Greenland, North America, and the Southern Hemisphere showing cold periods (Harvey 1980); the Cockburn Stade (Andrews and Ives 1972; Alley et al. 1997; Stager and Mayewski 1997); and the Younger Dryas event (Alley et al. 1993; Mayewski et al. 1993). A–E mark major periods of cold and intensified atmospheric circulation with A equal to the Little Ice Age (after O'Brien et al. 1995, 1963).

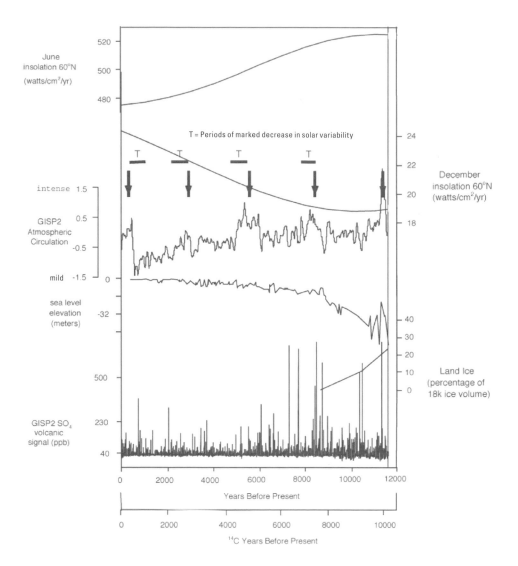

FIG. 7 There are several causes for change in climate during the Holocene. Several direct and indirect causes that may affect the changes in atmospheric circulation revealed by the GISP2 record are captured in this figure (modified from O'Brien et al. 1995). Changes in insolation are produced by Earth's changing position in relation to the sun. These changes in insolation differ according to time of the year (June and December values depicted) and latitude. As insolation changes, so does the distribution of heat over the surface of the Earth and as a consequence the position and the intensity of the storm systems that are recorded in the GISP2 record. Periods during which solar energy output may have fluctuated coincide with the timing of increased solar activity revealed in the GISP2 record (see Ts [after Stuiver and Braziunas 1989]). Changes in land ice volume, which directly affects sea level, caused the albedo and geography of the Earth's surface to change dramatically over the Holocene, further complicating the study of Holocene climatic change. Changes in the timing and magnitude of volcanic events (Zielinski et al. 1994) can affect climate because volcanic materials (dust and H_2SO_4) shield the Earth from incoming solar radiation (after Mayewski and White 2002).

(increased sea salt and dust) all vary, in general, together. However, the coherence between these variables weakens as the events become younger (this is not a sampling artifact in that sample resolution remains constant throughout the Holocene) and is particularly poor during periods between the cool events. This progressive decoupling of variables may be the manifestation of (1) increased regionalization of climate from Early to Late Holocene in response to decreased ice-sheet extent (reduced positive feedbacks) and increased northward penetration of mid-latitude circulation systems; and/or (2) the varying influence of several climate-forcing mechanisms (e.g., changes in total and season-to-season insolation, ice-sheet and sea-ice extent, solar variability, and volcanism; fig. 7).

ARCHAEOLOGICAL APPLICATION FOR ICE-CORE RECORDS

There are many potential archaeological applications for high-resolution paleoclimatic records such as the GISP2 ice-core record. In the preceding text, ice-core data have been translated into records of past environmental change that may have relevance to the interpretation of archaeological records. A few examples follow.

The GISP2 Cl series—a proxy for North Atlantic sea-ice extent (fig. 8)—can be used to identify two prominent periods that relate in timing to notable archaeological events. Increased levels characterize the Little Ice Age (LIA) (note rise in chloride as of AD 1400), a period of relatively greater

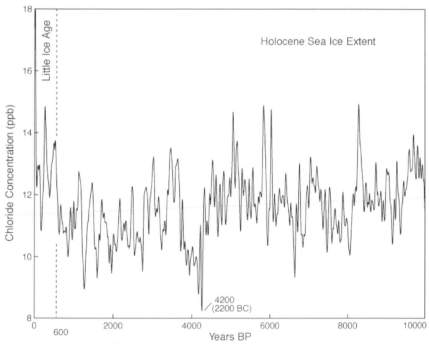

FIG. 8 Change in North Atlantic sea-ice extent over the last 10,000 years, determined from measurements of Cl in the GISP2 ice core (after Mayewski and White 2002, 117).

storminess over the ocean and land (figs. 3 and 6) plus more sea ice than during much of the Holocene. Note other LIA-like events during the Holocene (fig. 7) and potential controls on these events (fig. 8). The onset of the Little Ice Age is coincident with the decline of Norse settlements in Greenland (Buckland et al. 1996). Conversely, decreased values are characteristic of periods with less sea ice and more summer-like conditions over the North Atlantic. The period around 2200 BC on figure 8 shows the most dramatic decrease in chloride in the entire Holocene record and hence extreme summer-like conditions over the North Atlantic. Around 2200 BC much of modern-day Syria and possibly regions as far away as eastern China experienced drought-like conditions. Syria is characterized by a summer dry period and a winter wet period. Increase in summer-like conditions could have heralded conditions conducive to a longer dry season during the 2200 BC collapse period, contributing to one of the more dramatic disruptions in ancient civilization, the collapse of the Akkadian empire.

Associations between other archaeological events and the ice-core–derived climatic record can be developed by utilizing the figures in this paper that contain evidence of change in temperature, sea-level pressure, volcanic activity, aridity, biomass burning, and climate forcing. For more detailed examination of specific time periods of interest refer to the GISP2 data series and a variety of other paleoclimatic records available from www.ngdc.noaa.gov/paleo/.

ACKNOWLEDGMENTS

Support for the recovery, analysis, and interpretation of the GISP2 data presented in this paper came from the U.S. National Science Foundation Office of Polar Programs (OPP 8096305) and Earth System History Program (ESH 9808963).

REFERENCES CITED

Alley, Richard B., Debra A. Meese,
A. J. Shuman, Anthony J. Gow,
Karl C. Taylor, Pieter M. Grootes,
James W. C. White, Michael Ram,
Edwin D. Waddington, Paul A.
Mayewski, and Gregory A. Zielinski
1993 Abrupt Accumulation Increase at
the Younger Dryas Termination in the
GISP2 Ice Core. *Nature* 362: 527–529.

Alley, Richard B., Paul A. Mayewski,
Todd Sowers, Minze Stuiver,
Karl C. Taylor, and Peter U. Clark
1997 Holocene Climatic Instability:
A Large Event 8000–8400 Years Ago.
Geology 25: 482–486.

Andrews, John T., and Jack D. Ives
1972 Late and Post-glacial events
(<10,000 BP) in Eastern Canadian
Arctic with Particular Reference to the
Cockburn Moraines and the Break-up
of the Laurentide Ice Sheet. In *Climate
Changes in Arctic Areas during the Last
Ten-Thousand Years* (Yrjö Vasari, Hannu
Hyvärinen, and Sheila Hicks, eds.):
149–176. University of Oulu, Finland.

Barlow, Lisa K., James W. C. White,
Roger G. Barry, Jeffrey C. Rogers,
and Pieter M. Grootes
1993 The North Atlantic Oscillation
Signature in Deuterium and Deuterium
Excess Signals in the Greenland Ice
Sheet Project 2 Ice Core, 1840–1970.
Geophysical Research Letters 20:
2901–2904.

Behl, Richard J., and James P. Kennett
1996 Brief Interstadial Events in the
Santa Barbara Basin, NE Pacific, during
the Past 60 Kyr. *Nature* 379: 243–246.

Bender, Michael, Todd Sowers,
Mary-Lynn Dickson, Joseph Orchardo,
Pieter M. Grootes, Paul A. Mayewski,
and Debra A. Meese
1994 Climate Correlations between
Greenland and Antarctica during the
Past 100,000 Years. *Nature* 372: 663–666.

Blunier, Thomas, Jerome A. Chappellaz,
Jakob Schwander, Bernhard Stauffer,
and Dominique Raynaud
1995 Variations in Atmospheric Methane
Concentration during the Holocene
Epoch. *Nature* 374: 46–49.

Bond, Gerard, Wallace S. Broecker,
Sigfus J. Johnsen, Jerry McManus,
Laurent D. Labeyrie, Jean Jouzel, and
Gordon Bonani
1993 Correlations between Climate
Records from North Atlantic Sediments
and Greenland Ice. *Nature* 365: 143–147.

Brook, Edward J., Todd Sowers,
and Joseph Orchado
1996 Rapid Variations in Atmospheric
Methane Concentration during the Past
110,000 Years. *Science* 273: 1087–1091.

Buckland, Paul C., Thomas Amorosi,
Lisa K. Barlow, Andrew J. Dugmore,
Paul A. Mayewski, Thomas H.
McGovern, Astrid E. J. Ogilvie,
Jon P. Sadler, and Peter Skidmore
1996 Bioarchaeological Evidence and
Climatological Evidence for the Fate of
Norse Farmers in Medieval Greenland.
Antiquity 70: 88–96.

Chappellaz, Jerome A., Thomas
Blunier, Dominique Raynaud,
Jean Marc Barnola, Jakob Schwander,
and Bernhard Stauffer
1993 Synchronous Changes in
Atmospheric CH4 and Greenland
Climate between 40 and 8 Kyr B.P.
Nature 366: 443–445.

Cuffey, Kurt M., Richard B. Alley,
Pieter M. Grootes, and Sridhar
Anandakrishnan
1992 Toward using Borehole
Temperatures to Calibrate an Isotopic
Paleothermometer in Central Greenland.
*Palaeogeography, Palaeoclimatology and
Palaeoecology* 98: 265–268.

Cuffey, Kurt, M., Gary D. Clow, Richard B. Alley, Minze Stuiver, Edwin D. Waddington, and Richard W. Saltus
1995 Large Arctic-temperature Change at the Wisconsin-Holocene Transition. *Science* 270: 455–458.

Damon, Paul E., Songlin Cheng, and Timothy Linick
1989 Fine and Hyperfine Structure in the Spectrum of Secular Variations of Atmospheric ^{14}C. *Radiocarbon* 31: 704–718.

Damon, Paul E., and John L. Jirikowic
1992 The Sun as a Low-frequency Harmonic Oscillator. *Radiocarbon* 34: 199–205.

Dansgaard, Willi, Sigfus J. Johnsen, Henrik B. Clausen, Dorthe Dahl-Jensen, Niels S. Gundestrup, Claus U. Hammer, Christine S. Hvidberg, Jorgen P. Steffensen, Arny E. Sveinbjornsdottir, Jean Jouzel, and Gerard Bond
1993 Evidence for General Instability of Past Climate from a 250 Kyr Ice Core. *Nature* 364: 218–219.

Denton, George H., and Wibjörn Karlén
1973 Holocene Climatic Variations—Their Pattern and Possible Cause. *Quaternary Research* 3: 155–205.

Dibb, Jack E.
1992 The Accumulation of Pb-210 at Summit, Greenland since 1855. *Tellus* 44B: 72–79.

Domack, Eugene W., and Paul A. Mayewski
1999 A Bi-polar Signal in Late Holocene Paleoclimate: Comparison of the Marine and Ice Core Record. *The Holocene* 9: 247–251.

Grootes, Pieter M., Minze Stuiver, James W. C. White, Sigfus S. Johnsen, and Jean Jouzel
1993 Comparison of Oxygen Isotope Records from the GISP2 and GRIP Greenland Ice Cores. *Nature* 366: 552–554.

Harvey, L. D. Danny
1980 Solar Variability as a Contributing Factor to Holocene Climatic Change. *Progress in Physical Geography* 4: 487–530.

Johnsen, Sigfus, Henrik Clausen, Willi Dansgaard, Katherine Fuhrer, Niels Gundestrup, Claus Hammer, Peter Iversen, Jean Jouzel, Bernhard Stauffer, and Jorgen Steffensen
1992 Irregular Glacial Interstadials Recorded in a New Greenland Ice Core. *Nature* 359: 311–313.

Legrand, Michel, and Mayewski, Paul A.
1997 Glaciochemistry of Polar Ice Cores: A Review. *Reviews of Geophysics* 35: 219–244.

Mayewski, Paul A., and Michael Bender
1995 The GISP2 Ice Core Record—Paleoclimate Highlights. *Reviews of Geophysics*, Supplement, U.S. National Report to International Union of Geodesy and Geophysics 1991–1994: 1287–1296.

Mayewski, Paul A., and Frank White
2002 *The Ice Chronicles*. University Press of New England, Hanover, NH.

Mayewski, Paul A., W. Berry Lyons, Mary Jo Spencer, Mark S. Twickler, Bruce Koci, Willi Dansgaard, Cliff Davidson, and Richard Honrath
1986 Sulfate and Nitrate Concentrations from a South Greenland Ice Core. *Science* 232: 975–977.

Mayewski, Paul A., W. Berry Lyons, Mary Jo Spencer, Mark S. Twickler, Christopher F. Buck, and Sallie Whitlow
1990a An Ice Core Record of Atmospheric Response to Anthropogenic Sulphate and Nitrate. *Nature* 346: 554–556.

Mayewski, Paul A., Mary Jo Spencer, Mark S. Twickler, and Sallie Whitlow
1990b A Glaciochemical Survey of the Summit Region, Greenland. *Annals of Glaciology* 14: 186–190.

Mayewski, Paul A., Loren D. Meeker, Sallie Whitlow, Mark S. Twickler, Michael C. Morrison, Richard B. Alley, Peter Bloomfield, and Karl C. Taylor
1993 The Atmosphere during the Younger Dryas. *Science* 261: 195–197.

Mayewski, Paul A., Loren D. Meeker, Sallie Whitlow, Mark S. Twickler, Michael C. Morrison, Pieter M. Grootes, Gerard C. Bond, Richard B. Alley, Debra A. Meese, Anthony J. Gow, Karl C. Taylor, Michael Ram, and Mark Wumkes
1994 Changes in Atmospheric Circulation and Ocean Ice Cover over the North Atlantic during the Last 41,000 Years. *Science* 263: 1747–1751.

Mayewski, Paul A., Loren D. Meeker, Mark S. Twickler, Sallie I. Whitlow, Qinzhao Yang, W. Berry Lyons, and Michael Prentice
1997 Major Features and Forcing of High Latitude Northern Hemisphere Atmospheric Circulation over the Last 110,000 Years. *Journal of Geophysical Research* 102 (C12): 26,345–26,366.

Meeker, Loren D., and Paul Mayewski
2002 A 1400 Year Long Record of Atmospheric Circulation over the North Atlantic and Asia. *The Holocene* 12: 257–266.

Meeker, Loren D., Paul Mayewski, Mark S. Twickler, Sallie I. Whitlow, and Debra A. Meese
1997 A 110,000 Year Long History of Change in Continental Biogenic Source Strength and Related Atmospheric Circulation. *Journal of Geophysical Research* 102 (C12): 26,489–26,504.

Meese, Debra A., Richard B. Alley, R. Joseph Fiacco, Mark Germani, Anthony J. Gow, Pieter M. Grootes, Matthias Illing, Paul Mayewski, Michael Morrison, Michael Ram, Karl C. Taylor, Qinzhao Yang, and Gregory Zielinski
1994a Preliminary Depth-Age Scale of the GISP2 Ice Core, Specia. CRREL Report 94-1. U.S. Army Corps of Engineers, Hanover, NH.

Meese, Debra A., Richard B. Alley, Anthony J. Gow, Pieter M. Grootes, Paul A. Mayewski, Michael Ram, Karl C. Taylor, Edwin D. Waddington, and Gregory Zielinski
1994b The Accumulation Record from the GISP2 Core as an Indicator of Climate Change throughout the Holocene. *Science* 266: 1680–1682.

O'Brien, Suzanne R., Paul A. Mayewski, Loren D. Meeker, Debra A. Meese, Mark S. Twickler, and Sallie I. Whitlow
1995 Complexity of Holocene Climate as Reconstructed from a Greenland Ice Core. *Science* 270: 1962–1964.

Schott, Christine, Edwin D. Waddington, and Charles F. Raymond
1992 Predicted Time-scales for GISP2 and GRIP Boreholes at Summit, Greenland. *Journal of Glaciology* 38: 162–167.

Sowers, Todd, Michael Bender, Laurent Labeyrie, Doug Martinson, Jean Jouzel, Dominique Raynaud, Jean Jacques Pichon, and Yevgeniy Sergeevich Korotkevich
1993 A 135,000 Year Vostok-SPECMAP Common Temporal Framework. *Paleoceanography* 8: 737–766.

Stager, J. Curt, and Paul A. Mayewski
1997 Abrupt Mid-Holocene Climatic Transitions Registered at the Equator and the Poles. *Science* 276: 1834–1836.

Street-Perrott, F. Alayne, and R. Alan Perrott
1990 Abrupt Climate Fluctuations in the Tropics: The Influence of Atlantic Ocean Circulation. *Nature* 343: 607–612.

Stuiver, Minze, and Thomas F. Braziunas
1989 Atmospheric C-14 and Century-scale Solar Oscillations. *Nature* 338: 405–408.

1993 Sun, Ocean, Climate and Atmospheric $^{14}CO_2$: An Evaluation of Causal and Spectral Relationships. *The Holocene* 3: 289–305.

Stuiver, Minze, Thomas F.
Braziunas, Pieter M. Grootes,
and Gregory Zielinski
1997 Is There Evidence of Solar Forcing
of Climate in the GISP2 Oxygen Isotope
Record? *Quaternary Research* 48: 1–8.

Taylor, Karl C., Richard B. Alley,
R. Joseph Fiacco, Pieter M. Grootes,
Greg W. Lamorey, Paul A. Mayewski,
and Mary Jo Spencer
1992 Ice Core Dating and Chemistry by
Direct-current Electrical Conductivity.
Journal of Glaciology 38: 325–332.

Taylor, Karl C., Greg W. Lamorey,
Georgia Ann Doyle, Richard B. Alley,
Pieter M. Grootes, Paul A. Mayewski,
James W. C. White, and Lisa K. Barlow
1993 The "Flickering Switch" of Late
Pleistocene Climate Change. *Nature* 361:
432–436.

Taylor, Karl C., Paul A. Mayewski,
Richard B. Alley, Edward J. Brook,
Anthony J. Gow, Pieter M. Grootes,
Debra A. Meese, Eric S. Saltzman,
Jeffrey P. Severinghaus, Mark S.
Twickler, James W. C. White, Sallie I.
Whitlow, and Gregory A. Zielinski
1997 A Close Look at the Holocene/
Younger Dryas Transition Recorded
at Summit, Greenland. *Science* 278:
825–827.

White, James W. C., Lisa K. Barlow,
David Fisher, Pieter M. Grootes,
Jean Jouzel, Sigfus J. Johnsen, Minze
Stuiver, and Henrik Clausen
1997 The Climate Signal in Stable
Isotopes of Snow from Summit,
Greenland: Results of Comparisons
with Modern Climate Observations.
Journal of Geophysical Research 102 (C12):
26,425–26,440.

Wilson, Alexander T., and
Douglas J. Donahue
1990 AMS Carbon-14 Dating of Ice:
Progress and Future Prospects. *Nuclear
Instruments and Methods in Physics
Research* B52: 473–476.

Yang, Qinzhao, Paul A. Mayewski,
Sallie I. Whitlow, and Mark S. Twickler
1997 Major Features of Glaciochemistry
over the Last 110,000 Years in the GISP2
Ice Core. *Journal of Geophysical Research*
102: 23,289–23,299.

Zielinski, Gregory A., Paul A.
Mayewski, Loren D. Meeker, Sallie I.
Whitlow, Mark S. Twickler, Michael C.
Morrison, Debra A. Meese, Richard B.
Alley, and Anthony J. Gow
1994 Record of Volcanism since 7000
BC from the GISP2 Greenland Ice Core
and Implications for the Volcano-climate
System. *Science* 264: 948–952.

Zielinski, Gregory A., and
Grant R. Mershon
1997 Paleoenvironmental Implications
of the Insoluble Microparticle Record
in the GISP2 (Greenland) Ice Core
during the Rapidly Changing Climate
of the Holocene-Pleistocene Transition.
Geological Society of America Bulletin 109:
547–559.

EL NIÑO AND INTERANNUAL VARIABILITY OF CLIMATE IN THE WESTERN HEMISPHERE

Kirk Allen Maasch

Earth's climate is dynamic, changing on time scales ranging from months to hundreds of millions of years. Although the climate system is not yet fully understood, some aspects of climate change can be explained. In this paper I summarize what we know about annual-to-interannual variability of climatic conditions across the Western Hemisphere, with particular attention to El Niño/Southern Oscillation (ENSO).

The climate system comprises five basic components: atmosphere, hydrosphere, cryosphere, biosphere, and lithosphere. Each component is quite complex, containing many feedbacks, both positive and negative. Figure 1 summarizes the components of the climate system, along with their time constants. When a component of the climate system is perturbed, the time constant is the time required for that component to return to its original value. The entire system is forced externally, primarily by energy from the sun. Periodic annual and diurnal variations of solar forcing occur due to the rotation of Earth on an axis that is tilted 23.5 degrees relative to the plane of its elliptical orbit. Interactions, or feedbacks, between different parts of the climate system, geography, and this external forcing can lead to internally driven oscillations. The length of time, or periodicity, of such internally driven climate oscillations is related to the time constants of the specific components of the climate system that are involved in feedbacks that give rise to the oscillation.

BASIC CONCEPTS

Weather is the state of the atmosphere in terms of hot or cold, wet or dry, cloudy or clear, windy or calm. Instantaneous, or synoptic measurements of key meteorological variables are used to quantify the weather, namely temperature, humidity, pressure, winds, precipitation, and cloudiness. These meteorological elements are often shown on a map or chart at a given time for a particular region.

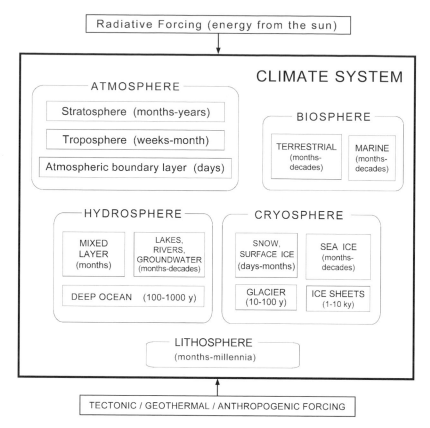

FIG. 1 Basic components of the climate system along with their typical time constants

Climate is the statistical collection of spatially- and/or time-averaged weather conditions at a given place. A 30-year interval is typically chosen to define climate. Thirty-year averages of meteorological variables are known as normals. At present, a normal refers to the 1971–2000 average for a particular variable. Note that the climate defined using different periods of time may be different (e.g., the normals defined by the 1931–60 average are different from those of 1961–90); spatial scale also affects the definition of normals.

The instrumental record of meteorological variables has been kept systematically for about 150 years. Stations where measurements have been made are not distributed uniformly in time or space. It is from this ~150-year instrumental record that climate variables can be calculated and examined in terms of any systematic climatic change that may have occurred. In particular, the climate variables that I use in this discussion of both the seasonal cycle and interannual climatic variability are from a shorter (about 50-year), but more homogeneous (spatially and temporally complete) record of instrumental measurements, the NCEP/NCAR Reanalysis Project, a description of which may be found in Kistler et al. (2001). These include:

1. Net radiation balance as a measure of the heat added to the climate system (energy in W/m²)
2. Near-surface air temperature (°C)
3. Sea-level pressure (force/area, in units of millibars, or mb [1 mb = 100 N/m²])
4. Near-surface winds (atmospheric motion, meters per second [m/s])
5. Precipitation rate in millimeters per day (mm/day)

These data are available online at http://www.cdc.noaa.gov. The NCEP/ NCAR dataset includes instantaneous values of each variable at four reference times: 0, 6, 12, and 18 GMT (Greenwich Mean Time). The spatial coverage is global on a 2.5-degree latitude by 2.5-degree longitude grid with 144 × 73 points spanning 90N–90S, 0E–357.5E. This dataset covers the period from 1948 to present. Monthly means were calculated by averaging 4 times daily data for each day of every month. Long-term means were obtained by averaging monthly means from 1968 to 1996, the most reliable part of the record. In this discussion, I reference long-term monthly means and anomalies (monthly means for individual years minus the long-term mean).

SEASONAL CYCLE: A PRIMER

In many places the magnitude of change over any particular year (a seasonal cycle) is greater than the interannual, or longer-time-scale variations of climate. Global patterns of atmospheric temperature, pressure, resultant winds, and precipitation are ultimately the result of uneven heating of Earth by the sun. More than 99% of the energy added to the Earth system comes from the sun in the form of electromagnetic radiation. This radiation is largely in the visible band (wavelengths ranging from 0.4–0.7 micrometer), but also in shorter (ultraviolet or UV) and longer (infrared or IR) wavelengths. Over a long time span, globally averaged incoming, short-wave radiant energy from the sun is balanced by outgoing, long-wave (IR) radiation from Earth back to space. At any given time and place, however, there are radiative imbalances.

Net Radiation

Net radiation added to the Earth system is equal to incoming short-wave minus reflected and backscattered short-wave minus outgoing long-wave. Figure 2 summarizes the seasonal cycle of net radiation. Earth Radiation Budget Experiment (ERBE) satellite measurements were used to calculate net radiation in January, April, July, and October. At low latitudes the amount of incoming short-wave radiation from the sun is greater than outgoing long-wave radiation, mainly because the sunlight is more direct (the sun is higher in the sky) and has less chance of being absorbed (passes through less atmosphere) before reaching Earth's surface. This surplus energy (red colors in fig. 2) leads to heating of Earth's surface and the lower atmosphere at these latitudes. At high latitudes in both hemispheres, outgoing radiation is greater

January

April

October

July

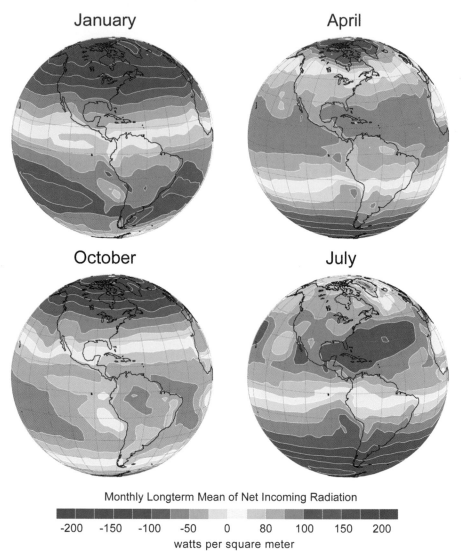

Monthly Longterm Mean of Net Incoming Radiation

-200   -150   -100   -50   0   80   100   150   200
watts per square meter

FIG. 2 Monthly mean January, April, July, and October net radiation over the Western Hemisphere based on 1985–89 ERBE data

than incoming. This deficit results in cold polar temperatures. The position of the belt of surplus radiation shifts from south of the equator in January (Southern Hemisphere summer) to north of the equator in July (Northern Hemisphere summer) tracking the most direct solar forcing through the seasonal cycle.

Temperature

The seasonally changing spatial pattern of near-surface temperature, shown in figure 3, is mainly controlled by latitude because of the dependence of net radiation (fig. 2) on the solar zenith angle, which is related to the inclination of Earth's axis of rotation. Incoming solar radiation is more direct at low latitudes, resulting in more energy/time/square meter (W/m^2)

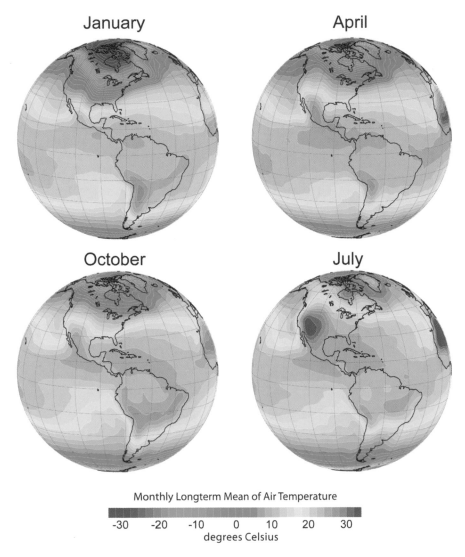

January April October July

Monthly Longterm Mean of Air Temperature

-30 -20 -10 0 10 20 30
degrees Celsius

FIG. 3 Long-term mean monthly near-surface air temperature for January, April, July, and October

available to heat the surface and ultimately the lower atmosphere. Additional controls of temperature are the amount of water vapor and other greenhouse gases (e.g., O_3, CO_2, CH_4), the distribution and type of cloud cover, albedo (reflectivity) of Earth's surface, composition of Earth's surface (i.e., land-sea distribution), position on the continent (i.e., east vs. west coast, coastal vs. mid-continent), ocean currents, and elevation. Inspection of the temperature patterns through a seasonal cycle (fig. 3) reveals the following:

1. Temperature decreases from the equator to the poles.
   The pattern of maximum temperature shifts seasonally to the summer hemisphere.

2. The rate of north-south temperature change, or temperature gradient (shown by the spacing of the isotherms), is greater in the winter hemisphere than in the summer hemisphere.

3. Seasonal equatorward and poleward migration of the isotherm patterns cover more latitude for land areas than for oceans.

4. Land heats and cools faster than do oceans, so continents are much colder in winter and warmer in summer than oceans at the same latitude.

5. Eastern boundary currents (e.g., Peru Current) are cold and flow equatorward, western boundary currents (e.g., Gulf Stream) are warm and flow poleward. Upwelling brings cold water to the surface of the ocean.

Pressure and Winds

The average air pressure (the force per unit area exerted by the weight of the atmosphere) at sea level (SLP) is equal to slightly more than 1000 millibars (mb). SLP is higher in some places than it is in others. The long-term monthly average spatial distribution of SLP over a seasonal cycle is shown in figure 4. The underlying cause for these SLP patterns is differential heating of Earth's surface, which also leads to the large-scale patterns of near-surface air and sea-surface temperature discussed above. The distribution of SLP over a seasonal cycle (fig. 4) is characterized by

1. A band of relatively low SLP centered near, but not directly on, the equator;

2. Subtropical regions (centered at around 30°N and 30°S) characterized by high SLP. Over the oceans subtropical centers of high pressure are very stable features throughout the entire year;

3. Low SLP at high-mid latitudes (around 60°N).

Horizontal differences in pressure cause atmospheric motion (winds). These pressure differences result in a pressure-gradient force that always points from high to low pressure and sets air in motion. The pressure-gradient force is proportional to the pressure difference between two locations divided by the distance between the two locations. The higher the pressure-gradient force, the faster the winds blow. Near Earth's surface two other forces also impact the magnitude and direction of the wind: coriolis and frictional forces. The coriolis force, due to Earth's rotation, deflects winds to the right in the Northern Hemisphere and to the left in the Southern Hemisphere. The frictional force is opposite the wind direction, thus decreasing wind speed.

Figure 4 shows long-term monthly average large-scale motions of the lower atmosphere as vector arrows laid over sea-level pressure (SLP). Wind patterns clearly shift seasonally along with the global pressure patterns. In

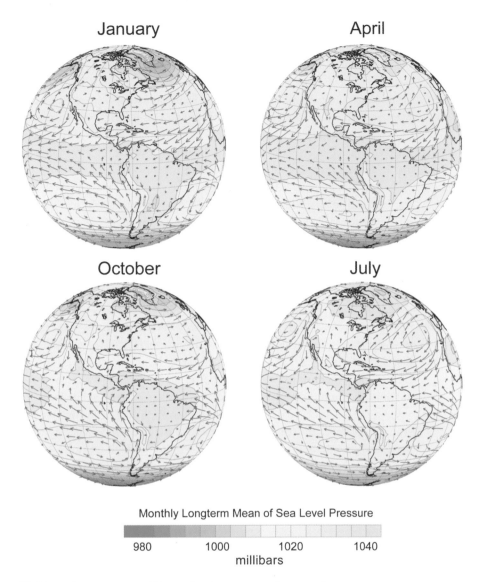

January

April

October

July

Monthly Longterm Mean of Sea Level Pressure

980   1000   1020   1040
millibars

FIG. 4 Long-term mean monthly sea-level pressure and near-surface winds for January, April, July, and October

the tropics the prevailing winds are easterly (from the east); in mid-to-high latitudes, they are westerly (from the west). Centers of subtropical high pressure, primarily over ocean basins, are characterized by divergent anticyclonic flow (clockwise in the Northern Hemisphere, counterclockwise in the Southern Hemisphere). These regions are also characterized by descending air (subsidence) and consequently skies are usually clear. Circulation around centers of low pressure is convergent and cyclonic (counterclockwise in the Northern Hemisphere, clockwise in the Southern Hemisphere). Low pressure is associated with rising motion and cloudy conditions.

Heat Transport

For equatorial regions *not* to get warmer and warmer and polar regions *not* to get correspondingly cooler and cooler, there must be a transport of heat from low to high latitudes. This heat transport is accomplished by both atmospheric and oceanic circulation. Transport in the atmosphere can take place as both sensible and latent heat. In the tropics heat is moved poleward by large-scale meridional circulation cells, known as Hadley cells (fig. 5). Surface heating in the equatorial region results in rising, poleward-moving air. As this air moves poleward in both hemispheres, it cools radiatively and sinks near 30°N and 30°S latitude, where it then returns at low levels toward the equator. In the Northern Hemisphere, the northeasterly trade winds blow equatorward from the subtropical high at around 30°N toward lower pressure near the equator, while in the Southern Hemisphere the south-easterly trade winds blow equatorward from the subtropical high at around 30°S. These wind fields converge near the equator in a region called the inter-tropical convergence zone (ITCZ), an area of convective rain showers and thunderstorms. The ITCZ is not stationary over the seasonal cycle, but tends to migrate to the warmest surface areas throughout the year. In January, the sunlight is most direct in the Southern Hemisphere, causing a southward displacement of the ITCZ. As the most direct solar radiation travels from the Southern Hemisphere to the Northern Hemisphere, the ITCZ migrates northward, attaining its maximum northward displacement in late June.

On a smaller scale, heat transport is also accomplished by hurricanes that evaporate water in the tropics and later release the latent heat acquired when condensation occurs at higher latitudes. At mid- to high latitudes heat is transported by frontal storms. These storms tend to track along the polar front (the boundary between cold polar air and warmer subtropical air).

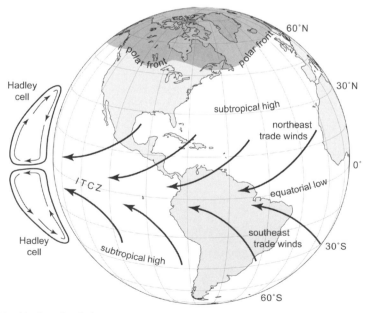

FIG. 5 Hadley circulation

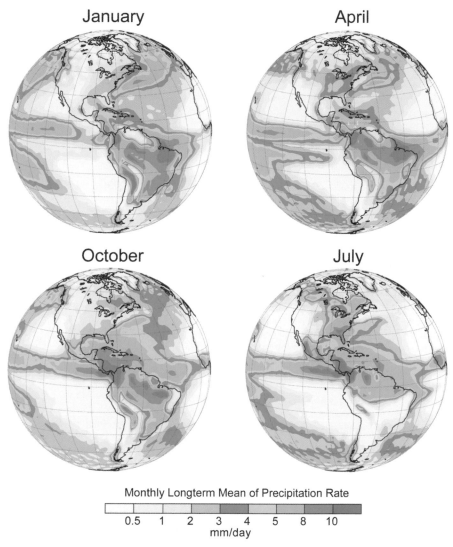

FIG. 6 Long-term mean monthly precipitation for January, April, July, and October

Precipitation

Figure 6 summarizes large-scale precipitation patterns, which are related to the atmospheric circulation discussed above. The most prominent feature is a band of maximum precipitation centered on the latitude where low-level convergence leads to rising motion (ITCZ). Precipitation is a function of latitude, elevation (due to orographic lifting, precipitation usually increases with elevation), distance from moisture sources (precipitation is usually lower at greater distances from the ocean), position within the continental land mass, prevailing wind direction, relation to mountain ranges (windward sides are typically cloudy and rainy, with leeward sides typically dry and sunny—the "rain shadow"), and relative temperatures of land and bordering oceans.

INTERANNUAL VARIABILITY AND ENSO

Seasonal cycles look different from year to year. While there is no single reason for these changes, El Niño/Southern Oscillation (ENSO) is an important coupled ocean-atmosphere phenomenon relating to climatic variability on interannual time scales.

El Niño and La Niña

The term *El Niño* was first used centuries ago by the coastal fishermen of Peru and Ecuador to describe a warm ocean current that typically appears in December (around Christmas) and lasts for several months (Philander 1990). It has long been recognized that when sea-surface temperatures warm during El Niño, rainfall also increases and the success of fisheries decreases along the north Peruvian coast. But not every summer sea-surface warming is the same. Some years are characterized by especially warm water that remains until May or even June. Through time, the term "El Niño" has come to refer only to these particularly strong warm intervals. Local warming of waters off Peru coincides with positive SST (sea-surface temperature) anomalies over a much larger domain, namely the eastern half of the equatorial Pacific. *La Niña* is the opposite phenomenon, referring to abnormally cold SST in the eastern half of the equatorial Pacific. We now know that these changes are part of a much larger climate pattern and that the atmosphere and ocean are coupled in the equatorial Pacific.

Southern Oscillation

The *Southern Oscillation*, originally named and described by Sir Gilbert Walker (1924), is a seesaw shift in surface air pressure differences between Darwin, Australia and Tahiti. Semipermanent, low SLP sits over Indonesia and northern Australia and the warm pool in the western Pacific. The southeastern Pacific is dominated by high SLP, as shown in figure 4. When SLP is higher than normal at Darwin, it is lower than normal at Tahiti and vice versa. The Southern Oscillation Index (SOI) is the measure of the monthly/seasonal fluctuations in surface air pressure differences at Tahiti and Darwin. El Niño and its sister event, La Niña, occur at the extreme phases of the Southern Oscillation. SOI has a negative value during an El Niño event. During "normal" years, low-level easterly trade winds blowing across the Pacific basin are accompanied by rising air in the western Pacific and descending air in the eastern Pacific (fig. 7). This large-scale zonal circulation is called the Walker Cell.

THE EL NIÑO/SOUTHERN OSCILLATION (ENSO) CYCLE

Together, both El Niño/La Niña and flips of the Southern Oscillation participate in the ENSO cycle. The warm events (El Niño) occur about every three to seven years, can last several months to over one year, and can also influence faraway places even outside of the tropics (teleconnections). Below I summarize in general terms what happens through an ENSO cycle, and then look in more detail at climate anomalies associated with the two largest El Niño events of the twentieth century.

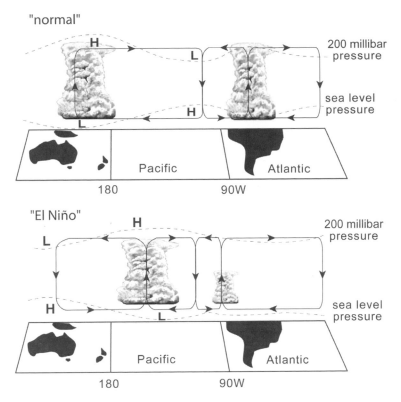

FIG. 7 Walker circulation

"Normal" Conditions in the Pacific Basin

In the equatorial Pacific, so-called normal conditions are characterized by trade winds (figs. 4 and 5) accompanied by slightly elevated sea level (~50 cm) on the western side of the basin, and a shallow thermocline (boundary between cold, nutrient-rich deep water and warm, nutrient-depleted surface water) on the eastern side. There is low atmospheric pressure and rising air over a pool of warm water in the western Pacific and high atmospheric pressure and sinking air over a tongue of cold, upwelling water off the coast of Peru and along the equator in the east (Walker cell shown in fig. 7, top; see fig. 10, top, for the cold tongue). The upwelling is caused by Ekman transport, as shown schematically in figure 8. Upwelling not only maintains cool SST, but also delivers nutrients to the surface waters, leading to the highly productive fisheries in this region. Clouds and precipitation occur where air rises, and so normally there is rainfall in the western equatorial Pacific and dry conditions in the eastern equatorial Pacific.

El Niño Conditions in the Pacific Basin

During an El Niño event, the difference in atmospheric pressure between the central and western Pacific decreases and the trade winds weaken. Sea level drops in the west and rises in the east. The thermocline becomes shallower in the west and deeper in the east. Upwelling weakens or shuts down, and eastern equatorial and coastal-surface waters warm and are depleted of nutrients. This in turn results in reduced productivity in this important fisheries area. In a

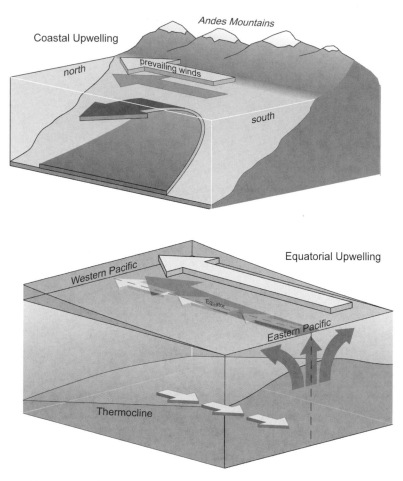

FIG. 8 Equatorial and coastal upwelling

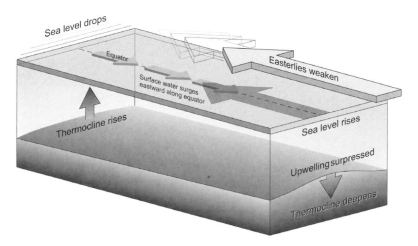

FIG. 9 Schematic diagram showing changes in
the Pacific basin during an El Niño event

strong El Niño, the effect in the ocean can reach as far north as the California coast. The large atmospheric convection cell over Indonesia moves eastward, resulting in dry conditions (drought) in the western Pacific including over Indonesia and Australia (fig. 7, bottom). Rainfall increases in the central and eastern Pacific (for example, in the Galapagos and coastal Ecuador and northern Peru). These changes are summarized in figure 9.

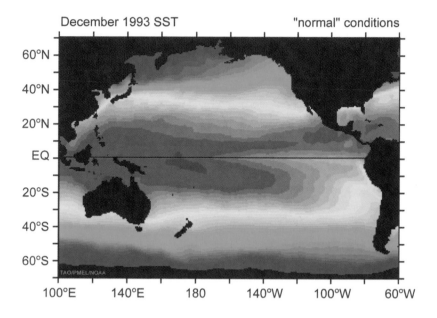

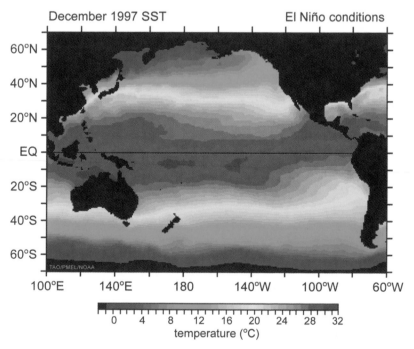

FIG. 10 Equatorial Pacific SST in December 1993 ("normal" year) and December 1997 (El Niño year)

Since 1950, thirteen El Niño events as defined above (1951–52, 1953–54, 1957–58, 1965–66, 1969–70, 1972–73, 1976–77, 1982–83, 1986–87, 1991–92, 1994–95, 1997–98, 2002–03) have affected the South American coast, each characterized by increased SST extending from the coast in a belt stretching some 8000 km across the equatorial Pacific. The weaker events raised eastern equatorial Pacific sea-surface temperatures by only around one to one and a half degrees Celsius and had only a minor impact on South American fisheries. But the much larger SST increases of the strong events, in particular 1982–83 and 1997–98, had a significant impact on both local weather and marine life, as well as on climatic conditions around the globe (teleconnections).

For comparison, figure 10 shows SST conditions during a "normal" year (December 1993) and during a very strong El Niño year (December 1997). Note that regions of cool SST along the Peruvian coast (~20°C, colored yellow) and along the equator in the east Pacific (~22–24°C, colored orange) during a normal year (1993) were some 5°C warmer during the 1997 El Niño event.

Measuring the Frequency and Strength of El Niño

The classic method for measuring the strength of the warm phase of the ENSO cycle (El Niño) is the SOI (fig. 11, top). The negative phase of the SOI (up on fig. 11, top) corresponds to times when the pressure difference between the eastern and western Pacific is less than normal, as occurs during an El Niño. Another diagnostic measure of El Niño strength is defined by the average SST in specific regions of the tropical Pacific Ocean. Figure 11 (middle) shows average SST in one such region, called Niño 3.4 (a box covering 5°S–5°N and 120°–170°W). More recently ENSO cycles have also been quantified using a multivariate ENSO index (MEI, Wolter and Timlin 1993). The MEI (fig. 11, bottom) monitors the coupled oceanic-atmospheric character of ENSO based on the main observed variables over the tropical Pacific Ocean. These surface marine observations, taken from the COADS (comprehensive ocean-atmosphere data set, available at http://icoads.noaa .gov/), are a weighted average of the main ENSO features contained in six different variables: sea-level pressure, the east-west and north-south components of the surface wind, SST, surface air temperature, and total amount of cloudiness (Wolter and Timlin 1993).

El Niño events since 1950 are labeled in figure 11. Based on the MEI, the El Niño events up to about 1972–73 were characterized by early warming in the far-eastern Pacific and reached their standardized peaks before the end of the first year. More recent El Niño events in 1982–83, 1986–87, and 1991–92 took longer to mature, typically reaching their peaks in the spring of the second year. The MEI peak in 1983 was the biggest of the century, while 1997–98 had two peaks just below the 1982–83 El Niño, one in July–August 1997, and one in February–March 1998. These 1982–83 and 1997–98 events will be examined in more detail below.

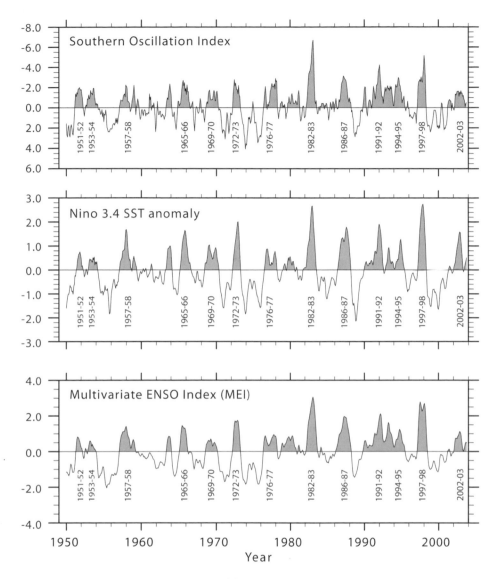

FIG. 11 Tropical Pacific indicators of ENSO from 1950 to 2003: Southern Oscillation Index (top), Niño 3.4 sea-surface temperatures (middle), and Multivariate ENSO Index (bottom). Note that each indicator is oriented such that peaks representing El Niño events point up.

FIG. 12 Regions where typical temperature and precipitation anomalies associated with El Niño have been assessed (based on last 10 events)

TELECONNECTIONS IN THE WESTERN HEMISPHERE

An "Average" ENSO Event

Defining a typical El Niño is not straightforward in terms of distant teleconnections. However, certain general patterns are consistent from one event to the next. Below I summarize the most basic patterns that characterize an "average" El Niño in the Western Hemisphere based on the past ten El Niño events: 1965–66, 1969–70, 1972–73, 1976–77, 1982–83, 1986–87, 1991–92, 1994–95, 1997–98, and 2002–03. I look specifically at regions that appear to have some correlation between El Niño and temperature and/or precipitation, or are relevant to the readers of this book. These regions, marked by numbers on figure 12, include 1) eastern equatorial Pacific; 2) coastal Ecuador and the north coast of Peru; 3) eastern Brazil; 4) central Chile; 5) Yucatan and the western Caribbean; 6) northern Mexico and southwestern United States; 7) southeastern United States; 8) northeastern United States; 9) Pacific northwest; and 10) Alaska. For the last ten events, I have characterized air temperature for each of these regions as warm, neutral, or cool and precipitation as wet, neutral, or dry. For air temperature, *warm* means that the region is more that 1°C higher than the long-term mean, *cool* is more than

FIG. 13 (right) Equatorial Pacific sea-surface temperature anomalies in July, October, January, and April for 1982–83 and 1997–98 El Niño events

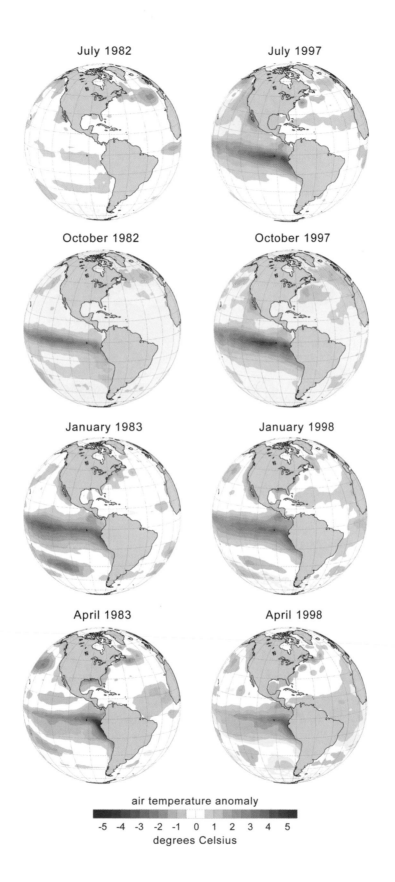

July 1982   July 1997

October 1982   October 1997

January 1983   January 1998

April 1983   April 1998

air temperature anomaly

-5 -4 -3 -2 -1 0 1 2 3 4 5

degrees Celsius

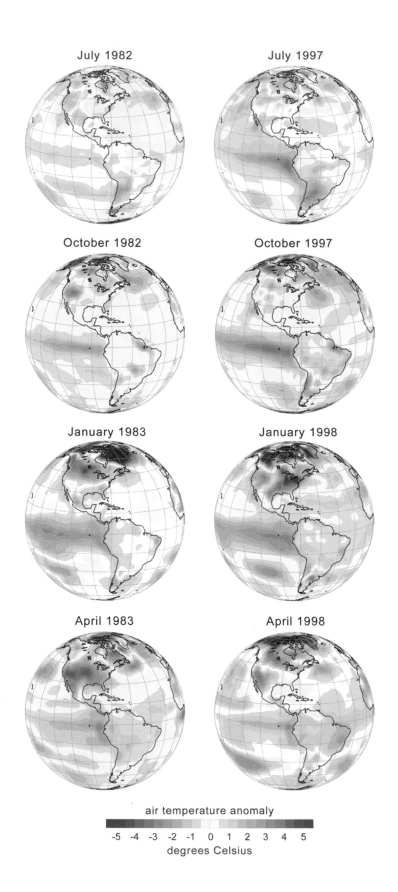

July 1982   July 1997

October 1982   October 1997

January 1983   January 1998

April 1983   April 1998

air temperature anomaly

-5 -4 -3 -2 -1 0 1 2 3 4 5
degrees Celsius

region	Warm (>1°C above normal)				Neutral (< ±1°C from normal)				Cool (>1°C below normal)			
	Jul	Oct	Jan	Apr	Jul	Oct	Jan	Apr	Jul	Oct	Jan	Apr
1	60%	70%	60%	40%	40%	30%	40%	50%	0%	0%	0%	10%
2	40%	20%	30%	40%	60%	80%	70%	60%	0%	0%	0%	0%
3	20%	40%	30%	30%	70%	30%	70%	70%	10%	30%	0%	0%
4	40%	10%	20%	30%	50%	20%	60%	60%	10%	70%	20%	10%
5	0%	0%	0%	20%	90%	100%	90%	50%	10%	0%	10%	30%
6	50%	10%	10%	20%	20%	70%	20%	0%	30%	20%	70%	80%
7	10%	20%	30%	20%	90%	60%	20%	40%	0%	20%	50%	40%
8	10%	20%	60%	20%	60%	30%	10%	40%	30%	50%	30%	40%
9	50%	40%	60%	40%	30%	30%	30%	30%	20%	30%	10%	30%
10	20%	40%	60%	30%	60%	20%	10%	70%	20%	40%	30%	0%

TABLE 1 Summary of the temperature anomalies (relative to long-term mean) for the last ten El Niño events in the Western Hemisphere in each of the ten regions labeled in figure 12

region	Wet (>1mm/day above normal)				Neutral (< ±1mm/day from normal)				Dry (>1mm/day below normal)			
	Jul	Oct	Jan	Apr	Jul	Oct	Jan	Apr	Jul	Oct	Jan	Apr
1	10%	10%	40%	40%	90%	90%	60%	40%	0%	0%	0%	20%
2	0%	30%	30%	30%	100%	70%	70%	70%	0%	0%	0%	0%
3	0%	10%	0%	10%	80%	30%	50%	30%	20%	60%	40%	60%
4	0%	0%	10%	0%	100%	100%	80%	100%	0%	0%	10%	0%
5	30%	20%	20%	40%	30%	40%	20%	40%	40%	40%	60%	20%
6	20%	10%	10%	10%	30%	80%	90%	90%	50%	10%	0%	0%
7	60%	30%	30%	50%	10%	50%	70%	40%	30%	20%	0%	10%
8	20%	20%	10%	20%	60%	70%	80%	50%	20%	10%	10%	30%
9	10%	30%	30%	20%	90%	60%	50%	40%	0%	10%	20%	40%
10	10%	50%	10%	20%	80%	40%	60%	70%	10%	10%	30%	10%

TABLE 2 Summary of the precipitation anomalies (relative to long-term mean) for the last ten El Niño events in the Western Hemisphere in each of the ten regions labeled in figure 12.

1°C lower than the long-term mean, and *neutral* is within ±1°C of the long-term mean. For precipitation rate, *wet* means that the region is more than 1 mm/day higher than the long-term mean, *dry* is more than 1 mm/day lower than the long-term mean, and *neutral* is within ±1 mm/day of the long-term mean. I describe the response in El Niño years for each of these ten regions, and tables 1 and 2 summarize temperature and precipitation anomaly data for a seasonal cycle centered on the peak of the event. I examine July and October for the year in which the El Niño begins, and January and April of the following year as the El Niño matures and ultimately ends.

Region 1: Eastern Equatorial Pacific

In general the tropics and subtropics throughout the Western Hemisphere are warmer than normal in most El Niño years. The most obvious changes during any given El Niño are found in the equatorial Pacific and consist of an increase in both SST and near-surface air temperature. In July, the eastern equatorial Pacific was warm for 6/10 events, October was warm in 7/10

FIG. 14 (left) Western Hemisphere air-temperature anomalies associated with the 1982–83 and 1997–98 El Niño events

events, and January was warm in 6/10 events. Most other El Niño events were neutral, tending toward warm. Only once were near-surface air temperatures cool in this region (one of the ten Aprils). There tends to be increased precipitation in January and April (each wet in this region for 4/10 El Niño events). Nearly all other El Niño events were neutral tending toward wet.

Region 2: Coast of Ecuador and Northern Peru

Precipitation rate can increase dramatically, producing floods in a normally very dry desert. Three of the ten El Niño events were wet in October, January, and April; all the rest were neutral tending toward wet. In this region, as in the eastern equatorial Pacific, normal precipitation is so low that even slightly elevated rainfall represents a significant increase in precipitation. This region experienced neutral to warm conditions in all four seasons during all ten El Niño events.

Region 3: Eastern Brazil

For the most part, El Niño events had warm (2/10 in July, 4/10 in October, and 3/10 in January and April) or neutral temperatures (7/10 in July, 3/10 in October, and 7/10 in January and April). Drought occurs in eastern Brazil during more than half of the last ten El Niño events. In October, January, and April, 6/10, 4/10, and 6/10 El Niño events are dry, respectively.

Region 4: Central Chile

Temperature in central Chile is cool in October 7/10 El Niño events and neutral in January and April in 6/10 El Niño events. Precipitation anomalies are small (neutral) in July, October, January, and April in 10/10, 10/10, 8/10, and 10/10 El Niño events, respectively.

Region 5: Yucatan, Northern Central America, and the Western Caribbean

This region experiences very little temperature change during El Niño events while nearly half of these events are dry in July, October, and January. For October (and perhaps July) this may in part be due to a tendency for fewer Atlantic tropical storms and hurricanes to form during El Niño events.

Region 6: Mexico and the Southwestern United States

El Niño events are characterized as cool in January (7/10) and April (8/10) in northern Mexico, Arizona, and New Mexico. Precipitation is close to normal during most El Niño events in October (8/10), January (9/10), and April (9/10). During July, 5/10 El Niño events are dry and 3/10 are neutral.

Region 7: Southeastern United States

Increased wintertime storm activity results in wetter than normal October and January conditions in the southeast along the Gulf Coast states from Louisiana to Florida. Above-average summer and spring are also typical in this region during El Niño events (July is wet 6/10 and April 5/10 events). Nearly half of the El Niño events are cool in January (5/10) and April (4/10).

FIG. 15 (right) Western Hemisphere precipitation anomalies associated with the 1982–83 and 1997–98 El Niño events

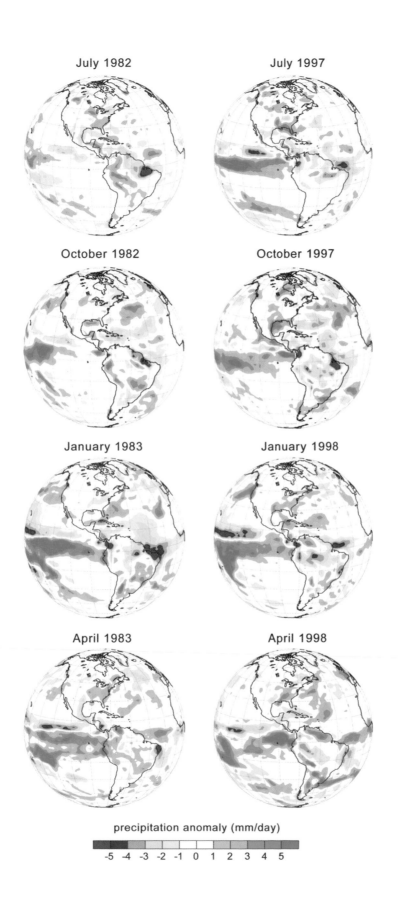

July 1982   July 1997

October 1982   October 1997

January 1983   January 1998

April 1983   April 1998

precipitation anomaly (mm/day)

-5 -4 -3 -2 -1 0 1 2 3 4 5

Region 8: Northeastern United States

October is more likely to be cool (5/10) and January more likely to be warm (6/10) during El Niño events (although January 2003 was one of the coldest in recent years). Most El Niño events are near normal in terms of precipitation. July, October, January, and April are neutral in terms of precipitation for 6/10, 7/10, 8/10, and 5/10 El Niño events, respectively.

Region 9: Pacific Northwest

During El Niño the Pacific northwest tends to be warmer than normal. July, October, January, and April are warm for 5/10, 4/10, 6/10, and 4/10 El Niño events, respectively. Three of ten events are neutral in each of these months. Most El Niño events have above- to near-average precipitation; however, 3/10 events in October and January are wet, and April is dry during 4/10 events.

Region 10: Coast of the Gulf of Alaska

October tends to be wetter than normal (5/10 El Niño events), while January is slightly drier (3/10 El Niño events). Air temperature in this region is about as likely to be above as below normal in any given El Niño year except in January, when 6/10 El Niño events are warm.

SUMMARY AND CONCLUSION

As is the case with the annual cycle, no two ENSO cycles are alike. While the general teleconnection patterns in the Western Hemisphere do have some persistent features, there are also differences from one El Niño to the next. In this paper I have compared and summarized interannual climate variability in terms of temperature and precipitation anomalies in ten different regions during the ten El Nino events that have occurred over the past forty years. For this time span, there are reliable globally distributed instrumental records available to make such comparisons. Figures 13, 14, and 15 demonstrate the variable expression of ENSO-related phenomena by comparing sea-surface temperature, near-surface air temperature, and precipitation anomalies for the 1982–83 and 1997–98 El Niño events. These two events, the strongest recorded in the twentieth century, demonstrate how ENSO events are alike in some ways, but different in others.

Although the record prior to around 1960 is not as complete, there are enough instrumental data to quantify annual-interannual climate variability back about 100–150 years. In addition, it is possible to extend such variability back many millennia using historical and proxy records of climate.

The strength and frequency of El Niño events has indeed varied over decadal (Chavez et al. 2003; Sandweiss et al. 2004) and even longer time scales (e.g., Sandweiss et al. 1996, 2001). From the longer instrumental records it is clear that El Niño events in the 1940s were strong, followed by several decades of weaker events. Then the 1980s and 1990s were again characterized by very strong El Niño events. Reconstruction of El Niño using historical records has also demonstrated changes in the character of the ENSO cycle over many centuries (e.g., Quinn 1992).

Many researchers who investigate time intervals prior to the instrumental record are interested in possible connections between societal or ecological change and climate. While it is my hope that the correlations drawn between El Niño and the climatic conditions in any specific region in the Western Hemisphere may be of use in this way, some caution must be taken considering the relatively short length of the meteorological data used in this study and the likelihood that teleconnections changed through time. In fact, because the ENSO cycle has evolved on time scales longer than this instrumental record, it is important to look also at the longer records of both climate change and El Niño.

REFERENCES CITED

Chavez, Francisco P., John Ryan, Salvador E. Lluch-Cota, and Miguel Ñiquen C.
2003 From Anchovies to Sardines and Back: Multidecadal Change in the Pacific Ocean. *Science* 299: 217–221.

Kistler, Robert, Eugenia Kalnay, William Collins, Suranjana Saha, Glenn White, John Woollen, Muthuvel Chelliah, Wesley Ebisuzaki, Masao Kanamitsu, Vernon Kousky, Huug van den Dool, Roy Jenne, and Michael Fiorino
2001 The NCEP–NCAR 50-Year Reanalysis: Monthly Means CD-ROM and Documentation. *Bulletin of the American Meteorological Society* 82 (2): 247–267.

Philander, S. George H.
1990 *El Niño, La Niña and the Southern Oscillation.* Academic Press, San Diego.

Quinn, William H.
1992 A Study of Southern Oscillation-Related Climatic Activity for A.D. 622–1900 Incorporating Nile River Flood Data. In *El Niño: Historical and Paleoclimatic Aspects of the Southern Oscillation* (Henry F. Diaz and Vera Markgraf, eds.): 119–149. Cambridge University Press, Cambridge.

Sandweiss, Daniel H., James B. Richardson III, Elizabeth J. Reitz, Harold B. Rollins, and Kirk A. Maasch
1996 Geoarchaeological Evidence from Peru for a 5000 Years B.P. Onset of El Niño. *Science* 273: 1531–1533.

Sandweiss, Daniel H., Kirk A. Maasch, Richard L. Burger, James B. Richardson III, Harold B. Rollins, and Amy Clement
2001 Variation in Holocene El Niño Frequencies: Climate Records and Cultural Consequences in Ancient Peru. *Geology* 29: 603–606.

Sandweiss, Daniel H., Kirk A. Maasch, Fei Chai, C. Fred T. Andrus, and Elizabeth J. Reitz
2004 Geoarchaeological Evidence for Multi-decadal Natural Climatic Variability and Ancient Peruvian Fisheries. *Quaternary Research* 61: 330–334.

Walker, Gilbert
1924 Correlations in Seasonal Variations in Weather, IX: A Further Study of World Weather. *Memoirs of the India Meteorological Department* 24: 275–333.

Wolter, Klaus, and Michael S. Timlin
1993 Monitoring ENSO in COADS with a Seasonally Adjusted Principal Component Index. In *Proceedings of the 17th Climate Diagnostics Workshop:* 52–57. NOAA/N MC/CAC, NSSL, Oklahoma Climate Survey, CIMMS and the School of Meteorology, University of Oklahoma, Norman, OK.

THE ANDES

CLIMATE CHANGE, EL NIÑO, AND THE RISE OF COMPLEX SOCIETY ON THE PERUVIAN COAST DURING THE MIDDLE HOLOCENE

James B. Richardson III
Daniel H. Sandweiss

For decades scholars have postulated that climatic and environmental change, El Niño, and tectonics were forcing mechanisms that have had both severe consequences and positive benefits for Andean coastal populations of the last thirteen thousand years. Nevertheless, the question remains open whether the early prehistory of coastal Peru shows such strong correlations between climatic, environmental, and cultural change that we can suspect a causal role for natural processes of change. We do not propose unilinear causation, much less environmental determinism, but we do believe that physical factors can and did play important roles in cultural change. Recent advances in the archaeology and paleoclimatology of the Peruvian coast provide new data on climatic change and related alterations of the distribution and abundance of natural resources available to the prehistoric inhabitants of Peru. In this context, we focus our discussion on Middle Holocene climatic change, El Niño, and the appearance of Late Preceramic monumental architecture on the Peruvian coast by considering two closely related questions:

1. In the 1960s, was Edward Lanning (1963, 1965, 1967) right to invoke climatic change to explain the cultural transformations that took place at the start of the Late Preceramic Period?
2. Given that maritime adaptations in the Andes began with the first inhabitants of the region and were not an invention of the Late Preceramic coast-dwellers (e.g., Sandweiss et al. 1998b, 1999), was Michael Moseley (1975, 1978, 1992) correct in the 1970s to identify intensive fishing as a critical new component of Late Preceramic socioeconomic organization?

We believe that both Lanning and Moseley correctly identified important factors in the initial development of complex societies of the Peruvian coast. However, only recently has paleoclimatic research provided the justification for including climatic change and intensive fishing in explanations of the early social transformations of coastal Peru. In brief, Lanning was right about increased marine productivity at the start of the Late Preceramic Period, and this increase supported the intensification of fishing identified by Moseley.

Though at first glance the questions raised above may appear unrelated to El Niño, the theme of this symposium, we will argue that changes in El Niño frequency—rather than the impact of single events—are the key to answering them. Therefore, we begin with a brief history of El Niño in Peruvian archaeology.

EL NIÑO IN PERUVIAN ARCHAEOLOGY

In the 1940s, Junius Bird (1948: 21) was the first archaeologist to state in print that Peruvian landscape alteration "can ultimately be correlated with the cultural record," in this case with the ceramics of Cupisnique through Chimú cultures of the north Peruvian coast. Bird based this assertion on evidence for five or six major floods from a Virú Valley river profile. Some years later, Larco Hoyle published similar observations; he may have influenced Bird's thinking on this issue, because the two interacted closely (Quilter, personal communication, 2005; Larco Hoyle 1963).

Based on his research around Ancón (see fig. 1 for location of sites mentioned in this chapter), Edward Lanning (1963, 1965) proposed that progressive dessication to 5800 cal yr BP[1] caused a shift from terrestrial *lomas* (fog-based plant community) hunting to a maritime way of life. We know now that maritime adaptations began long before the Late Preceramic Period (e.g., Keefer at al. 1998; Sandweiss et al. 1998b, 1999), but in Lanning's day, no one had reported earlier sites or considered bias in settlement patterns caused by sea-level change and the submergence of the continental shelf. Using Hutchinson's (1950) study of the rate of guano accumulation, Lanning (1963: 369–370) was the first to suggest that changes in the ocean circulation pattern of the Peru Current increased the productivity of marine resources, which he perceived as responsible for the establishment of coastal settlements by circa 5800 cal yr BP. Working at a time when few paleoclimatic records were available for coastal Peru, Lanning's use of climatic and oceanographic change to explain the archaeological record was a major, if little accepted breakthrough for the early 1960s. Indeed, in 1968 Craig and Psuty (1968: 126) referred to Lanning's ideas as a "… transparent adventure in environmental determinism."

In 1970 Mary Parsons argued that El Niño, rather than increasing aridity, had affected the lomas. She was the first archaeologist to present an in-depth discussion of El Niño and its potential for explaining cultural change.

---

1 Here and throughout the paper, we have converted radiocarbon years (yrs ^{14}C BP) to calendar years using Calib 4.2, based on datasets described by Stuiver et al. (1998a, 1998b) and available at http://www.calib.org. Our use of calendar years, which follows current practice, allows more effective cross-correlation of archaeological and paleoclimatic records, but differs from the radiocarbon chronology used in most of the earlier sources we cite.

FIG. 1 Location of sites mentioned in the text

In 1966 Michael Moseley (1975, 1978, 1992, n.d.) began working at Ancón on Lanning's and other Late Preceramic sites. He, too, incorporated El Niño into the explanatory framework that led to his Maritime Foundations of Andean Civilization hypothesis. This well-known idea credits intensive fishing and maritime gathering as the economic underpinnings of complex society on the Peruvian coast. From the beginning, however, Moseley also recognized the importance of agriculture, especially of "industrial" crops such as cotton and gourd that were important in the fishing technology.

Also stimulated by Lanning's research, James Richardson (1965, 1973, n.d.a; Lanning 1967: 54–55) began work in the Talara region in the mid-1960s. There, he found an extinct mangrove molluscan fauna that reflected wetter conditions prior to circa 5800 cal yr BP. Holocene beach-ridge studies also began at this time, and Richardson proposed that the ridges at

Chira and Colán on the north coast were the result of mega-Niños (Richardson 1981, 1983, n.d.b). In 1981, he also suggested that El Niño originated in the Holocene (Richardson 1981: 141).

In 1980 Moseley accepted a volunteer, Dan Sandweiss, whom he sent to investigate the Santa beach ridges. While at Santa, David Wilson told Sandweiss about a series of Preceramic sites he had discovered on the paleoshoreline fronting the Santa beach ridges. There, Sandweiss discovered more mollusks out of place (Sandweiss et. al. 1983). His continued research at the Ostra Base Camp and Ostra Collecting Station resulted in the first major discussion of a Mid-Holocene onset of El Niño. The debate on the origins of El Niño began in earnest in 1986 with the publication of "The Birth of El Niño: Geoarchaeological Evidence and Implications" in the new journal *Geoarchaeology*. In this article, Harold Rollins, Richardson, and Sandweiss proposed that El Niño was absent during the early millennia of human occupation in Peru, and that this climatic phenomenon began to affect cultural development only after 5800 years ago. Sandweiss, Rollins, and colleagues (Andrus et al. 2002; Reitz and Sandweiss 2001; Rollins et al. 1986, 1987, 1990; Sandweiss et al. 1983, 1996, 1997, 2001; Sandweiss and Rodriguez 1991; Sandweiss 1996a) also pioneered the use of mollusks and later fish to understand changes in ocean circulation and El Niño (see review in Sandweiss 2003). Moseley and his colleagues (Moseley et al. 1981, 1983; Nials et al. 1979) used their analyses of tectonics and El Niño flooding to understand Late Pre-Hispanic change in the Moche Valley on Peru's north coast. Since the 1980s, Moseley's group has been conducting research on El Niño flooding in the Ilo region on the southern Peruvian coast (Clement and Moseley 1991; deFrance et al. 2001; Keefer et al. 1998; Satterlee et al. 2001; Keefer and Moseley 2004; see Moseley and Keefer, this volume).

For many decades, then, archaeologists have used faunal displacements, beach-ridge studies, and flood deposits to incorporate El Niño into explanations of human adaptation and cultural change on the Peruvian coast. The precise dating, frequency, and severity of specific El Niño floods and droughts, however, remained elusive until coring of the Quelccaya Ice Cap by Lonnie Thompson and his colleagues provided a proxy record for the last 1500 years (Thompson et al. 1979, 1984, 1985, 1988). Theories on the collapse of Moche (Shimada et al. 1991) and Tiwanaku (Ortloff and Kolata 1993) drew causal inspiration from the Quelccaya ice-core record and, for the latter case, also from Lake Titicaca lake-level changes (Binford et al. 1997; Kolata et al. 2000; cf. Erickson 1999). Since the late 1980s there has been an "explosion" of archaeological research correlating El Niño with cultural shifts and collapse (e.g., Burger 1988; Fagan 1999; Moseley 1987, 1997; Richardson 1994; Sandweiss et al. 2001, Van Buren 2001; see also Billman and Huckleberry, Kiracofe and Marr, Roscoe, and Wilkerson, this volume).

THE MIDDLE AND LATE PRECERAMIC PERIODS OF THE PERUVIAN COAST

Dating between 8800 and 5800 cal yr BP, the Middle Preceramic Period of the Peruvian coast is poorly known. Sites such as Siches (e.g., Richardson 1973, 1978; Sandweiss et al. 1996), Ostra (e.g., Sandweiss 1996b; Sandweiss et al. 1983, 1996), Paloma (Quilter 1989), and the Ring Site (Sandweiss et al. 1989) were camps or villages of maritime hunters and gatherers without ceremonial architecture. Several highland sites on the western slopes of the Andes have small-scale public architecture, but nothing similar to the Late Preceramic sites of coast or highlands (Aldenderfer 1998; Dillehay 1992; Dillehay et al. 1989, 1997: 50). Although Middle Preceramic mounds may be located on the now-drowned continental shelf, we find this possibility unlikely because of the dearth of such structures elsewhere; where Middle Preceramic ceremonial activities have been identified, the scale and context are strikingly different from centers of the subsequent period.

In contrast, many more Late Preceramic Period sites are known, the sites are larger, and the scale of public architecture is much greater than in the Middle Preceramic Period. As Moseley (1992: 21–22) has pointed out, early monumental architecture is restricted to that part of the coast that today has the highest marine productivity as a result of the upwelling regime and the configuration of the continental shelf, from about 8° S latitude (Virú Valley) to just south of Lima at 12° S latitude.

The Late Preceramic Period dates from about 5800 to 4000 cal yr BP, and monumental construction probably began toward the beginning of the period (Feldman 1985; Haas et al. 2004). Consequently, the innovations of the Late Preceramic Period arose quite rapidly, with perhaps only centuries separating the earliest mounds from the massive constructions and complex organization of the recently excavated site of Caral (Shady Solís et al. 2001; Shady and Leyva eds. 2003; see Lanning 1967: 59; Burger 1992: 54; and Dillehay 1992: 57 for earlier assessments of the rapid development of the Late Preceramic temple centers). The apparent lack of Middle Preceramic antecedents for the early temple centers underscores the need to seek sources of change at the start of the Late Preceramic Period.

MIDDLE HOLOCENE CLIMATIC CHANGE

Today the north coast of Peru is washed by the cold Peru current to 4.0° S latitude, but in the Middle Holocene conditions between 10° and 4° S latitude were quite different (Reitz and Sandweiss 2001; Richardson 1973, 1978, 1981, 1998; Sandweiss 1986, 1996b, 2003; Sandweiss et al. 1983, 1996, 2001). Marine fauna from Middle Holocene, Middle Preceramic sites on the south coast of Ecuador (Las Vegas: Stothert 1985, 1988; Stothert et al. 2003) and the north coast of Peru as far south as the Casma Valley (Siches: Andrus et al. 2002; Sandweiss et al. 1996; Quebrada Chorrillos: Cárdenas Martín et al. 1993; Ostra: Andrus et al. 2002; Reitz and Sandweiss 2001; Sandweiss et al. 1996; Almejas: Pozorski and Pozorski 1995, 2003) provide faunal and geochemical evidence that we interpret to show a warmer-than-present ocean between 8800 and 5800 cal yr BP. At that latter date, mangrove mollusks disappeared from the archaeological record along the coast from Tumbes to Lambayeque (3.30° and 6.30° S latitude) and the central and north Peruvian coastal environment reverted to desert. Today, the southern extent of mangroves in southern Ecuador and northern Peru is at Puerto Pizarro (3.20° S latitude), but, between 11,000 and 5800 cal yr BP, the mangrove-bearing Preceramic sites were present as far south as the southern edge of the Llescus Peninsula at 6° S latitude (Cardenas et al. 1993; Richardson 1992, 1998; Sandweiss et al. 1996). In the Talara region, the Middle Preceramic Siches phase dates to 8800–5800 cal yr BP. The largest site of this phase, the Siches-type site at the head of Quebrada Siches, dates between 11,000 and 5500 cal yr BP and includes a component of the subsequent Honda phase (Richardson 1973, 1978; Sandweiss et al. 1996; unpublished data). This is the only known site from the northern coast of Peru that provides a stratigraphic sequence from the Middle to the early Late Preceramic Period. Our excavations in 1995 and 2001 have demonstrated that Siches maritime fisher-gatherers were adapted to a warm-water ocean regime. The following Late Preceramic Honda phase (5500–4200 cal yr BP) reflects the dramatic change from the warm-water fauna of Siches to the cold-water fauna of the present day Peru Current.

Just north of the Santa River near Chimbote, Peru, at 9° S latitude, Ostra Complex sites are situated on a paleoshoreline of a dry embayment, stranded from the ocean by a Holocene beach ridge set. Discovered in 1980 (Sandweiss 1986, 1996b; Sandweiss et al. 1983, 1996, 1998a; Reitz and Sandweiss 2001; Rollins et al. 1986), the Ostra Base Camp lies 5 km inland from the modern coast and dates between 7150 and 6200 cal yr BP. Both the fish and the mollusks are predominantly tropical species (Reitz and Sandweiss 2001). The mollusks were collected from the foreshore of the paleoshoreline, where many specimens are found in living position (Sandweiss et al. 1983). The Ostra complex provoked a debate as to whether the tropical mollusks lived in a closed shallow embayment warmed by the sun and therefore reflected local conditions (DeVries and Wells 1990; Sandweiss et. al. 1998a; Wells 1992, 1996; also Pozorski and Pozorski 2003), or whether the shellfish were living along the shore of the Mid-Holocene open ocean and therefore indicated oceanic conditions. Reitz and Sandweiss (Reitz and Sandweiss 2001; Sandweiss et al. 1998a) have now demonstrated that some of the tropical fish species came from the open ocean, thus proving that the Ostra people were exploiting resources that reflected open ocean conditions. Geochemistry of catfish otoliths for both Ostra and Siches showed that the ocean temperature was 3 to 4° C warmer than at present at circa 6800–7400 cal yr BP (Andrus et al. 2002), further supporting a warmer Mid-Holocene coastal current. However, Andrus and colleagues (2002) also show that coastal warming was seasonally structured, with warmer summers and similar to modern winters.

In addition to faunal elements from archaeological sites, several other lines of evidence support the hypothesis of a warm coastal current in northern Peru in the Mid-Holocene. Analysis of the soils and lomas plant community distributions indicates a marked difference in the environment north and south of 12° S latitude. The greater soil development and the lack of salt deposits reflect greater rainfall in the north (Noller n.d.). The endemism or geographic separation of plant communities in the southern lomas is the result of millennia of hyperaridity. In contrast, north of 12° S latitude, there are far greater similarities in the species composition of now-separate lomas clusters, suggesting that the northern lomas were continuous prior to 5800 cal yr BP as a result of increased precipitation (Rundel and Dillon 1998; Rundel et al. 1991; Sandweiss 2003; Dillon et al. 2003). Furthermore, the preservation of perishable artifacts and human burials is very poor north of Lima (12° S) until after 5800 cal yr BP, while even early sites in the south can show astounding preservation (e.g., Arriaza 1995; Sandweiss et al. 1997). Like the soil and plant data, this preservation pattern suggests long-term hyperaridity south of Lima (12° S latitude) and more frequent precipitation north of Lima and prior to 5800 cal yr BP.

Some geographic models suggest that the continental shelf exposed by lowered sea level until about 6000 cal yr BP may have been more amenable to human life than the adjacent coastal zones that are still above water (Faure et al. 2002; see also Roscoe's [this volume] consideration of the possible effects of meandering rivers on the exposed coastal margin). However, the sparse available data from terrestrial sites suggest that Middle Preceramic coastal populations consisted of small groups of fisher-gatherers who began settling in villages but who did not build monumental architecture (e.g., Benfer 1990). This characterization contrasts markedly with the Late Preceramic record.

Returning to the paleoclimatic record, after 5800 cal yr BP, the northern coast cooled and El Niño started up after a hiatus of at least three millennia (e.g., Sandweiss et al. 1996). However, the distribution of molluscan species in northern Peruvian sites suggests strongly that El Niño occurred much less frequently than today. El Niño frequency similar to today, with events as often as every three to seven years, did not begin until around 3200–2800 cal yr BP, at the end of the Initial Period (Sandweiss et al. 2001). In summary, then, we see the following history of El Niño frequency since the arrival of humans in Peru:

1) El Niño was present but of unknown frequency from 13,000–8800 cal yr BP during the Early Preceramic Period;
2) El Niño was absent or extremely rare from 8800–5800 cal yr BP during the Middle Preceramic Period;
3) El Niño was infrequent from 5800–3200 cal yr BP, during the Late Pre-Ceramic and Initial periods, with recurrence intervals of perhaps 50 to 100 years;
4) El Niño attained modern recurrence intervals only after 3200 cal yr BP.

Though derived initially from the faunal contents of Peruvian coastal sites, this reconstruction is now strongly supported by a variety of proxy records throughout the Pacific basin. These records include, among others, lake records from the Galapagos Islands (Riedinger et al. 2002), from mid-latitude Chile (Jenny et al. 2002; Maldonado and Villagrán 2002), from New Zealand (Shulmeister and Lees 1995), and from Australia (McGlone et al. 1992), and coral records from the western Pacific (Gagan et al. 1997, 1998; Tudhope et al. 2001). Recently, some climate models run for the Mid-Holocene in the eastern equatorial Pacific have produced results in accordance with the above interpretation of the paleodata (e.g., Clement et al. 2000; Liu et al. 2000; Sun 2000).

The major climatic transitions occurred at ca. 5800 and 3000 cal yr BP, and both are correlated with important cultural change: the beginning of Late Preceramic monumental construction shortly after 5800 cal yr BP and the abandonment of Initial Period temple centers shortly after 3000 cal yr BP. We focus on the first transition; Roscoe (this volume) discusses both of them.

MARITIME FOUNDATIONS

Returning to explanation of the archaeological record, the Maritime Foundations of Andean Civilization hypothesis (Moseley 1975, 1978, 1992) engendered heated discussions on the priority of maritime versus agricultural resources soon after its 1975 publication (Osborn 1977; Quilter and Stocker 1983; Raymond 1981; Wilson 1981). Although the excavation of numerous Late Preceramic sites had shown that domesticated plants provided food and raw material, it wasn't until Quilter's 1983 project at El Paraíso that the role of domesticated and wild plants and of marine animals could be quantified (Quilter 1991, 1992; Quilter et al. 1991). As expected, seafood provided virtually all of the animal protein consumed at the site. At the same time, the plants provided carbohydrates and essential nutrients, as well as raw materials such as cotton fiber. In 1992, Moseley reformulated the Maritime Hypothesis to take account of these new data.

From the mid-1980s to the late 1990s, few excavations were carried out on Late Preceramic coastal sites in the critical region of early temple centers. Recently, Ruth Shady's work at Caral (Shady and Leyva 2003, inter alia) has challenged the Maritime Hypothesis, and her findings have been supported by Haas et al.'s (2004) report of many more Late Preceramic monumental sites in the adjacent valleys of the north central coast.

Caral is located 23 km from the shore in the Supe Valley on the north central coast of Peru. It is the largest Late Preceramic site in South America, with six major pyramids, numerous smaller mounds, and various types of dwellings (Shady Solís et al. 2001; Shady and Leyva 2003). However, Caral is not the earliest monumental Late Preceramic center—on the basis of available data, that honor still goes to Aspero, a smaller site located next to the ocean in the same valley as Caral (Feldman 1985; Moseley and Willey 1973; Sandweiss and Moseley 2001), or possibly to some of the sites dated by Haas et al. (2004). Virtually all of the animal remains from the extensive excavations at Caral are marine, while cotton and gourds dominate the domesticated-plant assemblage (Shady Solís et al. 2001). We do not yet have equivalent data from the sites discovered by Haas et al. (2004), though their preliminary results show strong similarities to Caral. In short, the Late Preceramic monumental sites of the north central coast represent an early intensification of the pattern that led Moseley to formulate the Maritime Foundations hypothesis in the first place.

The Maritime Foundations hypothesis intends to explain the increasingly well-supported evidence for demographic growth and architectural intensification in the Late Preceramic Period. Implicit in all versions of the hypothesis is the idea that ancient Peruvians only began harvesting the rich ocean resources at the start of the Late Preceramic Period. In the 1990s, research in Early and Middle Preceramic sites such as Quebrada Jaguay (Sandweiss et al. 1998b), Quebrada Tacahuay (Keefer et al. 1998), the Ring Site (Sandweiss et al. 1989), Ostra (Sandweiss 1996b), and Siches (Richardson 1978; Sandweiss et al. 1996) demonstrated beyond doubt that even the first inhabitants of western South America included sophisticated fishermen who targeted particular species, built houses, and participated in long-distance exchange. Once again, the data force us to ask what changed between the Middle and Late Preceramic Periods. In other words, if marine-resource use was critical to the rise of civilization, why didn't earlier fishermen build temple mounds?

CONCLUSIONS

The northward emplacement of cold water that accompanied the onset of El Niño at 5800 cal yr BP must have brought increased nutrient upwelling along the central and north coast of Peru. Andrus et al. (2002: 1510) write that: "[t]his would alter the trophic structure of fin-fish populations, so that lower trophic level species [such as anchoveta and sardine] would be proportionally less abundant in the early Mid-Holocene than at present, and the diversity and equitability of the taxa would be diminished."

Despite the occasional presence of anchoveta at Middle Preceramic sites, Andrus et al.'s (2002) analysis of published zooarchaeological data supports this observation. Lanning (1963) was right: along the stretch of Peruvian coast where monumental architecture began, oceanographic change at the start of the Late Preceramic Period did lead to a more productive marine ecosystem. Furthermore, these changes favored intensification by enhancing the populations of small schooling fish such as anchoveta and sardine that can be netted in tremendous quantities. This new potential for intensification of the fishery helps explain why monumental architecture did not arise earlier, during the preceding millennia of sophisticated maritime adaptations. In short, the data remove a major objection to the Maritime Foundations Hypothesis and therefore support Moseley's proposition in a new way.

Do the above observations on physical and biotic change around 5,800 calendar years ago explain the apparent changes in human organization that began then? By themselves, certainly not. We can identify other physical factors that may also have intervened, such as the increased proximity of fishing and farming/gathering populations in the valleys as the result of rising sea level through the Middle Preceramic Period and of the dessication of the desert interfluves at the start of the Late Preceramic Period. Though virtually all animal protein in Late Preceramic coastal sites came from the ocean, the increase in the number of domesticated plants also surely played a role (Quilter 1991), perhaps in part by creating horizontal stratification between groups of fishing and farming specialists (Sandweiss 1996c). More important, as we recently wrote, "Technology, history, cultural practices, religion, perception, and individual and group idiosyncrasies can all affect the way a society and its members respond to change" (Sandweiss et al. 2001: 605). The onset of El Niño meant a significant change in interannual climatic variability that must have altered long- and short-term risk for different subsistence strategies and would most likely have affected the perceived relations between coastal societies and supernatural forces.

Regardless of the degree of social complexity implicit in the archaeological remains of Late Preceramic coastal Peru, any explanation for the transformations that took place at the beginning of that period will necessarily be complex, with a vast array of variables. Even then, unable to interview informants about causation, we can do no more than make models based on correlations. Our best strategy for the moment is to continue to identify important variables—such as change in climate and resource structure—and to refine our understanding of them.

REFERENCES CITED

Aldenderfer, Mark S.
1998 *Montane Foragers: Asana and the South-Central Andean Archaic.* University of Iowa Press, Iowa City.

Andrus C. Fred T., Douglas E. Crowe, Daniel H. Sandweiss, Elizabeth J. Reitz, and Christopher S. Romanek
2002 Otolith ∂¹⁸O Record of Mid-Holocene Sea Surface Temperatures in Peru. *Science* 295: 1508–1511.

Arriaza, Bernardo T.
1995 *Beyond Death: The Chinchorro Mummies of Ancient Chile.* Smithsonian Institution, Washington, D.C.

Benfer, Robert A., Jr.
1990 The Preceramic Site of Paloma, Peru: Bioindications of Improving Adaptation to Sedentism. *Latin American Antiquity* 1: 284–318.

Binford, Michael M., Alan L. Kolata, Mark Brenner, John W. Janusek, Matthew T. Seddon, Mark B. Abbott, and Jason H. Curtis
1997 Climate Variation and the Rise and Fall of an Andean Civilization. *Quaternary Research* 47: 235–248.

Bird, Junius B.
1948 Preceramic Cultures in Chicama and Virú. In *A Reappraisal of Peruvian Archaeology* (Wendell C. Bennett, comp.): 21–28. Memoirs of the Society for American Archaeology 4, Supplement to American Antiquity 13 (4), part 2. Society for American Archaeology and the Institute of Andean Research, Menasha, WI.

Burger, Richard L.
1988 Unity and Heterogeneity within the Chavin Horizon. In *Peruvian Prehistory: An Overview of Pre-Inca and Inca Society* (Richard W. Keatinge, ed.): 99–144. Cambridge University Press, Cambridge.

1992 *Chavín and the Origins of Andean Civilization.* Thames and Hudson, London.

Cárdenas Martín, Mercedes, Judith Vivar Anaya, Gloria Olivera de Bueno, and Blanca Huapaya Cabrera
1993 *Materiales arqueológicos del Macizo de Illescas Sechura-Piura: Excavaciones en Bayóvar, Nunura, Avic, Reventazón y Chorillos.* Pontificia Universidad Católica del Perú, Dirección Académica de Investigación, Lima.

Clement, Amy C., Richard Seager, and Mark A. Cane
2000 Suppression of El Niño during the Mid-Holocene by Changes in the Earth's Orbit. *Paleoceanography* 15: 731–737.

Clement, Christopher O., and Michael E. Moseley
1991 The Spring-Fed Irrigation System of Carrizal, Peru: A Case Study of the Hypothesis of Agrarian Collapse. *Journal of Field Archaeology* 18: 425–442.

Craig, Alan K., and Norbert P. Psuty
1968 *The Paracas Papers: Studies in Marine Desert Ecology,* vol. 1, no. 2: *Reconnaissance Report.* Occasional Publication 1 of the Department of Geography, Florida Atlantic University, Boca Raton.

deFrance, Susan D., David K. Keefer, James B. Richardson III, and Adán Umire Alvarez
2001 Late Paleo-Indian Coastal Foragers: Specialized Extractive Behavior at Quebrada Tacahuay, Peru. *Latin American Antiquity* 12: 413–426.

DeVries, Thomas J., and Lisa E. Wells
1990 Thermally-Anomalous Holocene Molluscan Assemblages from Coastal Peru: Evidence for Paleogeographic, Not Climatic Change. *Palaeogeography, Palaeoclimatology, Palaeoecology* 81: 11–32.

Dillehay, Tom D.
1992 Widening the Socio-economic Foundation of Andean Civilization: Prototypes of Early Monumental Architecture. *Andean Past* 3: 55–65.

Dillehay, Tom D., Patricia J. Netherly, and Jack Rossen
1989 Middle Preceramic Public and Residential Sites on the Forested Slope of the Western Andes, Northern Peru. *American Antiquity* 54: 733–759.

Dillehay, Tom D., Jack Rossen, and Patricia J. Netherly
1997 The Nanchoc Tradition: The Beginnings of Andean Civilization. *American Scientist* 85: 46–55.

Dillon, Michael O., M. Nakazawa, and Segundo Leiva González
2003 The Lomas Formations of Coastal Peru: Composition and Biogeographic History. In *El Niño in Peru: Biology and Culture over 10,000 Years* (Jonathan Haas and Michael O. Dillon, eds.): 1–9. Fieldiana Botany, n. s. 43. Field Museum of Natural History, Chicago.

Erickson, Clark L.
1999 Neo-environmental Determinism and Agrarian "Collapse" in Andean Prehistory. *Antiquity* 73: 634–642.

Fagan, Brian M.
1999 *Floods, Famines, and Emperors: El Niño and the Fate of Civilizations.* Basic Books, New York.

Faure, Hugues, Robert C. Walter, and Douglas E. Grant
2002 The Coastal Oasis: Ice Age Springs on Emerged Continental Shelves. *Global and Planetary Change* 33: 47–56.

Feldman, Robert A.
1985 Preceramic Corporate Architecture: Evidence for the Development of Non-Egalitarian Systems in Peru. In *Early Ceremonial Architecture in the Andes* (Christopher B. Donnan, ed.): 71–92. Dumbarton Oaks Research Library and Collection, Washington, D.C.

Gagan, Michael K., Linka K. Ayliffe, Sharon Anker, David Hopley, Malcolm T. McCulloch, Peter J. Isdale, John M. A. Chappell, and M. John Head
1997 Great Barrier Reef "Climatic Optimum" at 5800 y BP. *PAGES (Past Global Changes) Newsletter* 5: 15.

Gagan, Michael K., Linda K. Ayliffe, David Hopley, Joseph A. Cali, Graham E. Mortimer, John Chappell, Malcolm T. McCulloch, and M. John Head
1998 Temperature and Surface-Ocean Water Balance of the Mid-Holocene Tropical Western Pacific. *Science* 279: 1014–1018.

Haas, Jonathan, Winifred Creamer, and Alvaro Ruiz
2004 Dating the Late Archaic Occupation of the Norte Chico Region in Peru. *Nature* 432: 1021–1023.

Hutchinson, G. Evelyn
1950 *Survey of Contemporary Knowledge of Biogeochemistry,* vol. 3: *The Biogeochemistry of Vertebrate Excretion.* Bulletin of the American Museum of Natural History 96. American Museum of Natural History, New York.

Jenny, Bettina, Blas L. Valero-Garcés, Rodrigo Villa-Martínez, Roberto Urrutia, Mebus Geyh, and Heinz Veit
2002 Early to Mid-Holocene Aridity in Central Chile and the Southern Westerlies: The Laguna Aculeo Record (34° S). *Quaternary Research* 58: 160–170.

Keefer, David K., and Michael E. Moseley
2004 Southern Peru Desert Shattered by the Great 2001 Earthquake: Implications for Paleoseismic and Paleo–El Niño–Southern Oscillation Records. *Proceedings of the National Academy of Sciences of the United States of America* 101: 10,878– 10,883.

Keefer, David K., Susan D. deFrance, Michael E. Moseley, James B. Richardson III, Dennis R. Satterlee, and Amy Day-Lewis
1998 Early Maritime Economy and El Niño Events at Quebrada Tacahuay, Peru. *Science* 281: 1833–1835.

Kolata, Alan L., Michael W. Binford, Mark Brenner, John W. Janusek, and Charles R. Ortloff
2000 Environmental Thresholds and the Empirical Reality of State Collapse: A Response to Clark Erickson, Neo-environmental Determinism and Agrarian "Collapse" in Andean Prehistory [Erickson 1999]. *Antiquity* 74: 424–426.

Lanning, Edward P.
1963 A Pre-agricultural Occupation on the Central Coast of Peru. *American Antiquity* 28: 360–371.

1965 Early Man in South America. *Scientific American* 213 (4): 68–76.

1967 *Peru before the Incas.* Prentice-Hall, Englewood Cliffs, NJ.

Larco Hoyle, Rafael
1963 *Las épocas peruanas.* Museo Arqueológico, Lima.

Liu, Zhengyu, John Kutzbach, and Lixin Wu
2000 Modeling Climate Shift of El Niño Variability in the Holocene. *Geophysical Research Letters* 27: 2265–2268.

Maldonado, Antonio, and Carolina Villagrán
2002 Paleoenvironmental Changes in the Semiarid Coast of Chile (32°S) during the Last 6200 cal years Inferred from a Swamp-forest Pollen Record. *Quaternary Research* 58: 130–138.

McGlone, Matt S., A. Peter Kershaw, and Vera Markgraf
1992 El Niño/Southern Oscillation Climatic Variability in Australasian and South American Paleoenvironmental Records. In *El Niño: Historical and Paleoclimatic Aspects of the Southern Oscillation* (Henry F. Diaz and Vera Markgraf, eds.): 419–433. Cambridge University Press, Cambridge, England.

Moseley, Michael E.
1975 *The Maritime Foundations of Andean Civilization.* Cummings, Menlo Park, CA.

1978 *Pre-Agricultural Coastal Civilizations in Peru.* Carolina Biology Readers, Carolina Biological Supply Company, Burlington, NC.

1987 Punctuated Equilibrium: Searching the Ancient Record for El Niño. *Quarterly Review of Archaeology* 8: 7–10.

1992 Maritime Foundations and Multilinear Evolution: Retrospect and Prospect. *Andean Past* 3: 5–42.

1997 Climate, Culture and Punctuated Change: New Data, New Challenges. *The Review of Archaeology* 18 (1): 19–28.

n.d. Changing Subsistence Patterns: Late Preceramic Archaeology of the Central Peruvian Coast. Ph.D. dissertation, Department of Anthropology, Harvard University, 1968.

Moseley, Michael E., and Gordon R. Willey
1973 Aspero, Peru: A Re-examination of the Site and Its Implications. *American Antiquity* 38: 452–468.

Moseley, Michael E., Robert A. Feldman, and Charles R. Ortloff
1981 Living with Crises: Human Perception of Process and Time. In *Biotic Crises in Ecological and Evolutionary Time* (Matthew H. Nitecki, ed.): 231–267. Academic Press, New York.

Moseley, Michael E., Robert A. Feldman, Charles R. Ortloff, and Luis Alfredo Narváez
1983 Principles of Agrarian Collapse in the Cordillera Negra, Peru. *Annals of Carnegie Museum* 52: 299–327.

Nials, Fred L., Eric E. Deeds, Michael E. Moseley, Shelia G. Pozorski, Thomas G. Pozorski, and Robert A. Feldman
1979 El Niño: The Catastrophic Flooding of Coastal Peru. Part 1: *Field Museum of Natural History Bulletin* 50 (7): 4–14; part 2: *Field Museum of Natural History Bulletin* 50 (8): 4–10.

Noller, Jay S.
n.d. Late Cenozoic Stratigraphy and
Soil Geomorphology of the Peruvian
Desert, 3 Degrees–18 Degrees S: A
Long-Term Record of Hyperaridity
and El Niño. Ph.D. dissertation,
Department of Geological Sciences,
University of Colorado, Boulder, 1993.

Ortloff, Charles R., and Alan L. Kolata
1993 Climate and Collapse: Agro-
Ecological Perspectives on the Decline
of the Tiwanaku State. *Journal of
Archaeological Science* 20: 195–221.

Osborn, Alan J.
1977 Strandloopers, Mermaids,
and Other Fairy Tales: Ecological
Determinants of Marine Resource
Utilization—The Peruvian Case. In
For Theory Building in Archaeology
(Lewis Binford, ed.): 157–205.
Academic Press, New York.

Parsons, Mary Hrones
1970 Preceramic Subsistence on the
Peruvian Coast. *American Antiquity* 35:
292–304.

Pozorski, Shelia, and Thomas Pozorski
1995 Paleoenvironment at Almejas, a
Mid-Holocene Site in the Casma Valley,
Peru. *Society for American Archaeology:
Abstracts of the 60th Annual Meeting:* 154.
Society for American Archaeology Press,
Washington, D.C.

2003 Paleoenvironments at Almejas:
Early Exploitation of Estuarine Fauna
on the North Coast of Peru. In *El
Niño in Peru: Biology and Culture
over 10,000 Years* (Jonathan Haas
and Michael O. Dillon, eds.): 52–70.
Fieldiana Botany, n. s. 43. Field Museum
of Natural History, Chicago.

Quilter, Jeffrey
1989 *Life and Death at Paloma: Society
and Mortuary Practices in a Preceramic
Peruvian Village.* University of Iowa
Press, Iowa City.

1991 Late Preceramic Peru. *Journal of
World Prehistory* 5: 387–438.

1992 To Fish in the Afternoon: Beyond
Subsistence Economies in the Study of
Early Andean Civilization. *Andean Past*
3: 111–126.

Quilter, Jeffrey, and Terry Stocker
1983 Subsistence Economies and the
Origins of Andean Complex Societies.
American Anthropologist 85: 545–562.

**Quilter, Jeffrey, Bernardino
Ojeda E., Deborah M. Pearsall,
Daniel H. Sandweiss, John G.
Jones, and Elizabeth S. Wing**
1991 The Subsistence Economy of El
Paraíso, Peru. *Science* 251: 277–283.

Raymond, J. Scott
1981 The Maritime Foundations of
Andean Civilization: A Reconsideration
of the Evidence. *American Antiquity* 46:
806–821.

**Reitz, Elizabeth J., and
Daniel H. Sandweiss**
2001 Environmental Change at Ostra
Base Camp, A Peruvian Preceramic
Site. *Journal of Archaeological Science* 28:
1085–1100.

Richardson, James B., III
1965 Sitios precerámicos del extremo
norte del Perú. *Boletín del Museo
Nacional de Antropología y Arqueologia*
(Lima) 4: 2.

1973 The Preceramic Sequence and
the Pleistocene and Post-Pleistocene
Climate of Northwest Peru. In
Human Variation (Donald W. Lathrap
and Jody Douglas, eds.): 73–89.
University of Illinois Press, Urbana.

1978 Early Man on the Peruvian North
Coast, Early Maritime Exploitation and
Pleistocene and Holocene Environment.
In *Early Man in America from a Circum-
Pacific Perspective* (Alan L. Bryan, ed.):
274–289. Occasional Paper 1 of the
Department of Anthropology, University
of Alberta. Archaeological Researches
International, Edmonton.

1981 Modeling the Development of
Sedentary Maritime Economies on the
Coast of Peru: A Preliminary Statement.
Annals of Carnegie Museum 50: 139–150.

1983 The Chira Beach Ridges, Sea Level Change, and the Origins of Maritime Economies on the Peruvian Coast. *Annals of Carnegie Museum* 52: 265–276.

1992 Early Hunters, Fishers, Farmers and Herders: Diverse Economic Adaptations in Peru to 4500 B.P. *Revista de Arqueología Americana* 6: 71–90.

1994 *People of the Andes.* Smithsonian Institution/St. Remy, Washington, D.C.

1998 Looking in the Right Places: Pre-5,000 B.P. Maritime Adaptations in Peru and the Changing Environment. *Revista de Arqueología Americana* 15: 33–56.

n.d.a The Preceramic Sequence and Pleistocene and Post-Pleistocene Climatic Change in Northwestern Peru. Ph.D. dissertation, University of Illinois Urbana-Champaign, 1968.

n.d.b Holocene Beach Ridges between the Chira River and Punta Pariñas, Northwest Peru, and the Archaeological Sequence. Paper presented at the 39th Annual Meeting of the Society for American Archaeology, Washington, D.C.

Riedinger, Melanie A., Miriam Steinitz-Kannan, William M. Last, and Mark Brenner
2002 A ~6100 ^{14}C yr Record of El Niño Activity from the Galapagos Islands. *Journal of Paleolimnology* 27: 1–7.

Rollins, Harold B., James B. Richardson III, and Daniel H. Sandweiss
1986 The Birth of El Niño: Geoarchaeological Evidence and Implications. *Geoarchaeology* 1: 3–15.

Rollins, Harold B., Daniel H. Sandweiss, Uwe Brand, and Judith C. Rollins
1987 Growth Increment and Stable Isotope Analysis of Marine Bivalves: Implications for the Geoarchaeological Record of El Niño. *Geoarchaeology* 2: 181–187.

Rollins, Harold B., Daniel H. Sandweiss, and Judith C. Rollins
1990 Mollusks and Coastal Archaeology: A Review. In *Archaeological Geology of North America* (Norman P. Lasca and Jack D. Donahue, eds.): 467–478. Centennial Special Volume 4. Geological Society of America, Boulder, CO.

Rundel, Philip W., and Michael O. Dillon
1998 Ecological Patterns in the Bromeliaceae of the Lomas Formations of Coastal Chile and Peru. *Plant Systematics and Evolution = Entwicklungsgeschichte und Systematik der Pflanzen* 212: 261–278.

Rundel, Philip W., Michael O. Dillon, Beatriz Palma, Harold A. Mooney, Sherry L. Gulmon, and James R. Ehleringer
1991 The Phytogeography and Ecology of the Coastal Atacama and Peruvian Deserts. *Aliso* 13: 1–50.

Sandweiss, Daniel H.
1986 The Beach Ridges at Santa, Peru: Uplift, El Niño, and Prehistory. *Geoarchaeology* 1: 17–28.

1996a Environmental Change and Its Consequences for Human Society on the Central Andean Coast: A Malacological Perspective. In *Case Studies in Environmental Archaeology* (Elizabeth J. Reitz, Lee Newsom, and Sylvia Scudder, eds.): 127–146. Plenum Publishing, New York.

1996b Mid-Holocene Cultural Interaction on the North Coast of Peru and Ecuador. *Latin American Antiquity* 7: 41–50.

1996c The Development of Fishing Specialization on the Central Andean Coast. In *Prehistoric Hunter-Gatherer Fishing Strategies* (Mark Plew, ed.): 41–63. Boise State University, Boise, ID.

2003 Terminal Pleistocene through Mid-Holocene Archaeological Sites as Paleoclimatic Archives for the Peruvian Coast. *Palaeogeography, Palaeoclimatology, Palaeoecology* 194: 23–40.

Sandweiss, Daniel H., and
Michael E. Moseley
2001 Amplifying Importance of New
Research in Peru. *Science* 294: 1651–1652.

Sandweiss, Daniel H., and
María del C. Rodríguez
1991 Moluscos marinos en la prehistoria
peruana: Un breve ensayo. *Boletín de
Lima* 75: 55–63.

Sandweiss, Daniel H., Harold B.
Rollins, and James B. Richardson III
1983 Landscape Alteration and
Prehistoric Human Occupation on the
North Coast of Peru. *Annals of Carnegie
Museum* 52: 277–298.

Sandweiss, Daniel H., James B.
Richardson III, Elizabeth J. Reitz,
Jeffrey T. Hsu, and Robert A. Feldman
1989 Early Maritime Adaptations in
the Andes: Preliminary Studies at the
Ring Site, Peru. In *Ecology, Settlement,
and History in the Osmore Drainage,
Peru* (Don S. Rice, Charles Stanish,
and Phillip R. Scarr, eds.). BAR
International Series 545 (i): 35–84.
B.A.R., Oxford.

Sandweiss, Daniel H., James B.
Richardson III, Elizabeth J. Reitz,
Harold B. Rollins, and Kirk A. Maasch
1996 Geoarchaeological Evidence from
Peru for a 5000 Years B.P. Onset of El
Niño. *Science* 273: 1531–1533.

1997 Determining the Early History of El
Niño: Response [to comments]. *Science*
276: 966–967.

Sandweiss, Daniel H., Kirk A.
Maasch, Daniel F. Belknap, James B.
Richardson III, and Harold B. Rollins
1998a Discussion of "The Santa Beach
Ridge Complex," by Lisa E. Wells. In
Journal of Coastal Research 12 (1): 1–17
(1996). *Journal of Coastal Research* 14:
367–373.

Sandweiss, Daniel H., Heather
McInnis, Richard L. Burger,
Asunción Cano, Bernardino Ojeda,
Rolando Paredes, María del Carmen
Sandweiss, and Michael Glascock
1998b Quebrada Jaguay: Early South
American Maritime Adaptations in
South America. *Science* 281: 1830–1832.

Sandweiss, Daniel H., David K.
Keefer, and James B. Richardson III
1999 First Americans and the Sea.
Discovering Archaeology 1 (1): 59–65.

Sandweiss, Daniel H., Kirk A.
Maasch, Richard L. Burger, James B.
Richardson III, Harold B. Rollins,
and Amy Clement
2001 Variation in Holocene El Niño
Frequencies: Climate Records and
Cultural Consequences in Ancient Peru.
Geology 29: 603–606.

Satterlee, Dennis R., Michael E.
Moseley, David K. Keefer, and
Jorge E. Tapia A.
2001 The Miraflores El Niño Disaster:
Convergent Catastrophes and Prehistoric
Agrarian Change in Southern Peru.
Andean Past 6: 95–116.

Shady, Ruth, and Carlos Leyva (eds.)
2003 *La ciudad sagrada de Caral-
Supe: Los orígenes de la civilización andina
y la formación del Estado prístino en el
antiguo Perú.* Instituto Nacional de
Cultura, Proyecto Especial Arqueológico
Caral-Supe, Lima.

Shady Solís, Ruth, Jonathan
Haas, and Winifred Creamer
2001 Dating Caral, a Preceramic Site in
the Supe Valley on the Central Coast of
Peru. *Science* 292: 723–726.

Shimada, Izumi, Crystal B.
Schaaf, Lonnie G. Thompson,
and Ellen Mosley-Thompson.
1991 Cultural Impacts of Severe
Droughts in the Prehistoric Andes:
Application of a 1,500-year Ice Core
Precipitation Record. *World Archaeology*
22 (2): 247–270.

Shulmeister, James, and Brian G. Lees
1995 Pollen Evidence from Tropical Australia for the Onset of an ENSO Dominated Climate at Circa 4000 yrs BP. *The Holocene* 5: 10–18.

Stothert, Karen E.
1985 The Preceramic Las Vegas Culture of Coastal Ecuador. *American Antiquity* 50: 613–637.

1988 *La prehistoria temprana de la Península de Santa Elena, Ecuador: Cultura Las Vegas*. Miscelánea Antropológica Ecuatoriana, Serie Monográfica 10. Museos del Banco Central de Ecuador, Guayaquil, Ecuador.

Stothert, Karen E., Dolores R. Piperno, and Thomas C. Andres
2003 Terminal Pleistocene/Early Holocene Human Adaptation in Coastal Ecuador: The Las Vegas Evidence. *Quaternary International* 109–110: 23–43.

Stuiver, Minze, Paula J. Reimer, Edouard Bard, J. Warren Beck, Geoffrey S. Burr, Konrad A. Hughen, Bernd Kromer, Gerry McCormac, Johannes van der Plicht, and Marco Spurk
1998a INTCAL98 Radiocarbon Age Calibration 24,000–0. *Radiocarbon* 40: 1041–1083.

Stuiver, Minze, Paula J. Reimer, and Thomas F. Braziunas
1998b High-precision Radiocarbon Age Calibration for Terrestrial and Marine Samples. *Radiocarbon* 40: 1127–1151.

Sun, Dezheng Z.
2000 Global Climate Change and El Niño: A Theoretical Framework. In *El Niño and the Southern Oscillation: Multiscale Variability and Global and Regional Impacts* (Henry F. Diaz and Vera Markgraf, eds.): 443–463. Cambridge University Press, Cambridge, England.

Thompson, Lonnie G., Stefan Hastenrath, and Benjamin Morales Arnao
1979 Climatic Ice Core Records from the Tropical Quelccaya Ice Cap. *Science* 203: 1240–1243.

Thompson, Lonnie G., Ellen Mosley-Thompson, and Benjamin Morales Arnao
1984 El Niño–Southern Oscillation Events Recorded in the Stratigraphy of the Tropical Quelccaya Ice Cap, Peru. *Science* 226: 50–53.

Thompson, Lonnie G., Ellen Mosley-Thompson, John F. Bolzan, and Bruce R. Koci
1985 A 1500-Year Record of Tropical Precipitation in Ice Cores from the Quelccaya Ice Cap, Peru. *Science* 229: 971–973.

Thompson, Lonnie G., Mary Davis, Ellen Mosley-Thompson, and Kam-biu Liu
1988 Pre-Incan Agricultural Activity Recorded in Dust Layers in Two Tropical Ice Cores. *Nature* 336: 763–765.

Tudhope, Alexander W., Colin P. Chilcott, Malcolm T. McCulloch, Edward R. Cook, John Chappell, Robert M. Ellam, David W. Lea, Janice M. Lough, and Graham B. Shimmield
2001 Variability in the El Niño–Southern Oscillation through a Glacial-Interglacial Cycle. *Science* 291: 1511–1517.

Van Buren, Mary
2001 The Archaeology of El Niño Events and Other "Natural" Disasters. *Journal of Archaeological Method and Theory* 8: 129–149.

Wells, Lisa E.
1992 Holocene Landscape Change on the Santa Delta, Peru: Impact on Archaeological Site Distribution. *The Holocene* 2: 193–204.

1996 The Santa Beach Ridge Complex. *Journal of Coastal Research* 14: 367–373.

Wilson, David J.
1981 Of Maize and Man: A Critique of the Maritime Hypothesis. *American Anthropologist* 83: 93–120.

CATASTROPHE AND THE EMERGENCE OF POLITICAL COMPLEXITY: A SOCIAL ANTHROPOLOGICAL MODEL

Paul Roscoe

Catastrophes and disasters are commonly thought of as natural events, when really they are social, defined in reference to their negative consequences on human populations. A hurricane at sea is not a disaster, simply a hurricane, and even if it comes ashore it remains a "natural" phenomenon unless and until it harms human interests. The degree to which it becomes catastrophic is then determined by the technological and organizational characteristics of the communities it strikes. High seas may be little more than an interesting spectator event to foragers able to move at little cost to higher ground or to a society well protected by sea walls and overflow channels. Only if they strike permanent settlements in unprotected lowlands, as in the delta regions of Bangladesh, do they become catastrophic. Disasters, in short, are social as much as natural phenomena, defined by a society's technological, economic, social, and cultural structures.

As this volume recognizes, anthropology has paid only limited attention to the relationship between society and catastrophe, in particular to the influence of the latter on the evolution and devolution of the former. As van Buren (2001: 141) rightly laments, "we have learned much more about the geophysical parameters of prehistoric natural hazards than we have about their relation to and impact on past societies." To help fill this gap, this chapter develops a practice-based model of how political centralization is constructed in prestate society and of how these processes might be enabled or constrained by catastrophic events. It examines how political leaders create power; how prevailing environmental and cultural contexts shape the nature of the political hierarchies that result (whether they are more despotic than managerial, more secular than theocratic, and so on); and of how natural calamities affect these processes. To illustrate its utility, the model is then applied to two episodes from Peruvian prehistory. First, it is used to develop ideas that Richardson and Sandweiss (this volume, inter alia) and their colleagues have sketched about the role of sea-level rise and catastrophe in the origins of political complexity at the beginning of the Late Preceramic

Period. Second, it is used to generate a hypothesis of how the development of ENSO (El Niño Southern Oscillations) events on the coast could have been exploited by political entrepreneurs in the highlands, a hypothesis that makes some sense of the spatial chronology associated with the rise and fall of political centers in coastal and highland Peru.

POLITICAL EVOLUTION AND CATASTROPHE

To date, theoretical analysis of the role of catastrophe in prehistoric social process has been incidental rather than central to general theorizing about political evolution. Based on a premise that the emergence and development of political complexity is a social adaptation, adaptational (also, managerial, functional) models of political evolution argue that catastrophe promotes political complexity by stimulating the emergence or development of political institutions capable of preempting or managing the consequences. Moseley (1975), Osborn (1977), and Yesner (1980) have argued, for example, that political complexity emerged on the Peruvian coast as a means of buffering periodic declines of marine productivity caused by El Niño and other catastrophes.

In these adaptational models, the emergent organizations that supposedly "function" to increase community well-being are commonly represented as somehow developing independently of intentional human action through Darwinian or Darwinian-like processes, a stance that leaves the nature of the relevant political and social processes unclear. Over the last fifteen to twenty years, however, another line of theorizing has developed, sometimes dubbed the "political approach," that is in better accord with what social anthropology has learned about human political action. According to this approach, humans are self-interested agents who, within a culturally specific context of rules and resources, manipulate, strategize, and maneuver to advance their personal and group agendas. Political complexity is the product of power struggles among political entrepreneurs (variously called Accumulators, Aggrandizers, Achievers, or "Triple-A personalities") for control over people and resources.

According to this approach, and in contrast to the claims of adaptational models, a population does not necessarily benefit from the emergence of complexity: typically, in fact, it ends up exploited by its leaders and may have to be coerced into accepting their control. From this perspective, catastrophe might seem inimical to political evolution, but Arnold (1993: 102–107) has pointed out that it could be crucial to it. Conditions of environmental, military, or political stress, she argues, are the prime if not the only circumstances in which people can be induced to go along with an exploitative leadership because they recognize that otherwise they will suffer even worse consequences.

Accumulator, Aggrandizer, and Achiever models are rather crude in light of what we know about power and the construction of political dominance. For one thing, they explain almost nothing about the development of structure and its relationship to individual action: structure gets taken for granted as setting the "rules of the game," the values and rewards that motivate the action of political entrepreneurs, and the norms that regulate it. Practice theory (e.g., Bourdieu 1977; Giddens 1984) seeks to solve this problem with various models of the recursive nature of human practices, but the interaction of structure and action remains a difficult issue for political evolutionists (Weissner 2002).

A further problem with Accumulator, Aggrandizer, and Achiever models is that they tend to reduce power to economics: power becomes a derivative of economic indebtedness, the expropriation of foods and/or prestige goods, or storage (e.g., Arnold 1996: 93, 108; Brumfiel and Earle 1987; Cobb 1993; Hayden 1995; Testart 1982; Upham 1990). It is certainly the case that command of economic resources such as food, land, and wealth is a source of power, but there is overwhelming ethnographic evidence that political entrepreneurs make extensive use of still other types of resources. These include ideological resources (e.g., beliefs in spirits or sorcery), coercive resources (e.g., physical strength, advanced weaponry), social resources (e.g., membership in a powerful clan), cultural resources (e.g., a prestigious speech form, beauty), and symbolic resources (e.g., prestige, honor).

Finally, the political approach largely neglects the mechanics of how political entrepreneurs build power relations, the way in which environmental and cultural context affects the strategies and resources they deploy, and the consequences for the nature of the hierarchies they build. By focusing analysis on these mechanics and contexts, however, we can generate more realistic and productive models of human political action and of how catastrophe impacts it.

HOW TO BUILD A HIERARCHY: POPULATION DENSITY, POLITICAL INTERACTION, AND POWER

In the context of a practice theory of political evolution, I have dealt at length elsewhere with general issues concerning the construction of power relations and political hierarchy (Roscoe 1993, 2000). To begin with, it is important to differentiate prestige and hierarchies of value from power and political hierarchies. Prestige and its converse, opprobrium, are positive or negative appraisals of individuals that derive from evaluations of their actions in the context of a group's moral expectations. A hierarchy of value, which every society seems to construct, is the ranking of individuals in terms of their relative prestige. In contrast, power (or relational power), as Giddens puts it (1979: 88–94, 1984: 14–16, 257), is "the capacity to secure outcomes where the realization of those outcomes depends on the agency of others" (1979: 93). In more manageable terms, power is the capacity to get things done through the agency of others. Power, it should be emphasized, is a capacity, *not* a resource. Resources are the bases or vehicles of power, media such as sacredness, mental acuity, weaponry, land, and wealth that political actors deploy to achieve ends. To the extent that political entrepreneurs—would-be leaders—can habitually use the resources at their disposal to achieve their aims through the agency of specific others, they can be said to have constructed relations of power or a political hierarchy.

Political hierarchies vary significantly in "breadth" and "depth"—that is, in the *extent* of a leader's following (the *number* of followers controlled) and in its *effectiveness* (the *degree* of control over followers). At one end of the scale, leaders may control just a few people yet hold them in abject obeisance. Matoto, the pseudonymous leader of a Tairoran community in the eastern highlands of New Guinea, commanded no more than a couple of hundred people but exercised a despotic control over them, expropriating at his pleasure property, labor, and the sexual services of women, and occasionally even killing those in the community who offended him (Watson 1973). At the other end of the scale, leaders may command enormous followings yet exercise only slight control over any one member. The leaders of the 1967–72 Peli cult, a millenarian movement based in the Yangoru subdistrict of New Guinea (May 1982), secured tens of thousands of followers, but their control over any one individual was significantly less than that exercised by the average Yangoru Big-Man.

The extent, effectiveness, and nature of a political hierarchy depend crucially on some obvious but far from trivial conditions. To begin with, time budgets are crucial to building political hierarchies. If political entrepreneurs are to extend and exercise power over others, they must be able to interact with them: otherwise, they would be unable to deploy the resources at their disposal in order to manipulate others. It follows that the more time entrepreneurs can invest in these political interactions, the more followers they can recruit and/or the more effectively they can control them. The hours that an entrepreneur can invest in political action are limited by many factors, not least the attention he or she must devote to other, nonpolitical tasks such as subsistence, child rearing, and the like. Of the time that *can* be devoted to political action, however, the most critical constraint on the capacity to construct and extend power is the time that must be spent on bringing about political interactions.

Modes of communication and locomotion are pivotal in determining these interaction costs and hence the degree to which entrepreneurs are able to increase and expand their power. In advanced industrial states, technology radically reduces the costs of bringing about political interactions. Via mechanical forms of transport, leaders can use campaign bus tours or barnstorming rallies at airports to interact with tens of thousands of people spread over vast distances. Via electronic and print media—through TV political ads, telephone push-polls, and the like—they can manipulate millions more remotely. Absent these technologies, however, as in the early stages of political evolution, interaction is necessarily face-to-face and can only be brought about on foot. Under these circumstances the distribution of human population across the landscape—its density, degree of nucleation, and extent—crucially constrains or enables the manipulative ambitions of its leaders. In small, low-density, dispersed populations, political entrepreneurs must spend so much time traveling between interactions that they are unable to build significant hierarchies of power. Would-be leaders among the !Kung, for example, are faced with enormous travel costs if they are to interact with and build even a small following from a people scattered in small bands across the Kalahari at densities of about one person per square kilometer. The effort is hardly worthwhile, and it is no surprise that the !Kung are so egalitarian.

Things are very different where populations are larger, denser, and more nucleated. Under these conditions people live on an entrepreneur's doorstep, so to speak, facilitating interaction, manipulation, and the construction of power relations. On contact-era Tahiti, for instance, political entrepreneurs were surrounded by potential followers living at densities between 370 and 510 people/km². As we would expect, early European visitors to the island encountered powerful chiefs, some of whom commanded followings of seven thousand to fifteen thousand people (Roscoe 1993: 119).

RESOURCES, POLITICAL STRATEGIES, AND THE NATURE OF HIERARCHY

The characteristics of the resources that entrepreneurs can bring to political interactions are a second, critically important factor in the development of hierarchy. They affect the capacity of entrepreneurs to build a following, the way in which entrepreneurs must go about constructing power relations, and the nature of the hierarchy that results. The connections between resource characteristics and hierarchy are complex in the extreme, but two properties stand out for their consequences on political action. It matters greatly a) whether a resource functions to *attract and reward* followers or to *coerce and punish* them, and b) whether it is *natural* (i.e., material) or *supernatural* (i.e., magical or religious; table 1).

Whether a resource is attractive or coercive is crucial to the strategies political entrepreneurs must adopt and the interaction costs they must bear in building hierarchy. Attractive resources (e.g., land, food, sanctity, the promise of salvation) bolster power by functioning as rewards; coercive resources (e.g., physical strength, weaponry, sorcery, the threat of damnation) function as punishments. Attractive resources have the enormous advantage of permitting a leader to shift the costs of bringing about political interactions onto followers. Rewards—bread and circuses or millenarian promises—not only buy the support of the masses, they also typically induce them to converge on a central point where leaders or their agents can then interact with and manipulate them to consolidate or extend their power. A second advantage is that followers generally perceive rewards as enhancing their well-being. Since they can also refuse the reward should they wish, they are likely to view this sort of political action as moral and, if they accede to it, they will do so without resentment. Indeed, power exercised through reward traditionally has been defined as "authority"—as the "legitimate" use of power.

The principal advantage of coercive resources is that, in comparison to attractive resources, which followers are at liberty to refuse, coercive resources can be used to target and *compel* the action of others. The major disadvantages stem from precisely this compulsion: it generates distress in the target, and its use is therefore less likely than reward to be viewed as just or legitimate. (In contrast to "authority," some early definitions of power defined it as purely

or seizing by force sites already associated with ritual power, or by constructing monuments with the aesthetic power to conjure the presence of divinity.

To summarize the preceding discussions, four main conclusions can be drawn concerning the power-building enterprise in the early phases of political evolution, those stages in which hierarchy is most reliant on exogenous circumstances and that are of principal interest to this paper:

> a) Given the presence in a population of political entrepreneurs, increases in density and nucleation are sufficient in themselves to generate increases in political hierarchy because they reduce the costs of political interaction.
>
> b) To the extent that prevailing conditions tie people down, the power of leaders will rise and take on a coercive edge.
>
> c) To the extent that people are not tied down, entrepreneurs will prefer to deploy resources that are ideological and attractive rather than material and coercive; attractive material resources and coercive religious resources will occupy the middle ground.
>
> d) In comparison to political hierarchies built on coercion, those built on attractive religious resources are likely to be broad in their extent but limited in their effectiveness; hierarchies built on coercion are likely to be more effective in their control but more limited in their extent.

CATASTROPHE AND POLITICAL ACTION

Because of their suddenness and magnitude, natural disasters are usually presumed to generate radical social ruptures—switching a social system from one state to another (e.g., from a chiefdom back to an egalitarian system), or worse, simply annihilating it. It is common also to think of the social systems they hit as "societies," as discrete social units. In fact, societies as such are all but nonexistent: rather, humans live—and have lived for millennia—enmeshed in relational networks of exchange, trade, and warfare that transcend cultural gradients and extend across large parts of the planet. It is therefore useful to think of catastrophes as striking a few, localized links in a chain of social relationships that extends far beyond the site of destruction. It then becomes apparent that what may be a devastating event at some points in the chain may be advantageous at others.

In his remarkable ethnography of the hurricane that struck a large swath of the Mexican Gulf Coast, Wilkerson (this volume) graphically describes the effects at the disaster's epicenter. But he also points to the beneficial consequences for groups beyond the immediate sites of destruction, who were able to take advantage of the disruption in their neighbors' surveillance systems to rob and loot them of what was left of their property. Likewise, disasters can have both positive and negative effects on the creation of political hierarchy. At the center of devastation, the effects are usually negative. The cost of bringing about political interactions rises abruptly as people are scattered, die, or their time budgets get taken up with survival activities; material resources that entrepreneurs may have stored up for future political action may be destroyed; leadership claims to be favored by the gods may suffer a setback. Even at the center of destruction, though, all may not be lost. Material resources may be destroyed, but the disaster itself may be susceptible to cosmological exploitation. A good plague of locusts can be invaluable to an entrepreneur skillful enough to link it to claims of religious control. Arnold (1993) suggests that climatic disruptions were critical to the ability of Chumash leaders to consolidate control over their communities.

Most likely, though, political entrepreneurs beyond the immediate locus of destruction are the ones who profit most from a catastrophe. This was plausibly the case for the emergence and subsequent development of political complexity in Peru. The onset of ENSO events around 5800 cal yr BP may have interfered with the power-building activities of political entrepreneurs at the centers of devastation but the events offered new political resources and opportunities to entrepreneurs located further afield.

THE EMERGENCE OF POLITICAL COMPLEXITY IN COASTAL PERU

The Late Preceramic Period marked a time of striking social change on the coast of Peru (Burger 1995; Quilter 1991; Richardson 1994; Sandweiss 1996). Communities that had been low density, mobile, egalitarian hunter-fisher-gatherers in the Middle Preceramic Period now began to increase in population density, absolute size, and sedentism. As the foregoing arguments would predict, these changes in demography and settlement were associated with markers of increased political complexity, including an expansion of trade networks along the coast and, most conspicuously, the rise of monument building (e.g., Burger 1995; Haas et al. 2004; Moseley 1975; Richardson 1994; Shady Solís et al. 2001). At sites like Aspero (Feldman 1985) and others recently reported in the same region (Haas et al. 2004), monument construction may have been underway just a couple of centuries after the beginning of the Late Preceramic Period.

Sandweiss and Richardson (1999; Richardson 1981; Sandweiss 1996: 56–58; see also Raymond 1981) have suggested that the emergence and distribution of these early monuments were the product of the geomorphic effects of rising sea level on ancient coastal environments, coupled with the onset of ENSO events at the beginning of the Late Preceramic Period. Modern bathymetry suggests that 15,000 years ago the Peruvian coastal plain was as much as 100 km across. By the start of the Middle Preceramic Period, rising sea levels had reduced this width considerably, but it may still have been 10–15 km across in many places and perhaps 50 km wide in a few. Sandweiss and Richardson have pointed out that there is a rough correlation between variations in the width of this Middle Preceramic plain and the dates in the Late Preceramic Period when monument building appeared in different sections of the coast (fig. 1). The areas where monument building first emerged, between Salinas de Chao at 8° S and El Paraíso at around 12° S, were characterized in the Middle Preceramic Period by relatively

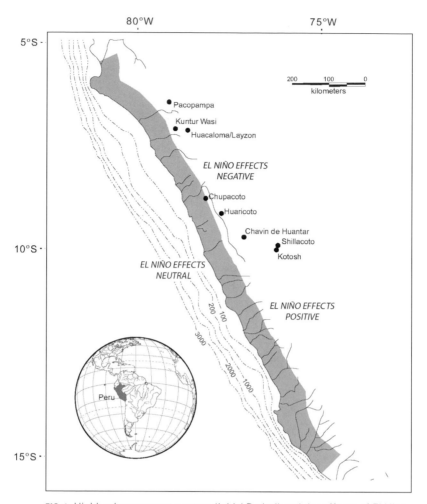

FIG. 1 Highland monument centers (Initial Period) and the effects of El Niño

wide coastal plains by Peruvian standards. In contrast, from the Rimac and Lurín valleys to the south, where the coastal plain was narrower, monument building did not appear until much later.

Sandweiss hypothesizes that the effects of the marine transgression up to 5800 cal yr BP on these geomorphological differences resulted in different historical trajectories at different locations along the coast. Where the coastal plain was narrow, the prevailing economy was limited to maritime hunting-gathering along the shore, coupled perhaps with terrestrial gathering farther inland; its transecting river valleys were too short and steep, and the rivers too fast flowing, to support significant levels of farming on the valley bottoms. Where the coastal plain was broad and the gradient of its transecting rivers shallow, however, valley farming was more feasible, and these valleys therefore may have supported two separate economies: a maritime-adapted economy on the coast and an incipient farming economy farther inland, on the coastal valley bottoms. If this was so, Sandweiss argues, then as sea levels rose through the Middle Preceramic Period and the coastal valleys were slowly inundated, these two economies eventually would have been brought together, producing the mixed valley-sea subsistence economy associated with early political complexity on the coast.

While it is plausible that the integration of fishing and agricultural economies would have synergistic political effects, how precisely would this work? The model advanced above suggests two chains of causation linked to increases in density and community size. First, rising sea levels would have forced a demographic packing sufficient in itself to generate political complexity. Where the coastal plain was broad, transecting river valleys were capable of supporting considerably greater populations than where it was narrow. The rivers themselves were longer, allowing more communities to exploit their banks, floodplains, and waters; and because their gradients were shallower, they were slower flowing, a condition conducive to the support of both aquatic faunal resources and alluvial farming. Where the plain was exceptionally broad and river gradients especially shallow, in fact, the development of meanders may have increased river lengths and decreased flows even further, allowing for the sustenance of even greater population levels.

As rising ocean levels gradually inundated the coast, the populations in these valleys would be forced to migrate landwards, resulting in demographic packing. Because refugee fluxes would be considerably greater among the large populations of the broader coastal plains than among the small populations of the narrower sections, however, this demographic packing would be more intense in the former sections than in the latter. As a result, political entrepreneurs in narrow sections of the plain would still be living among relatively small populations in which the costs of political interaction had changed only slightly over the Middle Preceramic. But those in broader sections would now find themselves living among large, high-density populations—the crucial feature in reducing interaction costs and facilitating the capacity to build power relations and mobilize labor for monument construction.

The second important consequence of the marine transgression was that to which Sandweiss draws attention, the integration in broader sections of the coastal plain of previously separated maritime and farming economies. Whether significant food cultivation of the valley floors was already underway at the beginning of the Late Preceramic Period is uncertain. Although there is little evidence to support the contention, Raymond (1981) makes the point that cultivation could have a much greater antiquity in coastal Peru than is commonly supposed because preservation in the material record is conspicuously biased against root crops. If this is correct and root crops were already being cultivated at the beginning of the Late Preceramic Period, then the sea-level rise that had caused demographic packing in broader sections of the coastal plain would have simultaneously provided the wherewithal to support it: bringing the procurement of marine resources into conjunction with incipient farming would have combined an extremely rich source of high-quality protein with a capacity for abundant carbohydrate production. Even were incipient root-crop cultivation not yet underway, this economic integration would at least allow for steep future increases in population. By the beginning of the Late Preceramic Period, in other words, political entrepreneurs in the broader sections of the coast may have been advantaged not only by a demographic packing that had already significantly reduced the costs of their political interaction but also by an economic integration capable of supporting additional population increases through intensification and thus of reducing these costs even further.

In sum, changes in population distributions during the Middle Preceramic Period may have been responsible for the differential fortunes of coastal entrepreneurs as they entered the Late Preceramic Period. Climate change then seems to have affected the political events that followed. To begin with, rainfall patterns began to change. In the Middle Preceramic Period, sea-surface temperature (SST) patterns (Andrus et al. 2002) indicate seasonal warming that would have caused summer rainfall on the north and north-central coast. Under these conditions, the populations of the coastal valleys could have dispersed to forage temporary florescences of vegetation on the intervalley uplands. Around 5800 cal yr BP, however, summer SSTs began to cool, ending these seasonal rains (Daniel H. Sandweiss, personal communication, October 2002). As a result, populations that had dispersed seasonally from the river valleys would now have been more permanently tied to them, increasing the manipulative and coercive capacities of valley entrepreneurs and their ability to mobilize labor.

With the Late Preceramic Period, though, catastrophe in the form of ENSO events also appeared as a variable of political consequence. Using molluscan and other data sources, Sandweiss and his colleagues (1996, 2001) find that, for several millennia prior to 5800 cal yr BP, El Niño events were either absent or very different from today. At about 5800 cal yr BP, however, major ENSO events begin to appear in the geoarchaeological record at a frequency of once every two or three generations. (Lesser episodes that did not affect marine biota as severely may have occurred even more often.)

These events affected sections of the coast in different ways (fig. 1). North of about the Casma and Supe valleys, they were probably devastating. If recent and historic ENSO events are any guide (e.g., Alcocer 1987 [1580]), they would have caused destructive flooding, declines in normal marine biota, plagues of insects, and disease. As Sandweiss (1996: 56–57) has noted, this may explain why monument building failed to get underway north of the Chao Valley until the Initial Period, even though the coastal plain in this region was broad, its river valleys presumably supported large populations in the Middle Preceramic Period, and its communities would have experienced a marked demographic packing as sea levels rose. Catastrophes of this scale would have eroded the ability of political entrepreneurs to construct and maintain hierarchy. Not only would their monument building and other political projects be precarious enterprises in areas subject to large-scale flooding, but the toll on population densities as followings were thinned by deaths and scattered by disaster would increase the costs of their political interactions, thereby decreasing their capacity to mobilize labor.

This detriment to political entrepreneurs in the north, however, served as an advantage to those farther south. In the Casma and Supe regions, the effects of ENSO events shade from devastating toward neutral, and from there on south they become increasingly positive. Rains are rarely intense enough to have much effect on irrigation systems, and changes in water temperature drive anchoveta and other species inshore, at least at the onset of the event, where they can be more easily harvested (Arntz 1986; Sandweiss 1995: 71, 1996: 56–57).

This pattern of periodic resource scarcity in the north and comparative abundance in the south may have produced north-to-south migration patterns in the wake of El Niño events (Sandweiss 1996: 57). If so, political entrepreneurs in the south would have gained political capacity as those in the north lost it. As refugees swelled local densities and populations, the costs of political interaction would have declined, while the scale of the available labor pool increased. Dispossessed of resources and dependent on the generosity of others, moreover, these immigrants would have been especially vulnerable to political exploitation, a phenomenon with ethnographic analogs among the Big-Men societies of highland New Guinea, where recurrent warfare rather than climatic catastrophe periodically put refugees to flight. To survive, these refugees had little option but to seek sanctuary with a neighboring Big-Man. Among the Melpa, for example, Big-Men provided refugees with land and piglets with which to start anew but, in return, those they took in were obliged to supply their labor "for years," looking after and tilling their gardens, fighting on their side as necessary, and sometimes becoming their bodyguards (Vicedom and Tischner n.d.: 68, 74–75; Ross, quoted in Mennis 1982: 177).

The appearance of monuments on the Peruvian coast was followed by a remarkable period of cultural florescence that lasted for over two and a half millennia, through the Late Preceramic Period to the end of the Initial Period. As the model above would predict, these political developments coincided with further increases in community size and density, fueled by events such as a rise in the abundance of fish stocks that accompanied a northward shift of the Humboldt Current (Richardson and Sandweiss, this volume).

Toward the end of the Initial Period, however, the trajectory of political development on the coast came to an abrupt end: the elaboration of coastal monuments halted and at least some sites were abandoned (Burger 1995: 183–184). As Sandweiss and his colleagues observe, these disruptions coincided with a shift in the frequency of major El Niño events. Where before they had occurred once every two or three generations, now they began to strike each decade, sometimes in a series of massive, closely spaced disasters (Richardson 1994: 87–89; Sandweiss et al. 2001: 605–606). By regularly cutting into population levels, damaging or destroying subsistence and residential sites, and scattering populations up and down the coast, this intensification of catastrophe may have been sufficient to increase the costs of political interaction to a level where they overwhelmed the capacity of political entrepreneurs to sustain hierarchy and to mobilize labor for monument construction.

EL NIÑO, RELIGIOUS EXPLOITATION, AND THE RITUAL CENTERS OF HIGHLAND PERU

The Late Preceramic and Initial periods saw the emergence and development of monument-building polities not only on the coast of Peru but also in the highlands, and using the model presented above I hypothesize that these were not independent events but rather the product of a ritual articulation fostered by the appearance of ENSO events on the coast. With the exception of Chavín de Huántar, our knowledge of highland monument building during these periods is limited, but the spatial distribution of construction activity shows an intriguing correlation with variations in the impact of El Niño along the coastline (fig. 1).

Several highland monument sites, including Pacopampa, Huacaloma, Layzón, and Kuntur Wasi, lie in the mountains directly above the sections of the coast that seem to have been most badly affected by ENSO events. Farther south, sites such as Chavín de Huántar, Shillacoto, and Kotosh were located above coastal sections where ENSO effects shaded from negative to neutral—though Chavín at least had ready access to the more seriously impacted sections farther north via the Santa Valley. Although El Niño effects were neutral *on average* along this section of the coast, it is unlikely that they were always neutral. Maasch (this volume) draws attention to significant temporal variations in temperature and precipitation anomalies that El Niño events have exhibited over the past forty years, and there are similar spatial variations from one event to the next (see for instance Billman and Huckleberry, this volume). In other words, this section of the coast doubtless experienced neutral effects during some El Niño events, but it was likely subjected to negative effects during others. Highland monument centers appeared above those sections of the coast subject to negative ENSO effects at greater or lesser frequencies. This contrasts to the situation yet farther south, along those sections of the coast where ENSO effects become positive. Here, no Late Preceramic monumental architecture has yet been discovered in the highlands, nor is there even evidence of large nucleated centers (Burger 1995: 124–127).

Many factors might account for this curious correlation, but we should not overlook the opportunities that El Niño events on the northern and central coast would have provided highland entrepreneurs in the mountains above. Prevailing scholarly opinion inclines to a view that Late Preceramic- and Initial-Period polities in the highlands—indeed everywhere in Peru at this time—were theocratic or ritual in nature. Richardson (1994: 90) has pointed to the absence of military fortification at highland centers, a feature to be expected if power were based on force. In the case of Chavín de Huántar, the best-known site of the Initial Period, there appears to have been no particular economic advantage or privileged access to mineral deposits or other natural resources that might account for its rise (Burger 1995: 129). Rather, the architecture, iconography, and internal artifacts of Chavín's monumental complex indicate that the site was primarily a religious center that had attracted pilgrims for hundreds of years (e.g., Burger 1995: 130, 163, 181; Kembel n.d.; Patterson 1971: 41–42; Richardson 1994: 80–81, 85–86; Rick 2004: 77).

The religious nature of highland monument centers accords with the argument advanced earlier that the optimal political strategy for securing a following in the early phases of political evolution is to deploy attractive religious resources rather than attractive or coercive material resources. That argument also specified, however, that in these early stages the success of political strategies based on ritual resources depends on exogenous circumstances that allow entrepreneurs to legitimate their claims to ritual power. If highland polities were religious in nature, then what circumstances might have allowed their entrepreneurs to deploy religious ideology to political effect?

Burger (1988: 141–142, also 1995: 190) has suggested that, at the end of the Initial Period, the expansion from the highlands of the symbolic culture known as the Chavín Horizon was connected to a "crisis cult" among coastal polities as they succumbed to the intensification of ENSO events on the coast. Whatever the merits of this argument as an explanation of the Chavín Horizon, the suggestion that catastrophes on the coast had political implications in the highlands is provocative. ENSO events had by this time been striking sections of the coast for over three millennia, albeit less frequently for most of that time, and the possibility that they constituted the exogenous conditions that facilitated the rise of Chavín is worth pursuing.

Consider how ENSO events and the destruction they wrought would have appeared to people on the coast. Like nonindustrial societies everywhere, they presumably attributed disruptive weather patterns to supernatural beings or magical action. In this case, an important diagnostic feature of the agency responsible would be the apparent provenience of the disruption. El Niño events originate far offshore to the west, but to people without a developed meteorological knowledge, the torrential rains and floodings that cause much of the damage and disorder would appear to originate in the highlands to the east. During El Niño years, rains that normally fall in the foothills or mountains appear to move down and out onto the coastal plain, while the destructive torrents of floodwater that course through the coastal valleys to the sea clearly emanate from the sierra side, not the ocean side.

For astute political entrepreneurs in the highlands, a perception that El Niños and their damage emanated from the mountains would be an exceptional opportunity to legitimate an ideological claim to power over the coastal people. By asserting that they embodied or possessed ritual power, they might present themselves as authors of ENSO events in order to leverage tribute or labor from the coastal population. The problem with coercion, however, as already noted, is that it generates resistance in its target: in this case, to claim responsibility for ENSO-related destruction would be to risk violent retribution from those affected. The preferable strategy, therefore, would be to manipulate coastal dwellers with an attractive resource. By shifting responsibility for events onto capricious cosmological powers in the mountains, for example, highland entrepreneurs might claim instead the ritual capacity to interpret the will of these powers and/or an ability to intercede with them on others' behalf.

Referring to the case of Chavín, Burger (1995: 156) has sketched how such a strategy could be made to work. Ethnographic evidence coupled with Chavín's iconography suggests that ancient Peruvians may have viewed the Amazonian lowlands as a site of ritual power.[2] If this were so, then highland entrepreneurs could profit politically by representing these eastern powers as the architects of ENSO disasters and then convincing others that they were capable, for a suitable consideration, of calming or diverting the wrath of these powers. The challenge would be to legitimate such a claim, to persuade others that one indeed possessed privileged knowledge of how to manage such a power.

An ethnographic analogy from the Yangoru subdistrict of New Guinea indicates that the location of highland sites, commanding the routes across the mountains from the Amazon to the ocean, may have been key. When European contact came to Yangoru in 1912, it was the good fortune of political entrepreneurs in Ambukanja, a village located in the high foothills of the Prince Alexander range, that they commanded the principal trade route across the coastal mountains to the main sites of European settlement at Boiken and Wewak. Capitalizing on this strategic position, Ambukanja quickly set about presenting itself to Yangoru as having privileged access to, knowledge of, and control over the European world. This legitimating claim allowed the village to construct at least six major millenarian movements over the next three decades, and it was not until the late 1940s, when a permanent European presence was established in Yangoru itself, that Ambukanja finally lost its monopoly and its ability to construct these followings (Roscoe 1988). Situated at natural choke points on major routes connecting the coast to the interior, highland entrepreneurs at places like Chavín may have been similarly positioned to legitimate claims to superior knowledge of, access to, and control over the powers of the Amazon and thus to exploit the peoples of the coast (Burger 1995: 128–129; Richardson 1994: 82).

As ENSO events began to affect long stretches of the coast during the Late Preceramic Period, therefore, the stage was set for strategically located polities in the highlands to develop parasitically on the coastal polities. For one thing, they were advantageously placed to exploit for their own gain the fears and misfortunes of those coastal populations subject to threatening or destructive ENSO events. For another, if ENSO events did indeed periodically scatter populations from the northern Peruvian coast, as suggested earlier, some displaced migrants may have moved east into the highlands as well as along the coast, swelling the size and density of northern highland settlements and facilitating the hierarchy-building efforts of their entrepreneurs. A whole string of political dominions may have linked the northern

2 Burger's assertion of the tropical lowland origins of Chavín religion has not gone unchallenged: Bischof (1994), Elera (1993), and Richardson (1994: 93–96) have argued that Chavín's architecture and iconography derive from the coast rather than the Amazonian lowlands. These arguments certainly cast doubt on theories that Chavín religion had its origins to the east of the Andes, but they do not thereby refute the proposition that early Peruvians viewed the Amazon as a source of ritual power. Indeed, the very fact that coastal culture incorporated tropical forest elements suggests that these elements were associated with power, and though these Peruvians may once also have inhabited the western Andean slopes, they might still have viewed the Amazon as their cosmological home.

highlands to the coast, with a ritual center in the mountains exerting influence over a number of polities on the coast below. If the arguments advanced earlier concerning the properties of ritual power are correct, these centers would have exerted limited control over any one coastal individual, no more perhaps than in the typical Melanesian millenarian movement, but the extent of their control may have been broad, drawing on the populations of several coastal valleys to which they had access.

Evidence for how long such polities might have been in place is scant. There is evidence of long-distance exchange between the coast and the highlands dating back to the Late Preceramic Period (e.g., Shady Solís 2005), and monument building was clearly underway at inland La Galgada well before the Initial Period (Grieder et al. 1988). In the case of Chavín, Kembel (n.d.) suggests, the earliest stages of temple construction may have extended well back into the Initial Period or even earlier. In any case, archaeological signatures of the earliest ritual domination may be difficult to discern given that major ENSO events were occurring only once every two or three generations and initial attempts at manipulation may have produced rather short-lived movements. Only after time and successive bursts of exploitation had institutionalized political resources to a degree and provided a greater permanence to the architecture of power might such signatures have become readily visible in the form of ceremonial spaces and monument centers. At Huaricoto, for example, Burger and Salazar-Burger's (1985) excavations revealed an episodic development sequence in which structures became larger, more permanent, and more numerous in periodic bursts through time.

Aldenderfer (2004: 30) has suggested that at La Galgada, one of the more precocious of the highland monument centers, leadership became institutionalized at the beginning of the Initial Period by imitating what had already occurred at coastal sites, 1,000 m below and three- to four-days' walk away. More than emulating leadership at coastal sites, however, the foregoing arguments suggest that highland polities like La Galgada may have owed their very existence to those sites, parasitically siphoning resources out of them by ritual artifice. If this were so, then with the intensification of ENSO events around 3000 cal yr BP and the subsequent decline of the coastal polities, we would expect a similar decline of the highland monument centers not long after. Significantly, the highland centers did begin to disintegrate around this time, though the chronology of these events is poorly known. According to Burger's influential chronology of events at Chavín, perhaps the best-known sequence, Chavín's decline did not start until the end of the Early (Chavín) Horizon, some 450 years after most of the coastal centers had ceased to function (Burger 1995: 284). This is a significant lag but not implausible if the highland centers were able to persist for a while after the coastal collapse by deploying institutionalized resources built up over previous centuries. More recently, however, Kembel (n.d.) and Rick (2004: 73) have suggested that the decline of Chavín was "roughly contemporaneous" with the decline and collapse of the coastal centers, occurring around 600 to 400 BC. Whatever the precise timing, the monuments at Chavín, Kotosh, Huaricoto, Kuntur

Wasi, Pacopampa, and other sites began to crumble and were not repaired; mundane housing was built over erstwhile sacred places; and radical changes in material culture suggest that whatever had once sustained the highland polities had now passed (Burger 1995: 228; see also Kembel n.d.; Rick 2004).

CONCLUSION

In comparison to social anthropologists such as myself, archaeologists have a tough row to hoe. From little more than incompletely preserved material remains, they are challenged to understand social processes that even those able to quiz humans directly in the field find difficult to penetrate. Nor have social anthropologists done much to help. We are awfully good at confounding archaeology with cautionary tales but, as prehistorians quite rightly protest, we seldom offer much in the way of solutions.

Most of us recognize, though, the value of a collaborative anthropology that seeks to explain both the ethnographic record and the prehistoric remains of *Homo sapiens sapiens*' activities on earth. This paper has attempted to develop a practice-based social theory of political evolution and how it is affected by calamity. What it advocates is the analytical value of locating ourselves as political entrepreneurs in an ecological and social landscape—of making explicit, in effect, what entrepreneurs have to know and do if they are to build a hierarchy. What it promises, at least for the early stages of political evolution, are predictions about the strategies that entrepreneurs must deploy if they are to be successful in a particular landscape and about the nature of the hierarchies they will build. Applying the argument to the Middle Preceramic/Late Preceramic–Period transition in Peru, it is possible to model social processes behind the connections that several scholars have drawn among geophysical events, climate-induced catastrophes, and the emergence and development of political complexity on the coast. In the more speculative case of Late Preceramic– and Initial-Period monument building in highland Peru, the argument generates a hypothesis to explain spatial- and temporal-event correlations between the highlands and the coast that hardly seem coincidental. Whether these efforts have value, of course, remains to be judged. If nothing else, though, it is to be hoped that they illustrate how theories distilled from the ethnographic record have the potential to be as "hard" as material remains in paring down the multiple possibilities that confront us as we try to understand the human past.

ACKNOWLEDGMENTS

My deepest thanks to Dan Sandweiss for stimulating my interest in early Peru several years ago, for directing me to relevant literature for this paper, and for so patiently answering my numerous, usually hopelessly unsophisticated questions. I also greatly appreciate discussions with or comments from Silvia Kembel, Jeff Quilter, Jim Richardson, and Ulrike Claas. Given that attempts at cross-disciplinary work are especially prone to mistakes and naiveties, I must forcefully stress that none of the above bears responsibility for any such failing.

REFERENCES CITED

Alcocer, Francisco de
1987 [1580] *Ecología e Historia: Probanzas de indios y españoles referentes a las catastróficas lluvias de 1578, en los corregimientos de Trujillo y Saña* (Lorenzo Huertas Vallejos, transcriber). CES Solidaridad, Chiclayo, Peru.

Aldenderfer, Mark
2004 Preludes to Power in the Highland Late Preceramic Period. In *Foundations of Power in the Prehispanic Andes* (Kevin J. Vaughan, Dennis Ogburn, and Christina A. Conlee, eds.): 13–35. Archaeological Papers of the American Anthropological Association 14. University of California Press, Berkeley.

Andrus, C. Fred T., Douglas E. Crowe, Daniel H. Sandweiss, Elizabeth J. Reitz, and Christopher S. Romanek
2002 Otolith $\partial^{18}O$ Record of Mid-Holocene Sea Surface Temperatures in Peru. *Science* 295: 1508–1511.

Arnold, Jeanne E.
1993 Labor and the Rise of Complex Hunter-Gatherers. *Journal of Anthropological Archaeology* 12: 75–119.

1996 The Archaeology of Complex Hunter-Gatherers. *Journal of Archaeological Method and Theory* 3: 77–126.

Arntz, Wolf
1986 The Two Faces of El Niño 1982–1983. *Meeresforschung* 31: 1–46.

Bischof, Henning
1994 Toward the Definition of Pre- and Early Chavín Art Styles in Peru. *Andean Past* 4: 169–228.

Bourdieu, Pierre
1977 *Outline of a Theory of Practice* (R. Nice, trans.). Cambridge University Press, Cambridge, England.

Brumfiel, Elizabeth M., and Timothy K. Earle
1987 Specialization, Exchange, and Complex Societies: An Introduction. In *Specialization, Exchange, and Complex Societies* (Elizabeth M. Brumfiel and Timothty K. Earle, eds.): 3–13. Cambridge University Press, Cambridge, England.

Burger, Richard L.
1988 Unity and Heterogeneity within the Chavín Horizon. In *Peruvian Prehistory: An Overview of Pre-Inca and Inca Society* (Richard W. Keatinge, ed.): 99–144. Cambridge University Press, Cambridge, England.

1995 [1992] *Chavín and the Origins of Andean Civilization.* Thames and Hudson, London.

Burger, Richard L., and Lucy Salazar-Burger
1985 The Early Ceremonial Center of Huaricoto. In *Early Ceremonial Architecture in the Andes* (Christopher B. Donnan, ed.): 111–138. Dumbarton Oaks Research Library and Collection, Washington, D.C.

Burridge, Kenelm
1969 *New Heaven, New Earth: A Study of Millenarian Activities.* Basil Blackwell, Oxford.

Carneiro, Robert L.
1970 A Theory of the Origin of the State. *Science* 169: 733–738.

1981 The Chiefdom: Precursor of the State. In *The Transition to Statehood in the New World* (Grant D. Jones and Robert R. Kautz, eds.): 37–75. Cambridge University Press, New York.

Cobb, Charles R.
1993 Archaeological Approaches to the Political Economy of Nonstratified Societies. In *Advances in Archaeological Method and Theory* (Michael B. Schiffer, ed.): 5, 43–100. University of Arizona Press, Tucson, AZ.

Elera, Carlos
1993 El Complejo cultural Cupisnique: Antecedentes y desarrollo de su ideología religiosa. In *El Mundo Ceremonial Andino* (Luis Millones and Yoshio Onuki, eds.): 229–257. Senri Ethnological Studies 37. National Museum of Ethnology, Osaka.

Feldman, Robert A.
1985 Preceramic Corporate Architecture: Evidence for the Development of Non-Egalitarian Social Systems in Peru. In *Early Ceremonial Architecture in the Andes* (Christopher B. Donnan, ed.): 71–92. Dumbarton Oaks Research Library and Collection, Washington, D.C.

Giddens, Anthony
1979 *Central Problems in Social Theory: Action, Structure, and Contradiction in Social Analysis.* University of California Press, Berkeley.

1984 *The Constitution of Society: Outline of the Theory of Structuration.* University of California Press, Berkeley.

Gilman, Antonio
1981 The Development of Social Stratification in Bronze Age Europe. *Current Anthropology* 22: 1–23.

Grieder, Terence, Alberto Bueno Mendoza, C. Earle Smith, Jr., and Robert Malina (eds.)
1988 *La Galgada, Peru: A Preceramic Culture in Transition.* University of Texas Press, Austin.

Haas, Jonathan, Winifred Creamer, and Alvaro Ruiz
2004 Dating the Late Archaic Occupation of the Norte Chico Region in Peru. *Nature* 432: 1020–1023.

Hayden, Brian
1995 Pathways to Power: Principles for Creating Socioeconomic Inequalities. In *Foundations of Social Inequality* (T. Douglas Price and Gary M. Feinman, eds.): 15–86. Plenum Press, New York.

Kembel, Silvia Rodriguez
n.d. Architectural Sequence and Chronology at Chavin de Huantar, Peru. Ph.D. dissertation, Department of Anthropological Sciences, Stanford University, Stanford, CA, 2001.

Lindstrom, Lamont
1984 Doctor, Lawyer, Wise Man, Priest: Big-Men and Knowledge in Melanesia. *Man*, n.s. 19: 291–309.

May, Ronald J.
1982 The View from Hurun: The Pele Association. In *Micronationalist Movements in Papua New Guinea* (Ronald J. May, ed.): 31–62. Australian National University, Canberra.

Mennis, Mary R.
1982 *Hagen Saga: The Story of Father William Ross, First American Missionary to Papua New Guinea.* Institute of Papua New Guinea Studies, Boroko.

Moseley, Michael E.
1975 *The Maritime Foundations of Andean Civilization.* Cummings, Menlo Park, CA.

Osborn, Alan J.
1977 Strandloopers, Mermaids, and Other Fairy Tails. In *For Theory Building in Archaeology: Essays on Faunal Remains, Aquatic Resources, Spatial Analysis, and Systemic Modeling* (Lewis R. Binford, ed.): 157–205. Academic Press, New York.

Patterson, Thomas C.
1971 Chavín: An Interpretation of Its Spread and Influence. In *Dumbarton Oaks Conference on Chavín, October 26th and 27th, 1968* (Elizabeth P. Benson, ed.): 29–48. Dumbarton Oaks Research Library and Collection, Washington, D.C.

Quilter, Jeffrey
1991 Late Preceramic Peru. *Journal of World Prehistory* 5: 387–438.

Raymond, J. Scott
1981 The Maritime Foundations of Andean Civilization: A Reconsideration of the Evidence. *American Antiquity* 46: 806–821.

Richardson, James B., III
1981 Modelling the Development of Sedentary Maritime Economies on the Coast of Peru. *Annals of Carnegie Museum of Natural History* 50: 139–150.

1994 *People of the Andes.* Smithsonian Books, Washington, D.C.

Rick, John W.
2004 The Evolution of Authority and Power at Chavín de Huántar, Peru. In *Foundations of Power in the Prehistoric Andes* (Kevin J. Vaughan, Dennis Ogburn, and Christina A. Conlee, eds.): 71–89. Archaeological Papers of the American Anthropological Association 14. University of California Press, Berkeley.

Roscoe, Paul B.
1988 The Far Side of Hurun: The Management of Melanesian Millenarian Movements. *American Ethnologist* 15: 515–529.

1993 Practice and Political Centralization: A New Approach to Political Evolution. *Current Anthropology* 34: 111–140.

2000 Costs, Benefits, Typologies, and Power: The Evolution of Political Hierarchy. In *Hierarchies in Action: Cui Bono?* (Michael W. Diehl, ed.): 113–133. Occasional Paper 27. Center for Archaeological Investigations, Southern Illinois University, Carbondale, IL.

n.d. Religious Thought and Scientific Experiment: An Anthropological View. Master's thesis, Department of Social Anthropology, University of Manchester, Manchester, England, 1977.

Sandweiss, Daniel H.
1995 Cultural Background and Regional Prehistory. In *Pyramids of Túcume: The Quest for Peru's Forgotten City* (Thor Heyerdahl, Daniel H. Sandweiss, and Alfredo Narváez): 56–77. Thames and Hudson, New York.

1996 The Development of Fishing Specialization on the Central Andean Coast. In *Prehistoric Hunter-Gatherer Fishing Strategies* (Mark G. Plew, ed.): 41–63. Department of Anthropology, Boise State University, Boise, ID.

Sandweiss, Daniel H., and James B. Richardson III
1999 Las fundaciones precerámicas de la etapa Formativa en la costa peruana. In *El Formativo Sudamericano* (Paulina Ledergerber, ed.): 179–188. Editorial Abya-Yala, Quito.

Sandweiss, Daniel H., James B. Richardson III, Elizabeth J. Reitz, Harold B. Rollins, and Kirk A. Maasch
1996 Geoarchaeological Evidence from Peru for a 5000 Years B.P. Onset of El Niño. *Science* 273: 1531–1533.

Sandweiss, Daniel H., Kirk A. Maasch, Richard L. Burger, James B. Richardson III, Harold B. Rollins, and Amy Clement
2001 Variation in Holocene El Niño Frequencies: Climate Records and Cultural Consequences in Ancient Peru. *Geology* 29: 603–606.

Shady Solís, Ruth
2005 *Caral Supe, Perú: La civilización de Caral-Supe; 5000 años de identidad cultural en el Perú.* Instituto Nacional de Cultura, Proyecto Especial Arqueológico Caral-Supe, Lima.

Shady Solís, Ruth, Jonathan Haas, and Winifred Creamer
2001 Dating Caral, a Preceramic Site in the Supe Valley on the Central Coast of Peru. *Science* 292: 723–726.

Testart, Alain
1982 The Significance of Food Storage among Hunter-Gatherers: Residence Patterns, Population Densities, and Social Inequalities. *Current Anthropology* 23: 523–537.

Upham, Steadman
1990 Analog or Digital? Toward a Generic Framework for Explaining the Development of Emergent Political Systems. In *The Evolution of Political Systems: Sociopolitics in Small-Scale Sedentary Societies* (Steadman Upham, ed.): 87–115. Cambridge University Press, Cambridge, England.

Van Buren, Mary
2001 The Archaeology of El Niño Events and Other "Natural" Disasters. *Journal of Archaeological Method and Theory* 8: 129–149.

Vicedom, Georg F., and Herbert Tischner
n.d. The Mbowamb: The Culture of the Mount Hagen Tribes in East Central New Guinea, vol. 2 (1): *Social Organisation*; vol. 2 (2): *Religion and Cosmology* (F. E. Rheinstein and E. Klestadt, trans). Menzies Library, Australian National University, Canberra.

Watson, James B.
1973 Tairora: The Politics of Despotism in a Small Society. In *Politics in New Guinea: Traditional and in the Context of Change; Some Anthropological Perspectives* (Ronald Murray Berndt and Peter Lawrence, eds.): 224–275. University of Washington Press, Seattle.

Wiessner, Polly
2002 The Vines of Complexity: Egalitarian Structures and the Institutionalization of Inequality among the Enga. *Current Anthropology* 43: 233–252.

Yesner, David R.
1980 Maritime Hunter-Gatherers: Ecology and Prehistory. *Current Anthropology* 21: 727–735.

DECIPHERING THE POLITICS OF PREHISTORIC EL NIÑO EVENTS ON THE NORTH COAST OF PERU

Brian R. Billman

Gary Huckleberry

In recent years, archaeologists have frequently cited El Niño as an important catalyst of social transformation in the prehistoric central Andes (see, for example, Fagan 1999; Moseley et al. 1983; Pozorski 1987; Sandweiss et al. 1999, 2001; Shimada et al. 1991). The importance of these events is recognized by many archaeologists, although the long-term impact of thousands of years of frequent, disastrous El Niño events on the production of political institutions, ideologies, and day-to-day social practices has yet to be fully addressed. At the same time, scholars are only just starting to investigate the specific ways in which El Niño events may have affected particular societies (e.g., Satterlee et al. 2001; Moseley and Keefer, this volume).

In searching for evidence of environmental change many archaeologists have relied upon distant sources of data, such as pollen and other materials trapped in layers of Andean glaciers. We believe, however, that it is better to use local proxy records of past El Niño events (see, for instance, Sandweiss 2003 and sources therein) that can be more directly related to specific environmental impacts on individuals and households. In this paper, we explore both the theory and data needed to construct models of human–El Niño interactions for the central Andes in the prehistoric era.

A model of El Niño–induced, prehistoric social change should be based on local information on the environmental impacts of El Niño, rather than on generalized region-wide data. It should explain how El Niño impacted individuals and households and examine the range of options available to those individuals and households. And finally, the range of options should be historically contingent and socially constructed. In other words, options were constrained by the historical and social context and varied according to the social, economic, and political position of each individual and household. We will present our approach in this chapter, concluding with a specific case study on how individuals and families in the Chimú empire coped with disastrous El Niño events.

DEVELOPING LOCAL PROXY RECORDS
IN THE MOCHE AND CHICAMA VALLEYS

Before we can understand the links between El Niño and sociopolitical transformation, we must first demonstrate a correlation between a specific environmental event and a specific sociopolitical event or process. Previously, investigators have made links to distant but fine-grained records, that is, continuous, high-resolution recording systems such as ice cores, tree rings, or corals (Baumgartner et al. 1989). For instance, Izumi Shimada and colleagues (Shimada et al. 1991) relied on data from the Quelccaya ice cores to infer the occurrence of a severe drought and strong El Niño events during the proposed collapse of Moche polities in the AD 500s (see also Fagan 1999). Although an excellent fine-grained proxy record for precipitation in the southern Peruvian Andes (14° S lat.), the Quelccaya ice cores are located far from the north coast of Peru on the east slope of the Andes. As Michael Moseley observed, "Andean ice cores record precipitation derived from the Amazon Basin and Atlantic Ocean whereas the heavy rainfall that drives flooding along the Peruvian coast is derived from warm, moist air from the tropical Pacific" (Moseley 1987: 7).

Unfortunately, the use of distant proxy records to reconstruct local events is highly problematic. El Niño events are circum-Pacific phenomena, but the severity of the local expression of a specific El Niño can vary significantly within and between regions (see Caviedes 2001: chaps. 4, 5). Although strong El Niño events often result in droughts in the southern Andes of Peru, where the Quelccaya Glacier is located, not all events result in such droughts, and not all southern highland droughts are caused by El Niño.

On the north and central coast of Peru, Waylen and Caviedes (1986, 1987) have demonstrated that the severity of coastal flooding varies considerably from valley to valley. Spatial variability of El Niño flooding is the result of varying hydrologic-basin characteristics (e.g., slope, drainage density, and surficial geology) and the spatial scale of localized tropical convective storms (Waylen and Caviedes 1986, 1987). Their study indicates that river valleys with large, low-altitude, tributary catchment areas, such as the Santa and Chicama valleys, are more likely to experience severe flooding during very strong El Niño events than valleys with small, low-altitude, tributary catchment areas, such as the Moche Valley. The reason is that most El Niño–related precipitation falls on the west slope of the Andes below 1000 m.a.s.l. (meters above sea level).

For the Moche River, mean annual discharges can vary by over a factor of ten from year to year (e.g., Barrena 1994 for Moche River discharge at Quirihuac). However, most of this annual variation is not driven by El Niño rains, which concentrate below 1000 m.a.s.l., but rather by variation in seasonal rainfall in the uppermost reaches of the river basin (Nials et al. 1979, 1: 10). Although anecdotal evidence indicates that the largest floods documented for the Moche River occurred during the very strong El Niño years of 1925, 1941, 1972, 1983, and 1998, the historic discharge data do not reveal a direct relationship between El Niño and flooding (Barrena 1994). For example, the maximum monthly mean discharge, sum of mean monthly discharge, and mean annual discharge for the strong La Niña years of 1950–51, 1955–56, 1970–71, 1973–74, and 1974–75

La Niña

Year	Max. Month (m³/s)	Sum of Mean Monthly Q (m³/s)	Mean Annual Q (m³/s)
1950–51	13	47	4
1955–56	70	178	15
1970–71	55	129	11
1973–74	17	79	7
1974–75	60	135	11
mean	43.0	113.6	9.6

Strong El Niño

Year	Max. Month (m³/s)	Sum of Mean Monthly Q (m³/s)	Mean Annual Q (m³/s)
1931–32	26	110	9
1940–41	39	101	8
1957–58	35	75	6
1972–73	61	164	14
1982–83	55	182	15
mean	43.2	126.4	10.4

TABLE 1 Selected La Niña and El Niño years and associated discharge values for the Moche River at Quirihuac. Discharge data for Quirihuac from Barrena (1994), La Niña years from NOAA (http://www.cpc.ncep.noaa.gov/products/analysis_monitoring/ensostuff/ensoyears.shtml), strong El Niño years from Quinn et al. (1987).

are comparable to those for the very strong El Niño years of 1931–32, 1940–41, 1957–58, 1972–73, and 1982–83 (table 1). That the three largest maximum mean monthly discharges (1932–33, 1947–48, and 1966–67) do not correspond with very strong El Niño years is striking. This supports Waylen and Caviedes's (1986) observation that the Moche Valley with its high-elevation watershed is less responsive to El Niño events than most other coastal rivers. The river can still experience catastrophic flooding during El Niño events, as the destructive inundations of 1925, 1983, and 1998 remind us, but we can nonetheless expect that specific histories of flooding will vary among valleys on the coast of Peru (see, for instance, Netherly n.d.: 290).

Given variation within the coastal region, use of distant proxy records, such as the Quelccaya ice cores, should be done with extreme caution. Although Quelccaya, and other distant proxy records, might give us some indication of changes in the relative frequency of El Niño events over the course of centuries, linking specific droughts in the Quelccaya ice cores to floods in a specific north coastal river valley seems highly problematic.

In sum, the contrasting flood behavior of Peru's coastal rivers and the spatially limited regional paleoflood database for the coast makes local proxy records of El Niño history imperative. Stimulated by paleo–El Niño research initially performed over twenty-five years ago (Nials et al. 1979, 1, 2; Moseley et al. 1983), we have developed two new local proxy records of El Niño flooding in the Moche and Chicama valleys.

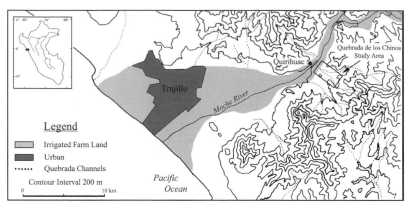

FIG. 1 Map of study area

THE QUEBRADA DE LOS CHINOS (QC) STRATIGRAPHIC LOCALITY

The first local proxy record is the Quebrada de los Chinos (QC) Stratigraphic Locality, which is located on a small tributary of the Moche River approximately 40 km from the Pacific coast (fig. 1). The quebrada drains an area of approximately 65 km² with a basin relief of 2200 m; approximately 60 percent of the catchment area is below 1000 m.a.s.l. Mean annual precipitation at nearby Quirihuac is 30 mm (Barrena 1994) and, according to local farmers, streamflow at QC in the 1900s occurred only during the very strong El Niño events of 1925, 1983, and 1998. Unlike the Moche River, which derives

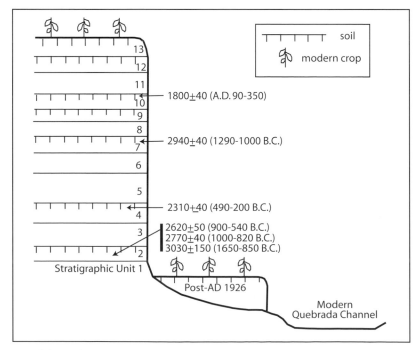

FIG. 2 Schematic cross-section and column of the Quebrada de los Chinos Stratigraphic Locality with selected ¹⁴C ages. Calibrated ages in parentheses (see Table 1).

Stratum	Ceramic Age	^{14}C Sample	Uncorrected ^{14}C Ages	Calibrated (2σ) ^{14}C Ages	^{14}C Sample Material
Historic chacra soil		CAMS-74943	730±30	AD 903–914	sediment: humic acids
10		CAMS-74942	1800±40	AD 90–350	charcoal
9		CAMS-68221	2670±40	900–790 BC	sediment: humic acids
8					
7		CAMS-68220	2940±40	1290–1000 BC	charcoal
5	Early Salinar (400–200 BC)				
4	Early Salinar (400–200 BC)	CAMS-68217 CAMS-68222 CAMS-68218 CAMS-68219 CAMS-68223	modern 2310±40 4210±40 4390±40 4910±40	490–460; 420–340; 320–200 BC 2900–2830; 2820–2020 BC 3270–3230; 3110–2900 BC 3780–3640 BC	seed corn sediment: humic acids sediment: decalcified seds. sediment: humins
2	Late Guañape (800–400 BC)	CAMS-68214 CAMS-68215 CAMS-68216	2620±50 3030±150 2770±40	900–750; 690–660; 640–540 BC 1650–850 BC 1000–820 BC	seed wood charcoal

TABLE 2 Radiocarbon and ceramic ages from Quebrada de los Chinos Stratigraphic Locality. Calibrated ages were determined using OxCal v. 3.3 based on Stuiver et al.'s (1998) atmospheric data.

most of its flow from seasonal rainfall in the highlands above 2500 m.a.s.l., QC supports flow only during low-elevation heavy rainstorms associated with El Niño. Hence, flood deposits at QC represent a discontinuous record of past El Niño events: not every El Niño event results in a flood at QC, but every flood is related to El Niño.

The QC Stratigraphic Locality is a 4 to 5 m high streamcut that exposes a vertical sequence of 13 flood and debris flow deposits (fig. 2; Huckleberry and Billman 2003, n.d.). Strata 2 and 4 contain coarse bedload and abundant cultural material (sheet midden) suggesting high-energy inundation of habitation areas. Also present are buried agricultural soils (A horizons) evidenced by organic dark colors and silty textures. These sediments are derived from irrigation waters supplied by the Moche River.

Based on surficial geologic mapping of the lower quebrada (Huckleberry and Billman 2003) and the analysis of more than 100 diagnostic sherds and 6 AMS (accelerator mass spectrometry) dates (table 2) from the QC Stratigraphic Locality, we were able to determine that 12 of 13 flood events occurred between 800 BC and AD 700 (fig. 2) (Huckleberry and Billman n.d.).[1] These floods were probably large enough to inundate much of the lower quebrada and associated agricultural communities at the edge of the Moche River floodplain. These QC floods probably correlate with large floods on the lower

1 All radiocarbon dates derived from organic fractions in soil A horizons yielded ages approximately 700 to 2,000 years too old based on associated ceramics. At present, we do not know the source of old carbon causing the erroneous ages. Consequently, we disregard all radiocarbon dates from soil organic fractions. We also disregard the modern date on a seed from an Early Salinar context.

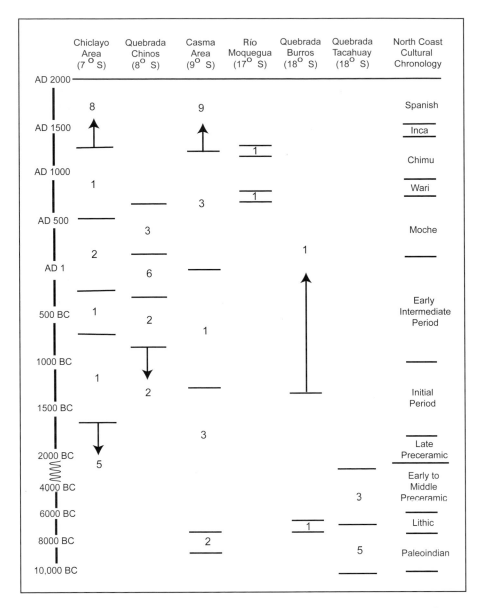

FIG. 3 Regional paleoflood record and cultural chronology. Chiclayo and Casma record from Wells and Noller (1999) and Wells (1990); Río Moquegua record from Magilligan and Goldstein (2001); Quebrada Burros from Fontugne et al. (1999); and Quebrada Tacahuay from Keefer et al. (1998). Numbers refer to flood events between dated horizons (horizontal bars).

Moche River, as evidenced by El Niño flooding in 1925 and 1998, although it is possible for storm cells to be small in area and isolated, resulting in relatively large instantaneous discharges at QC and other tributary valleys and more moderate flow along the Moche River.

The 1500-year paleoflood chronology at QC does not reflect the absolute frequency of very strong El Niño events because not every very strong El Niño generates a flood at QC. There also may be erosional unconformities within the stratigraphic record. As Wells (1990) noted, stratigraphic paleoflood records from these high-energy fluvial systems are best interpreted as recording a minimum number of flood events within a given period. Although discontinuous, these geologic records nonetheless provide insight into past changes in flood frequency and can be compared with each other to infer flood frequency changes in time and space (fig. 3). It is important to note that different stratigraphic paleoflood records from the Peruvian coast are derived from different geomorphic contexts including small quebradas (e.g., Quebrada Chinos, Muerto River, Quebrada Burros, and Quebrada Tacahuay) and larger drainages (e.g., Casma River). As previously noted, large and small drainages may respond to (and record) El Niño events differently. Also, these localities cover 11° of latitude and, given that the source of moisture for El Niño rain along the coast is the equatorial Pacific, one can expect more frequent storm events to the north. Consequently, there is considerable variability in the age and number of debris flow and flood deposits preserved at these paleoflood sites (fig. 3).

Because not all very strong El Niño events necessarily result in flooding at QC, what is important is not the total number of floods recorded at the site, *but rather any evidence for changes in the relative frequency of those large flood events.* Changes through time in the frequency of debris flows and large overbank floods in lowland drainages should reflect changes in the frequency of El Niño events. This hypothesis assumes that the processes that cause differential preservation of flood deposits do not vary with changes in El Niño frequency. Although changes in the frequency of flooding within the QC drainage basin were probably correlated with changes in the frequency of very strong El Niño events, geomorphic changes within the fluvial system, such as changing bank height and lateral channel shifts, could have altered the relationship between flooding and overbank sedimentation (Baker 1989; House et al. 2002). Indeed, a QC channel shift likely occurred after AD 700, causing the loci of deposition to shift away from the QC stratigraphic locality. Such limits are common to all stratigraphic-based paleoflood records and should not be ignored. Nonetheless, we believe that past changes in the frequency of flood deposition at QC and other paleoflood stratigraphic sites reflect changes in El Niño activity (Wells and Noller 1999), considering that other factors can influence when a locale will accumulate and preserve a flood signature.

The QC record suggests changes in the frequency of flooding, particularly during the period represented by Strata 4 through 10. If we take the midpoint values of the 2-sigma range age estimates of the most reliable ^{14}C sample from Stratum 4 (corn) and the charcoal sample from Stratum 10, the sequence of Strata 4 through 10 represents an approximately 600-year span (345 BC–AD 220), during which six large floods occurred. This period corresponds to the end of the Early Horizon and the early part of the Early Intermediate Period (EIP; the Salinar and Gallinazo phases in the Moche Valley sequence). No other paleoflood sequence along the Peruvian coast records such a high frequency prior to AD 1000.

FIG. 4 Aerial photograph of Quebrada del Oso study area

QUEBRADA DEL OSO STRATIGRAPHIC LOCALITIES

Seeking additional proxy records of past El Niño events, we conducted a reconnaissance of the La Cumbre Intervalley Canal (Huckleberry and Billman n.d.) constructed by the Chimú during the LIP in an effort to divert water from the Chicama River over the divide into the lower Moche Valley. At Quebrada del Oso (QO) (fig. 4), we found two sequences, QO Stratigraphic Locality 1 (QOSL1) and Locality 2 (QOSL2), where a canal aqueduct dammed two small quebradas. The quebradas are first-order tributaries with small, low-altitude catchment basins, draining barren rocky hillslopes. Unlike most aqueducts associated with the canal, these aqueducts have never been breached by runoff or other activities. Except for a small "spillover" located ~100 m upslope from QOSL2, runoff is entirely trapped where the quebrada intersects the canal. Behind the canal aqueducts are sandy and silty laminated deposits that record discrete runoff events.

As at QC, runoff on low-elevation quebradas in the Chicama Valley is linked to El Niño. Consequently, these deposits represent another local proxy record for past ENSO (El Niño/Southern Oscillation) activity. Moreover, because these first-order tributaries are steep with little soil cover, there is a lower rainfall threshold necessary to generate runoff, and smaller rainfall events are likely to be recorded. This contrasts with QC, where rainfall events have to be large and intense for infiltration capacity to be exceeded and runoff to reach the terminus of the quebrada system. Deposits at QO are therefore likely to provide a more continuous record of past El Niño events than at QC.

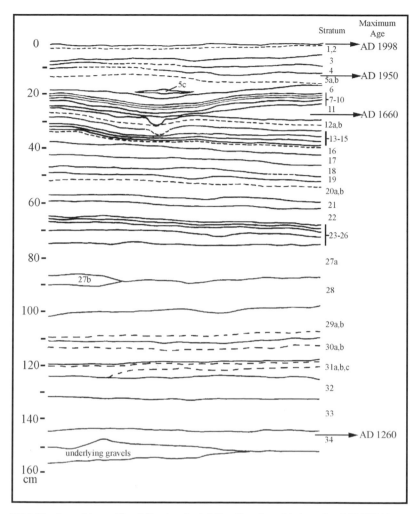

FIG. 5 Stratigraphic profile of Quebrada del Oso Stratigraphic Locality 1 (QOSL1) with dates based on archaeological, ^{14}C, and ^{137}Cs age control

We did not reach the bottom of the QOSL2 laminated sequence. We therefore focus on QOSL1, where we excavated a 1.5 by 1 m stratigraphic pit to the bottom of the laminated deposits at 1.6 m depth and encountered 33 discrete, coarse-to-fine couplets (fig. 5). The sequence displays a general trend of decreasing stratum thickness with time, reflecting the gradual filling of the inadvertent reservoir with sediment. Deposition was initially concentrated close to the berm, but, with time, the reservoir filled with sediment and the locus of deposition shifted farther upslope toward the mouth of the channel. Sediments increasingly became deposited over a larger area, and the thickness of the strata near the berm decreased with time. This may provide a bias in the record such that the minimum discharge necessary to inundate the entire ponded area and leave a deposit at QOSL1 may have increased through time. That is, smaller runoff events are more likely to be registered in the lower stratigraphy, and the frequency of couplet deposition should have decreased through time under uniform climate and ENSO activity. Nonetheless, given the steep gradient of the quebrada and the small size of the ponded area at QOSL1 (< ~800 m²), only the smallest runoff events would not be recorded in the upper profile.

Unlike the QC Stratigraphic Locality, our limited test excavations at QOSL1 yielded very little datable material and no artifacts. The base of the sequence is dated by the construction of the Intervalley Canal. Radiocarbon dates from the Quebrada del Oso section of the Intervalley Canal reported by Pozorski and Pozorski (1982) suggest that the canal was completed between approximately AD 1175 and 1265.[2] Upper deposit ages are constrained in part by a single ^{14}C date and the beginning of radioactive ^{137}Cs production associated with nuclear testing in 1950 (Huckleberry and Billman n.d.). The uppermost deposit dates to the 1997–98 El Niño, when the area last received heavy rainfall.

Consequently, the QO sequence contains a depositional history of between 735 and 825 years. With 33 depositional events recognized, the average runoff frequency is one event approximately every 25 years. Given that El Niño periodicity over the last 5,000 years is estimated to have been ~ 2–9 years (Rodbell et al. 1999), QOSL1 records only those El Niño events during which a storm cell passed over the small catchment area and generated enough rainfall to produce streamflow in the quebrada.

Historical research by Quinn et al. (1987) on El Niño frequency provides a means of evaluating the completeness of the QOSL1 record. Quinn and his colleagues documented 47 strong, quite strong, or very strong El Niño events from 1526 to 1987. In addition, they identified 32 near-moderate to moderate events from 1800 to 1987. These data suggest the following conclusions about the frequency of El Niño on the north coast of Peru since European contact:

1. The average frequency of near-moderate to very strong events from 1800 to 1987 was one every 3 to 4 years.
2. Strong to very strong events had an average frequency during the 460 period of one every 9 to 10 years.
3. Quite strong to very strong events during the 460 period had an average frequency of one every 25 years.
4. Very strong events over the same period had an average frequency of one every 57 years (just 8 events).

Although archival records have limitations in reconstructing El Niño events, it is clear from the research by Quinn et al. (1987) that not every historically known El Niño is recorded at the QOSL locality; more than 50 events have been documented for the last 200 years, whereas the locality evidences just 33 events

2 Of the twelve dates reported by Pozorski and Pozorski (1982), we selected only those dates from the Quebrada del Oso section of the canal that were analyzed at the radiocarbon laboratory at Washington State University. After adjustment of these dates to the 5568 Libby half-life, the pooled mean of the seven dates is 817±24, which yielded a two-sigma calibrated date of AD 1177–1265 (Southern Hemisphere calibration curve) using Calib 5.0.1 (Stuiver et al. 1998). An unresolved issue is the delta C-13 values of the dated materials. Because of this uncertainty, we offer *a provisional date of between AD 1175 and 1265* for the Intervalley Canal and associated ponded deposits. Further dating is required to test this provisional date.

over the last approximately 750 years. This is not surprising given that historically moderate El Niño events typically do not cause heavy rainfall in the Moche and Chicama valleys. Apparently, even many strong events may not have resulted in significant rainfall at Quebrada del Oso. However, the historical frequency of events classified as quite strong to very strong is similar to the record at QOSL1, circa one per 25 years. This may indicate that the Quebrada del Oso localities and other similar laminated deposits associated with archaeological features have the potential to yield highly detailed local proxy records of the strongest El Niño events. With careful dating, these laminated sediments may provide insight into decadal-to-century–scale variation in ENSO activity (e.g., Chavez et al. 2003), thus improving the temporal precision necessary to better construct and evaluate human behavioral-ecological models.

DISCUSSION

Our discussion of the QO and QC stratigraphic records as well as more distant proxies underscores two crucial points relevant to reconstructing past El Niño events. First, different types of proxy records have varying degrees of preservation and resolution (Ortlieb and Macharé 1993). Laminated sediment records from hillslopes and first-order tributaries will record more El Niño events than overbank quebrada deposits, due to the greater threshold of rainfall necessary for quebrada flooding and a reduced potential for erosional unconformities. However, if laminated sediments are associated only with Ceramic-Period archaeological structures, they will not have the time depth to evaluate models of millennial-scale changes in ENSO during the Holocene (e.g., Anderson 1992; Moy et al. 2002; Rodbell et al. 1999). Both types of stratigraphic paleoflood records need to be evaluated.

Second, the apparent incongruity among global (Cole 2001) and regional paleoflood proxy records (e.g., fig. 3) of Late Holocene El Niño events demonstrates that (1) stratigraphic-based proxy records of El Niño are discontinuous and large samples are needed to better assess flood-regime changes in time and space, and (2) the severity of local impacts of any given El Niño event varies considerably within and between regions. Only by developing local proxy records can we both increase the spatial and temporal sample of paleoenvironmental data and demonstrate possible correlations of changes in El Niño to periods of sociopolitical transformation.

Although the two new records presented here have limitations, they provide greater temporal resolution for El Niño flood events than previous stratigraphic paleoflood data from the coast. Moreover, they are the first proxy records of prehistoric El Niño frequency for the Moche and Chicama valleys where many important sociopolitical transformations took place during the prehistoric era. Further work is needed on laminated sediments associated with the La Cumbre Intervalley Canal as well as other well-dated archaeological features. As noted many years ago by Nials and his colleagues, long, linear structures that transect drainages are well suited for studying past El Niño events (Nials et al. 1979, 1: 13).

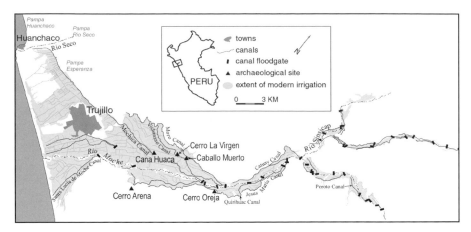

FIG. 6 Map of Mochica Canal in the Moche Valley

ALL POLITICS ARE LOCAL: UNDERSTANDING THE IMPACT OF EL NIÑO AT THE LOCAL LEVEL

One crucial piece of information for interpreting the impacts of El Niño events identified in our local proxy records is that moderate and perhaps even some strong El Niño events typically result in little or no damage to agricultural production in the Moche and Chicama valleys. For instance, the 2002–3 event was characterized as a moderate El Niño, yet this event resulted in only a few brief episodes of light rainfall in the lower Moche Valley. Precipitation was higher than average in Trujillo; however, no flooding occurred in either valley. If anything the rainfall during 2002–3 was advantageous to farming and grazing in the lower and middle valleys. Our research suggests that only very strong events cause significant flooding and damage to irrigated fields and canals in the Moche and Chicama valleys. In the 1900s, only the 1925–26, 1982–83, and 1997–98 events caused extensive damage.

Although very strong El Niño events can cause considerable damage to irrigation canals, prehistoric canals could have been rehabilitated relatively quickly (Billman 2002). With the exception of the massive La Cumbre Inter-valley Canal, the longest prehistoric canals in the Moche Valley could have been completely reconstructed in six months by the group of people supported by the canal. For instance, the Moche-phase Mochica Canal (fig. 6), which irrigated 4850 ha of irrigated land, could have supported 12,610 people or perhaps 8827 people if one third of the land was fallow (Billman 2002: 384). This canal could have been constructed from scratch by approximately 600 people working full-time, six days a week for six months (Billman 2002: 383). Therefore, even after a very strong El Niño, it could have been completely rebuilt before the start of the next growing season by a small subset of the people using the canal.

In contrast to repairing damaged canals, damage to fields is much more difficult to repair (Huckleberry and Billman n.d.). In order to reclaim damaged fields, overbank deposits of silts must be broken up and reworked, which requires considerable effort especially before the introduction of mechanized farm equipment (Moore 1991). Areas stripped of soil require even greater amounts of effort and time to restore to production. Boulders and cobbles must be removed, and yields remain low for years as soil is gradually restored by sediment born by irrigation water.

The extent of damage to fields by a very strong El Niño varies considerably both between and within valleys. The 1997–98 flood caused much more severe damage in the Chicama Valley than in the Moche Valley. Floods covered large areas of the lower Chicama Valley, burying irrigated fields with thick deposits of silts and stripping soil from fields on or adjacent to old and new flood channels. In the middle Chicama Valley, bottomland was scoured by massive flooding that stripped off soil, leaving behind cobble- and boulder-covered fields.

Flood damage was much more limited in the Moche Valley. In the lower Moche Valley, the course of the river was widened by flooding; however, no new or old flood channels were cut by flooding, and deposits of silt from overbank flooding were rare. In the middle Moche Valley, old flood channels were opened, and the river channel was widened; however, the extent of damage was far less than in the middle Chicama Valley. Our field observations confirm predictions by Waylen and Caviedes (1987) that drainages, such as the Moche, with small catchment areas below 1,000 m.a.s.l. will be less affected by El Niño flooding than valleys with large, low-altitude catchment areas, such as the Chicama Valley.

As a result, during the prehistoric era, people in the Moche Valley usually would have experienced less agricultural damage from El Niño events than people living in the Chicama Valley. This point had profound implications for prehistoric people living on the coast, especially for the ability of leaders to respond to El Niño damage. It may help to explain why centralized polities in the Moche Valley often dominated the Chicama Valley, even though the Chicama contains several times more arable land than the Moche.

Another important point is that flood damage is not evenly distributed throughout a valley. In the Chicama and Moche Valleys, flood damage was much more severe in the middle than in the lower valley. In middle valleys, gradients are steeper and rivers are confined by the foothills of the Andes, thus concentrating and increasing the speed of floodwater. In 1997–98, damage to agriculture in the middle valleys was most severe on the floodplains of the rivers and on old flood channels, where flooding destroyed fields by stripping away soil.

In the middle Moche Valley, we documented several cycles of flooding, floodplain stabilization, agricultural encroachment, and subsequent floodplain destruction during the twentieth century. Aerial photographs from 1942, 1987, and 1997 document some of these changes (fig. 7). The channel in 1942 is relatively narrow (50–70 m wide; table 3) and bounded by heavily vegetated bars. This photograph was taken approximately sixteen years after the large floods of 1926, allowing adequate time for riparian vegetation and agricultural fields to colonize gravel bars. The 1987 aerial photograph reveals

FIG. 7 Repeat aerial photography (1942, 1987, 1997) of the Moche River at Quirihuac. QC marks the Quebrada de los Chinos active channel.

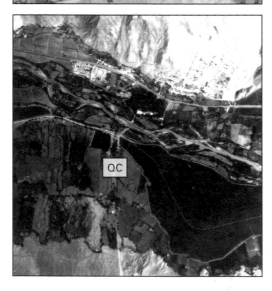

Year	1942	1987	1997	2000
Upstream	50 m	270 m	70 m	150 m
Bridge	50 m	150 m	60 m	180 m
Downstream	70 m	170 m	40 m	170 m
Years after strong El Niño	16	4	14	2

TABLE 3 Time series of channel widths for the Moche River at Quirihuac, from aerial photographs

a wider Moche River channel with less agricultural and riparian vegetation than in 1942. This picture was taken only four years after the El Niño of 1982–83, and although gauged records of mean monthly discharge do not record large floods during that year (Barrena 1994), the wider channel (150–270 m) suggests that recent flooding occurred and stripped away previous vegetation from the gravel bars. Ten years later in 1997, prior to floods associated with the 1997–98 El Niño event, riparian plants and agricultural fields had expanded across gravel bars and encroached upon the thalweg; channel widths were reduced to ~40–~70 m (table 3). During our visit in 1997, the footbridge at Quirihuac spanned a channel ~60 m wide. Upon our return in 1999, the footbridge was gone, and the concrete embankment supporting an earlier bridge was entirely within the active channel. In 2000, the channel width at the location of the former bridge was ~180 m.

These aerial photographs document both Moche River channel changes and the tendency of humans to maximize arable area for crop production within the active floodplain. The floodplain is the easiest location to divert water for irrigation in the middle Moche Valley, but it is also the area most prone to flood destruction. In 1942, a large area of the floodplain north and east of the Quirihuac bridge was not cultivated. Aerial photos reveal that this land was transected by flood channels formed by the floods of 1926 and was possibly inundated during the high runoff of 1932–33. By 1987, half of this parcel of land was claimed as an ecological preserve; the other half was cultivated. Likewise, agricultural plots had expanded onto several gravel bars located downstream from the bridge between 1987 and 1997. Most of these fields were destroyed by flooding and channel widening in 1998.

These observations document a cycle of flooding, bar deposition, vegetation growth, and agricultural encroachment. This cycle is followed by an eventual large flood that destroys the agricultural fields, deposits coarse alluvium, and renews the cycle. Planting crops in ephemeral river channels or in the flood channels of perennial rivers represents an opportunistic strategy for food production in arid lands (e.g., Doolittle 1989); however, such plantings run a high risk of destruction by flood.

The Moche River floodplain is the most dynamic component of the landscape in the valley. Channel changes occur in relation to large-peak instantaneous discharges associated with very strong El Niño events. Hence, periods of increased frequency and magnitude of El Niño would reduce the amount of farmland available in the Moche Valley. Flood-induced channel widening would have more of an impact on farmers in the narrow middle valley, where arable land is confined by mountain slopes. In the lower Moche Valley, where the floodplain widens along the coastal plain, a smaller percentage of the arable land is affected by channel changes. However, extreme flooding could still have serious impacts on downstream farmers through damaged canal headgates and possible channel downcutting.

These cycles of agricultural expansion and flood-induced contraction are likely universal along the northern Peruvian coast, although the frequency and magnitude varies from valley to valley. After the 1998 flood, irrigation canals were restored in both the Moche and Chicama valleys in only a few months: however, recuperation of fields destroyed by erosions or overbank sediment deposits was still going on five years after the flood.

By this logic, periods of increased frequency and magnitude of El Niño events would have resulted in a reduction of available farmland on the north coast and a concomitant increase in conflict over arable land as well as an increase in landless farmers. The social and economic debts of households farming areas with the highest probability of flood damage would also have risen.

Another implication of this dynamic is that agricultural production would have been especially susceptible to damage in the Late Preceramic Period (2700–1800 BC), when farming was largely limited to fields on or adjacent to the floodplains, and in the Initial Period (1800–800 BC), when agriculture was focused on the middle valleys. This suggests that river valleys with small catchment areas below 1000 m.a.s.l., such as the north central and central coast valleys between Supe and Lurín, would have been especially advantageous for early agriculture. Interestingly, the Late Preceramic Period on the north central coast was characterized by the earliest monumental construction on the Peruvian coast (Haas et al. 2004; Shady Solís et al. 2001).

In sum, increased frequency and magnitude of El Niño events would not have impacted all people equally. Those living and farming on or adjacent to the floodplains of rivers and quebradas faced the greatest risk; those living in middle valleys faced greater risk than lower-valley dwellers; and those living in small valleys would have been less affected than those living in large valleys. In fact, recent experience suggests that in some sections of certain valleys, most farmers survive severe events with minimal damage to agricultural land, while a few face devastating losses. With information on local impacts, researchers can begin to examine how people might have responded to severe El Niño events or a change in frequency of events.

PEERING INTO THE BLACK BOX:
HUMAN RESPONSES TO EL NIÑO EVENTS

In the mid-1980s a remarkable social movement swept through northern Peru. Rural communities organized nightly patrols to protect private property and created local courts to punish thieves (see Starn 1999 for a detailed history of this social movement). These *ronda campesina* organizations were directed by locally elected ronda committees, which held regular public meetings, organized communal labor projects (such as construction of schools and ronda meeting houses), and distributed aid from nongovernmental organizations. These organizations were established without governmental approval or help from national political organizations. On many occasions, the national government tried to stop this movement and abolish ronda organizations—in some cases intervening with armed force. Nonetheless, by 1990 over 3,400 communities had formed ronda committees over an area of 60,000 sq m of northern Peru (Starn 1999: 3–4). Orin Starn, who chronicled the rise and fall of the ronda campesina, considers it to have been "one of the fastest-growing and biggest movements for change in Latin America during the second half of the twentieth century" (Starn 1999: 4).

According to ronda members and leaders interviewed by Starn, the rapid spread of this movement was, in part, due to the 1982–83 El Niño. He concludes that "it was an initiative born of desperation. In 1983, the flooding connected to the El Niño weather pattern had destroyed roads, crops, and houses and had set off a spiral of crime and lawlessness in the desperate, hungry countryside" (Starn 1999: 2). The ronda movement of Peru provides us with a contemporary case in which El Niño played a role in the spread of a social movement, allowing us to examine human–El Niño interaction.

El Niño is often seen by archaeologists and earth scientists as a sort of "black box." Human societies enter one end of the box, El Niño occurs, and societies exit the box transformed in some dramatic fashion. What is frequently lacking in these inferences of El Niño–induced sociopolitical transformations, however, is an explanation of how El Niño events led to specific human responses that in turn resulted in social change. If we are to understand how El Niño shaped the lives of people in the central Andes, we need a much better understanding of how people assess and respond to environmental disasters. The spread of ronda organizations in the 1980s illustrates three key concepts for understanding El Niño–human interaction.

Groups are Formed and Maintained through the Actions of Individuals

Our first concept is that individuals are the relevant unit of analysis for understanding human–El Niño interaction. Although conceptions of individuality, self, or personhood vary historically and between cultures, all people live within their own separate, unique biological body. Consequently, human social groups are not an organic whole but a dynamic aggregate of individuals, each with their own motives, temperaments, values, and life histories. Our point is not that groups are unimportant. On the contrary, we propose that the formation of social groups is an essential element of social change. However, in order to understand groups, we must understand the dynamics of individual action.

The implication of this is that models of environmental-induced social change should specify how individuals were impacted by environmental change and how they responded to that change. Implicit in this shift to individual-based models is the notion that the impact of a specific El Niño varies considerably across communities and regions. Individuals in farming households that were devastated by El Niño floods may have had few options. One option may have been to seek help from individuals in wealthy households that survived the floods with little loss of land, labor, or seeds. Differential impacts at the individual and household levels create opportunities for well-situated individuals and households to gain control over the actions of others through the manipulation of social and economic debts. Alternatively, individuals may engage in theft from well-off families or form social movements to resist existing sociopolitical power structures.

The case of the spread of ronda organizations illustrates that social change is the result of the aggregate of individual actions (Starn 1999: 36–69). Following the 1982–83 El Niño many rural families in northern Peru saw their small fields destroyed. Other historical circumstances further reduced or eliminated other sources of income for rural families, leaving them with few options. In response, some individuals turned to theft of property. Although not everyone turned to theft, enough people did that many rural families lost goods to robbery. At its root, this crime wave was the result of countless decisions made at the family and individual level to rob or not to rob.

In response, some families banded together to conduct nightly patrols (Starn 1999: 36–69). No one family could protect their private property; however, a group of families could protect each other. In the community of Tunnel Six at first only a few men went out on patrol, but as the patrols proved successful, others joined, until all the men in the community took part (Starn 1999: 2). One by one individuals joined rondas, until thousands of communities and perhaps hundreds of thousands of people were involved. This entire social movement resulted from innumerable personal decisions about whether or not to participate in rondas campesinos.

The spread of the ronda movement illustrates that social groups and social movements are a messy, often chaotic aggregate of heterogeneous individual actions. The 1982–83 El Niño played a role in the spread of the ronda movement by significantly reducing options for many people in northern Peru, while at the same time creating new opportunities for group action.

Individuals have Cognitive Agency

Our second point is that individuals have been endowed by biological evolution with *cognitive agency*, the ability to act purposefully to alter the external world. In the case of the ronda movement, many individuals in northern Peru perceived that they were facing extreme poverty or even starvation as a result of the 1982–83 El Niño (Starn 1999). As a result of this perceived danger, they acted to avoid the consequences of El Niño. Some turned to theft, and, in turn, some members of rural communities adopted a new form of protection, ronda organizations. In other words, individuals perceived threats to their well-being caused directly or indirectly by El Niño, weighed various options, and then acted purposefully.

From the perspective of behavioral ecology, the cognitive agency manifested in the spread of rondas is the result of millions of years of biological evolution. The biological basis of our cognitive capacities evolved during the Pleistocene, which was a particularly turbulent period of global climate change. Recent research indicates that many of the changes in global climate during the Pleistocene occurred over the course of a few decades, rather than a few centuries (Mayewski and White 2002: 80–110). Consequently, individuals were able to perceive the occurrence of rapid climate change over the course of their adult lives and had to respond to that change in order to survive, reproduce, and raise their children. Rapid global climate change, manifested at the local level, may well have been one of the most important selective pressures shaping changes in hominid cognition during the Pleistocene.

A key result of this selective pressure is that people have the ability to perceive and respond to climate change and natural disasters. If people were passive and incapable of adapting to climate change, our species would never have survived the Pleistocene. Many species did not. Approximately 200 genera, including our closest hominid ancestors, Neanderthals and *Homo erectus*, were extinct by the close of the Pleistocene.

Because of this selective pressure, we are endowed with capacities that are well suited for coping with climate change and environmental disasters. We are, in a sense, natural-born historical ecologists. Our remarkable capacities for long-term memory coupled with our rich verbal linguistic abilities allow us to create detailed, multigenerational records of climate change. Further, we have an extraordinary capacity to classify ecological phenomena such as species, plant communities, and predator-prey relationships. Most important, we have the capacity to infer causal relationships between events that are temporally correlated—an ability that is useful in many cases, but in others leads to false inferences.

Three conclusions can be derived from an understanding of human cognitive abilities informed by evolutionary theory. First, the extent to which people perceive El Niño as a danger is a crucial variable in model building. If severe El Niño events occurred only once every few generations, then people probably would not have perceived them as a significant risk. Therefore measuring changes in the frequency and severity of El Niño events is essential (Sandweiss et al. 2001: 605).

Second, in order for people to change their actions in response to a perception of risk, they also must believe that their actions can have a positive (or negative) impact on the problem. Because our cognitive capacities are an essential part of our adaptation to climate change, *how we react is tied to what we perceive and what we believe.*

Third, because selective pressure in biological evolution occurs at the individual level, not at the level of the species, community, or population, human beings have the evolved capacity to calculate self-interest in regard to their own survival, reproduction, and the survival of close kin. People also have the capacity to recognize that their self-interests may converge with the interests of others, even those not related by kinship.

These observations raise two questions: what historical circumstances promoted (or in other circumstances inhibited) collective actions by individuals, and what historical circumstances created opportunities for certain individuals to gain control over the actions of other individuals?

Human Actions are Historically Contingent and Socially Constrained

Although individuals are endowed with the ability to perceive climatic risk and act purposefully to reduce or avoid risk, the available options for action are constrained by historical circumstance, society, and the social position of individuals. This means that any explanation of culture change in the Andes that invokes El Niño as a catalyst must be a historical explanation. Although El Niño events have occurred for much of the span of human occuption in the Andes, each new El Niño impacts a new society, albeit in some cases a society that is only slightly different than that during a previous El Niño. Nonetheless, slight changes in population, economy, land-use, political institutions, and beliefs can lead to radically different human responses to El Niño. Because global climate is a complex system and human society a complex web of individual actions, we should expect nonlinear relationships between climatic perturbations and human responses.

Although El Niño played no part in the creation of the first rondas campesinas, the 1982–83 El Niño appears to have been a tipping point in the spread of this movement. The 1982–83 event arrived at just the right moment in time, when social, political, and economic conditions were just right in northern Peru to push people to adopt this new movement. By contrast, the 1925–26 event, which also had a severe impact on northern Peru, apparently did not result in sweeping social movements. The 1925–26 event occurred in a very different social and political world. Rural and urban populations were much lower; the Peruvian economy was in the middle of an international economic boom; and northern Peru was dominated by large haciendas. Hacienda owners, in collaboration with the government and police, maintained control over rural populations. By 1982, many of these mechanisms of rural control had been eliminated or weakened; the social and demographic landscape was profoundly different; the national political order had been changed by a military coup in the 1960s; and the national economy was in a deep depression. In this context, El Niño provided impetus for people to act in new ways.

EL NIÑO AND EMPIRE IN THE MOCHE VALLEY DURING THE LATE INTERMEDIATE PERIOD

The Chimú empire was a complex web of individuals and families with varying motivations and goals: rulers at different levels in the political hierarchy, crafting families, fishermen, administrators, elite retainers, and, most important, the farming households that constituted the vast majority of the population. Remarkably, the Chimú rulers in the Moche Valley were able to hold this complex political system together and conquer the north coast. Given the complex nature of Chimú social and political relations, why didn't the Chimú empire collapse in the face of disastrous El Niño events? To answer this question we need to explore four other questions.

What Was the Historical and Social Context of El Niño activity in the Moche Valley in the Late Intermediate Period?

The Late Intermediate Period on the north coast (AD 1000–1460) was a time of imperial conquest, population growth, and agricultural expansion. Originating in the Moche Valley, the Chimú empire formed by AD 1000 and conquered all of the north coast of Peru by the AD 1300s (Mackey 1989: 122), resulting in the largest prehistoric state on the Andean coast prior to the Inca conquest. For the first time, river valleys from Tumbes to Supe, a distance of 1000 km, were united in a single polity.

Two important social and political practices documented for the late prehistoric era in the Moche Valley were (1) patron-client relationships between hereditary rulers and groups of farmers, fishermen, and craft specialists; and (2) the belief in the divinity of rulers. Historical sources from the north coast indicate that a network of patron and client relationships existed between rulers and commoners during the reign of the Chimú empire (Cock 1986; Netherly 1984, n.d.; Ramírez 1996: 42–97; Rostworowski 1975, 1977). Known as *señores, curacas, caciques,* or *principales,* these rulers were arrayed in a political hierarchy ranging from local rulers to the king of the Chimú empire. Each curaca controlled a group of families known as a *parcialidad* and irrigated land (or some other resource such as fishing rights to a specific area). In exchange for protection and access to resources, each parcialidad of farmers, fishermen, or crafting households provided annual tribute payments, in the form of goods and labor, to the curaca.

The curacas of the Chimú empire were not only political leaders. They also were responsible for maintaining the natural order and for preventing the occurrence of severe El Niño events. Chimú kings and curacas were viewed as divine beings, capable of intervening on behalf of their subjects. John Rowe (1948: 47) observed that "evidently differences between social classes were great and immutable on the north coast, for the creation legend told at Pacasmayo relates that two stars gave rise to the kings and nobles and the other two to the common people." The often-cited story of Fempellec, the lord of the Lambayeque Valley, who was put to death after thirty days of rain (Rowe 1948: 38), indicates that prehistoric rulers on the north coast could be held responsible for a severe El Niño.

The historical context of high population densities, large-scale irrigation, and regional political integration meant greatly reduced options for some individuals (especially in terms of residential mobility) while others would have been empowered through the control of elements of the regional political economy (see Roscoe, this volume).

What Were the Local Impacts of Very Strong El Niño Events?

Based on our investigations in the Moche and Chicama valleys before and after the 1997–98 El Niño (Huckleberry and Billman n.d.), we can deduce some of the consequences for people in the LIP facing a very strong El Niño event. Certain parcialidades would have had their crops completely destroyed by rains and flooding during the first crop cycle of the agricultural year (January to June). In some parcialidades, fields would have been destroyed by erosion or deposition of sediment. Virtually all canal intakes would have been destroyed. Flooding of quebradas would have destroyed relatively short segments of canals. In some parcialidades, many family homes and storerooms would have been destroyed by rain or flooding. Parcialidades of fishermen would have had several months of low productivity as cold-water fish, shellfish, and sea mammals were killed off by warm water. The influx of warm-water fish and shellfish would have offset some of the loss of local species.

In essence, many families would have faced challenges at three different temporal scales. During and immediately after El Niño, there would have been an urgent need for food as many people faced famine because of the loss of crops during the first growing season, the destruction of stored foods and seeds, and the reduced yields from marine resources. Next, people would have had to restore irrigation canals, a task that could have been performed by canal user groups in less than a half year, well before the start of the next January–June growing season. Over the long-term of several years, the biggest challenge would have been reclaiming agricultural land destroyed by flooding.

Our previous discussion demonstrates that the impact of these events would have varied greatly: lower valleys would have fared better than middle valleys, and valleys with small low-altitude, tributary catchments would have been better off than valleys with large low-altitude, tributary catchments. While some farmers faced devastating loss of fields and canals, many farmers, especially those in the lower Moche Valley, would have survived very strong El Niño events with minimal damage to their fields. This uneven distribution is a key to understanding human responses to El Niño in the LIP.

What Options Were Available to Individuals and Families Faced With a Very Strong El Niño Event?

The post–El Niño political landscape would have been filled with individuals and social groups seeking divergent ends. The crucial challenge for the leadership of the Chimú empire was to mobilize food and labor for short-term relief of starving families and to restore canals and fields. Because the empire controlled the entire north coast, its political reach encompassed lower and middle valleys, large- and small-catchment valleys, and heavily damaged as well as intact areas. While mobilizing labor at the local level would have been sufficient to restore canals, action at the regional level would have been required to provision starving farmers and reclaim fields. In order to cope with these challenges and maintain their political legitimacy, the regional political leadership would have had to move labor and staples from areas of relative abundance to areas of devastation.

Jerry Moore (1991) documents such a scenario archaeologically. In the Casma Valley, a massive reclamation project involving the construction of ridged fields was undertaken after a particularly devastating El Niño. A labor camp was established adjacent to the ridged fields, and the laborers at the camp were provisioned with stored foodstuffs and craft goods. Apparently, the Chimú leaders at the regional level responded to El Niño by moving people to the Casma Valley and provisioning them with stored food and tools. The ridged-field project would have met the short- and long-term challenges by providing food to farmers in exchange for work on the project and by quickly reclaiming new areas to replace damaged fields.

In addition to the material demands, disastrous El Niño events represented a fundamental challenge to the political and ideological legitimacy of Chimú royalty. Because they were responsible for maintaining the natural order, rulers would have had to meet spiritual obligations embodied in their divine status. In other words, Chimú royalty had to work to maintain the faith of the people in the ideological system, which was the basis of royal legitimacy.

Over the long term, the regional movement of goods and labor after very strong El Niño events would have been asymmetric. Farmers in areas that rarely encountered disaster would have been periodically called upon to help farmers in areas that frequently faced disasters. Rulers would have had to apply both coercion (the threat of punitive sanctions backed up by the military power of the empire) and ideological legitimation of patron-client relationships and divine rulership in order to maintain this long-term system of asymmetric exchange (see Roscoe, this volume, for consideration of coercive options and sources of power). A belief system that granted divine leaders the right to the labor of others and a regional system of patrons and clients would have provided leaders the means and the legitimacy to cope with El Niño disasters. In essence, the practices and obligations associated with patron-client relationships and the concept of divine rulership were fundamental to the strategies used by curaca and commoner families to survive periodic, disastrous El Niño events.

A key to understanding the production and maintenance of these practices can be found in the long period between the first indications of El Niño and the occurrence of flooding (Quilter and Stocker 1983). With the first warning signs of El Niño it would have been imperative for rulers to organize stockpiling of foodstuffs, construct dams, and perhaps even relocate families out of harm's way. An impending El Niño would have been a tangible test of a ruler's competency and his ability to mobilize labor and resources through the patron-client system. Monuments in the Moche Valley, such as

Caballo Muerto, Huaca del Sol, and the *ciudadelas* at Chan Chan, were (and still are) tangible evidence of such abilities. To the populace, these massive monuments conveyed an unambiguous message that leaders had the power to mobilize goods and labor in times of crisis.

The warning signs of an impending El Niño would also have provided an opportunity for rulers to demonstrate their divine connection. With the first signs of an El Niño, rulers could have sought divine help through the performance of rituals. The recognition of the legitimacy of divine rulers capable of supernatural intervention seems at odds with the periodic occurrence of very strong El Niño events, because divine rulers would have been blamed for such events. However, because most such events do not result in severe flooding, *divine intervention of leaders would have appeared effective most of the time.* The differential destruction of El Niño coupled with early warning of the coming floods provided an ideal context for the creation and maintenance of both divine rulers and the formation of patron and clients.

Perhaps because the Moche Valley was less susceptible to flooding during very strong events, rulers there would have appeared more successful in their divine interventions than leaders in other valleys. Political leaders in the Moche Valley also may have found themselves uniquely positioned to exploit their northern neighbors in the years following a strong event. The conquest of the Lambayeque Valley by the Chimú empire is said to have occurred after a severe El Niño (Rowe 1948: 38–39). This process may, in part, explain why Moche Valley polities frequently dominated the Chicama Valley, although the Chicama had much a larger population and far more agricultural land.

How Often Did People Face the Challenge of Very Strong El Niño Events in the Chimú Empire?

A final key to understanding the resilience of the Chimú political system is that very strong El Niño events might not have been frequent in the LIP. The local proxy record at the QOSLs (Quebrada del Oso Stratigraphic Localities) indicates that after about AD 1175, during the later half of the Chimú empire, very strong events occurred at an *average frequency* of approximately one every twenty-five years. We italicize "average frequency" because, at this point, we lack the precise stratigraphic age control necessary to know how this frequency might have varied over the course of each century. Because the QOSLs encompass the onset and retreat of the Little Ice Age, changes in the frequency of El Niño might have occurred during the seven-century span of the locality. The relationship between global climate change and the frequency of very strong events, however, is poorly understood (Cole 2001; Tudhope et al. 2001). At an average frequency of one very strong event per generation, limits of divine rulers and patron-client relationships might not have been fully tested.

CONCLUSION

Would Chimú rulers have been able to maintain their political dominance had the frequency of very strong El Niño events increased dramatically in the LIP? What if very strong events occurred several times a lifetime? Such an increase would have severely strained patron-client relationships and tested people's faith in divine rulers. As we develop more detailed local proxy records of very strong El Niño events on the north coast of Peru, we can begin to examine the resilience of prehistoric political systems there.

Rather than a history of frequent political collapse, the prehistoric era in the Moche Valley is characterized by long periods of political stability. Our current picture of the cultural history of the north coast suggests that people were quite adept at creating durable political institutions that lasted for hundreds of years and through many El Niño–induced disasters. The complexities of human society and global climate change and the durability of Andean polities demand that we be extremely cautious in our leap from El Niño–political-change correlations to causal inference.

In our view, our collective challenge in the discipline of anthropology is to develop robust models of human-environmental relationships that are informed and enriched by both the methodological rigor of the natural sciences and the conceptual richness of contemporary social and biocultural theory. Because anthropology is situated at the boundaries of the natural sciences, social sciences, and humanities, anthropologists are in a unique position to rigorously examine the interaction of environmental and social change. In the Andes, future work should focus on how environmental perturbations like El Niño affected the production and transformation of political institutions, ideologies, and social practices. An approach grounded in contemporary social and biocultural theory in areas where the archaeological and paleoenvironmental record are well defined will likely be the key to detailed modeling of human response to El Niño. The Moche Valley is a strong candidate for such future studies.

ACKNOWLEDGMENTS

Funding for the research presented here was provided by the National Science Foundation and the University Research Council and Institute for Latin American Studies at the University of North Carolina at Chapel Hill. Commentary was provided by Daniel Sandweiss, Jeffrey Quilter, Jonathan Haas, Celeste Gagnon, Greg Wilson, Amber Vanderwarker, and anonymous reviewers. Any errors and misjudgments are our own. We thank Jeffrey Quilter and Daniel Sandweiss for their thoughtful stewardship of this volume.

REFERENCES CITED

Anderson, Roger
1992 Long-term Changes in the Frequency of Occurrence of El Niño Events. In *El Niño: Historical and Paleoclimatic Aspects of the Southern Oscillation* (Henry F. Diaz and Vera Markgraf, eds.): 193–200. Cambridge University Press, Cambridge and New York.

Baker, Victor R.
1989 Magnitude and Frequency of Palaeofloods. In *Floods: Hydrological, Sedimentological, and Geomorphological Implications* (Keith Beven and Paul Carling, eds.): 171–183. John Wiley, Chichester.

Barrena, Sadi Dávila
1994 *Estudio geodinámico de la cuenca del Río Moche.* Boletin 15, serie C: Geodinámica e Ingenieria Geológica. Instituto Geológico, Minero y Metalúrgico, Lima.

Baumgartner, Timothy R., Joel Michaelsen, Lonnie G. Thompson, Glen T. Shen, Andrew Soutar, and Richard E. Casey
1989 The Recording of Interannual Climatic Change by High-resolution Natural Systems: Tree-rings, Coral Bands, Glacial Ice Layers, and Marine Varves. In *Aspects of Climate Variability in the Pacific and the Western Americas* (David H. Peterson, ed.): 1–14. Geophysical Monograph 55. American Geophysical Union, Washington, D.C.

Billman, Brian R.
2002 Irrigation and the Origins of the Southern Moche State on the North Coast of Peru. *Latin American Antiquity* 13: 371–400.

Caviedes, César N.
2001 *El Niño in History: Storming through the Ages.* University Press of Florida, Gainesville.

Chavez, Francisco P., John P. Ryan, Salvador E. Lluch-Cota, and Miguel Ñiquen C.
2003 From Anchovies to Sardines and Back: Multidecadal Change in the Pacific Ocean. *Science* 299: 217– 221.

Cock, Guillermo A.
1986 Power and Wealth in the Jequetepeque Valley in the Sixteenth Century. In *The Pacatnamu Papers*, vol. 1 (Christopher B. Donnan and Guillermo A. Cock, eds.): 171–182. Museum of Culture History, University of California, Los Angeles.

Cole, Julia
2001 A Slow Dance for El Niño. *Science* 291: 1496–1497.

Doolittle, William E.
1989 Arroyos and the Development of Agriculture in Northern Mexico. In *Fragile Lands of Latin America: Strategies for Sustainable Development* (John O. Browder, ed.): 251–269. Westview Press, Boulder, CO.

Fagan, Brian
1999 *Floods, Famines, and Emperors: El Niño and the Fate of Civilizations.* Basic Books, New York.

Fontugne, Michel, Pierre Usselmann, Danièlle Lavallée, Michèle Julien, M., and Christine Hatté
1999 El Niño Variability in the Coastal Desert of Southern Peru during the Mid-Holocene. *Quaternary Research* 52: 171–179.

Haas, Jonathan, Winifred Creamer, and Alvaro Ruiz
2004 Dating the Late Archaic Occupation of the Norte Chico Region in Peru. *Nature* 432: 1020–1023.

House, P. Kyle, Phillip A. Pearthree, and Jeanne E. Klawon
2002 Historical Flood and Paleoflood Chronology of the Lower Verde River, Arizona: Stratigraphic Evidence and Related Uncertainties. In *Ancient Floods, Modern Hazards: Principles and Applications of Paleoflood Hydrology* (P. Kyle House, Robert H. Webb, Victor R. Baker, and Daniel R. Levish, eds.): 267–293. Water Science and Application 5. American Geophysical Union, Washington, D.C.

Huckleberry, Gary, and
Brian R. Billman
2003 Geoarchaeological Insights
Gained from Surficial Geologic
Mapping, Middle Moche Valley, Peru.
Geoarchaeology: An International Journal
18: 505–521.

n.d. Impacts of El Niño Flooding on
Sociopolitical Organization in the
Moche Valley, Peru. Final report
submitted to the National Science
Foundation, Washington, D.C., 2002.

Mackey, Carol J.
1989 Chimu Administration in
the Provinces. In *The Origins and
Development of the Andean State*
(Jonathan Haas, Shelia Pozorski,
and Thomas Pozorski, eds.): 121–129.
Cambridge University Press, Cambridge,
England.

Mayewski, Paul A., and Frank White
2002 *Ice Chronicles: The Quest to
Understand Global Climate Change.*
University Press of New England,
Hanover, NH.

Moore, Jerry D.
1991 Cultural Responses to
Environmental Catastrophes: Post-El
Niño Subsistence on the Prehistoric
North Coast of Peru. *Latin American
Antiquity* 2: 27–43.

Moseley, Michael E.
1987 Punctuated Equilibrium:
Searching for the Ancient Record
of El Niño. *Quarterly Review of
Archaeology* Fall: 7–10.

Moseley, Michael E., Robert A.
Feldman, Charles R. Ortloff,
and Afredo Narváez
1983 Principles of Agrarian Collapse
in the Cordillera Negra, Peru. *Annals
of the Carnegie Museum* 52: 299–327.

Moy, Christopher M., Geoffrey O.
Seltzer, Donald T. Rodbell, and
David M. Anderson
2002 Variability of El Niño/Southern
Oscillation Activity at Millennial
Timescales during the Holocene Epoch.
Nature 420: 162–165.

Netherly, Patricia J.
1984 The Management of Late Andean
Irrigation Systems on the North Coast of
Peru. *American Antiquity* 49: 227–254.

n.d. Local Level Lords on the North
Coast of Peru. Ph.D. dissertation,
Department of Anthropology, Cornell
University, Ithaca, NY, 1977.

Nials, Fred, Eric. E. Deeds, Michael E.
Moseley, Shelia G. Pozorski, Thomas
Pozorski, and Robert Feldman
1979 El Niño: The Catastrophic Flooding
of Coastal Peru. *Field Museum of Natural
History Bulletin* 1, 50 (1): 4–10 and 2, 50
(2): 4–14.

Ortlieb, Luc, and José Macharé
1993 Former El Niño Events: Records
from Western South America. *Global
and Planetary Change* 7: 181–202.

Pozorski, Thomas
1987 Changing Priorities in the Chimu
State: The Role of Irrigation Agriculture.
In *The Origins and Development of the
Andean State* (Jonathan Haas, Shelia
Pozorski, and Thomas Pozorski, eds.):
111–120. Cambridge University Press,
Cambridge, England.

Pozorski, Thomas, and Shelia Pozorski
1982 Reassessing the Chicama–Moche
Intervalley Canal: Comments on
"Hydraulic Engineering Aspects of the
Chimu Chicama–Moche Intervalley
Canal." *American Antiquity* 47: 851–868.

Quilter, Jeffrey, and Terry Stocker
1983 Subsistence Economies and the
Origins of Andean Complex Societies.
American Anthropologist 85: 545–562.

Quinn, William H., Victor T. Neal,
and Santiago E. Antunez de Mayolo
1987 El Niño Occurrences over the Past
Four and a Half Centuries. *Journal of
Geophysical Research* 92: 14,449–14,461.

Ramírez, Susan E.
1996 *The World Upside Down: Cross-
Cultural Contact and Conflict in Sixteenth-
Century Peru.* Stanford University Press,
Stanford, CA.

Rodbell, Donald T., Geoffrey O. Seltzer, David M. Anderson, Mark B. Abbott, David B. Enfield, and Jeremy H. Newman
1999 An ~15,000-Year Record of El Niño-Driven Alluviation in Southwestern Ecuador. *Science* 283: 516–520.

Rostworowski de Diez Canseco, María
1975 Pescadores, artesanos, y mercaderes costeños el Perú prehispánico. *Revista del Museo Nacional* 41: 311–349.

1977 *Costa peruana prehispánica.* Instituto de Estudios Peruanos, Lima.

Rowe, John H.
1948 The Kingdom of Chimor. *Acta Americana* 6 (1–2): 26–55.

Sandweiss, Daniel H.
2003 Terminal Pleistocene through Mid-Holocene Archaeological Sites as Paleoclimatic Archives for the Peruvian Coast. *Palaeogeography, Palaeoclimatology, Palaeoecology* 194: 23–40.

Sandweiss, Daniel H., Kirk A. Maasch, and David G. Anderson
1999 Transitions in the Mid-Holocene. *Science* 283: 499–500.

Sandweiss, D. H., Kirk A. Maasch, Richard L. Burger, James B. Richardson III, Harold B. Rollins, and Amy Clement
2001 Variation in Holocene El Niño Frequencies: Climate Records and Cultural Consequences in Ancient Peru. *Geology* 29: 603–606.

Satterlee, David R., Michael E. Moseley, David K. Keefer, and Jorge Tapia A.
2000 The Miraflores El Niño Disaster: Convergent Catastrophes and Prehistoric Agrarian Change in Southern Peru. *Andean Past* 6: 95–116.

Shady Solís, Ruth, Jonathan Haas, and Winifred Creamer
2001 Dating Caral: A Preceramic Site in the Supe Valley on the Central Coast of Peru. *Science* 292: 723–726.

Shimada, Izumi, Crystal Barker Schaaf, Lonnie G. Thompson, and Ellen Mosley-Thompson
1991 Cultural Impacts of Severe Droughts in the Prehistoric Andes: Application of a 1,500-Year Ice Core Precipitation Record. *World Archaeology* 22: 247–270.

Starn, Orin
1999 *Nightwatch: The Politics of Protest in the Andes.* Duke University Press, Durham, NC.

Stuiver, Minze, Paula J. Reimer, Edouard Bard, J. Warren Beck, G. S. Burr, Konrad A. Hughen, Bernd Kromer, Gerry McCormac, Johannes van der Plicht, and Marco Spurk
1998 INTCAL98 Radiocarbon Age Calibration 24,000–0 cal B.P. *Radiocarbon* 40: 1041–1083.

Tudhope, Alexander W., Colin P. Chilcott, Malcolm T. McCulloch, Edward R. Cook, John Chappell, Robert M. Ellam, David W. Lea, Janice M. Lough, and Graham B. Shimmield
2001 Variability in the El Niño–Southern Oscillation through a Glacial-Interglacial Cycle. *Science* 291: 1511–1517.

Waylen, Peter R., and César N. Caviedes
1986 El Niño and Annual Floods on the North Peruvian Littoral. *Journal of Hydrology* 89: 141–156.

1987 El Niño and Annual Floods in Coastal Peru. In *Catastrophic Flooding* (Larry Mayer and David Nash, eds.): 57–78. Allen and Unwin, Boston.

Wells, Lisa E.
1990 Holocene History of the El Niño Phenomenon as Recorded in Flood Sediments of Northern Coastal Peru. *Geology* 18: 1134–1137.

Wells, Lisa E., and Jay Noller
1999 Holocene Coevolution of the Physical Landscape and Human Settlement in Northern Coastal Peru. *Geoarchaeology: An International Journal* 14: 755–789.

DEADLY DELUGES IN THE SOUTHERN DESERT: MODERN AND ANCIENT EL NIÑOS IN THE OSMORE REGION OF PERU

Michael E. Moseley
David K. Keefer

Torrential rainfall associated with El Niño events in 1982–83, 1992–93, and 1997–98 produced floods and debris flows in the lower Osmore drainage and adjacent coastal watersheds in southern Peru. Although these events caused local damage to infrastructure in this region, deposits left by the events are typically thin and fine-grained and are confined to incised drainage channels, with only local overtopping. Flood and debris-flow deposits that are much thicker, coarser-grained, and more laterally extensive dominate the sequence of Holocene and Late Pleistocene sediments in this region, implying that the events that produced them were more severe and the likely damage more catastrophic than anything known from the recent record. As El Niño is the main phenomenon producing significant rainfall in this region in modern times and as environmental indicators in the older sediments suggest that the climate was similar when they were deposited, we infer that these older sediments, too, were the products of El Niño events. The domination of the older sequence by large-scale deposits suggests, among other effects, that most deposits the scale of those from 1982–83, 1992–93, and 1997–98 do not survive in the geologic record, but rather are obfuscated, eroded, and reworked during more severe events.

Historically, scholars infer that one set of large-scale flood and debris-flow deposits in this region resulted from a combination of events—a great earthquake generating exceptional quantities of sediment, subsequently entrained during a severe El Niño–generated flood. They further surmise that an earlier set of massive deposits resulted from exceptional quantities of rainfall and runoff produced by an exceptionally severe El Niño event that exceeded all historical precedents. Archaeological studies indicate that this event decimated the local population and infrastructure, substantially contributing to significant local cultural collapse. In this paper we detail the evidence for this event and its consequences.

EL NIÑO AND COASTAL PERUVIAN FLOODS

Under normal conditions, the coast of Peru is in the rain shadow of the Andean cordillera, which blocks the flow of the moisture-laden easterly winds that bring precipitation to the Amazon Basin, the high Andes, and the altiplano of southern Peru and Bolivia. Normal atmospheric conditions associated with the cold, north-flowing Peru Current bring persistent fog but little precipitation to the coastal plain or low Coastal Cordilleras west of the Andes, which are thus part of one of earth's driest deserts.

Weather along this desert coast changes dramatically during El Niño, herein taken to be the warm phase of the El Niño–Southern Oscillation (ENSO) phenomenon (see Maasch, this volume). During El Niño, moisture-laden air sweeping onshore from the tropical eastern Pacific typically brings heavy rainfall and severe flooding to this region. At the same time, El Niño typically brings drought to the Andes and altiplano of southern Peru and Bolivia. Effects of historical El Niño events have varied considerably in geographic extent and intensity, and the events are variously rated as weak, moderate, strong, or very strong. Weak events affect northern Peru and southern Ecuador relatively frequently but their consequences are generally limited. Strong and very strong events are less frequent but far more consequential because, among other effects, they can cause heavy rainfall and flooding along the entire coastal watershed of the central Andes. In the Osmore region along the far south coast of Peru (approximately 17°–18°S; fig. 1), modern sedimentary deposits from flood events were produced by the very strong (Quinn and Neal 1995; Davis 2001) El Niños of 1982–83 and 1997–98 and the strong (Davis 2001), multiyear El Niño of 1991–95, which caused local flooding in 1992–93.

Because of the close historical correlation between El Niño and flooding along the central Andean coast, prehistoric flood deposits found there have been widely interpreted as evidence of earlier El Niño events (Nials et al. 1979; Moseley et al. 1981, 1992a, 1992b; Craig and Shimada 1986; Pozorski 1987; Moseley 1987, 1990; Wells 1987, 1990, n.d.; Donnan 1990; McClelland 1990; Shimada 1990; McGlone et al. 1992; Satterlee n.d.; Grodzicki 1994; Keefer et al. 1998, 2003; Fontugne et al. 1999; Satterlee et al. 2001). Both historical and prehistoric El Niños have had devastating effects for coastal populations, in the Osmore region (Moseley et al. 1992a; Reycraft 2000; Satterlee n.d.; Satterlee et al. 2001) and elsewhere (Nials et al. 1979; Moseley et al. 1981; Caviedes 1984; Moseley 1987, 1990; Donnan 1990; McClelland 1990; Shimada 1990; Wells 1990; Moseley and Richardson 1992; Bawden 1996). Indeed, calamitous erosion and burial of prehistoric cultural landscapes led in the 1970s to the first geoarchaeological detection of an ancient El Niño incident (Nials et al. 1979). This prehistoric flood and a number of others (Moseley et al. 1981; Wells 1990; Keefer et al. 2003) are posited to be the products of "mega-Niños" because the scale of the associated deposits substantially exceeds that of the most severe twentieth-century Andean coastal floods.

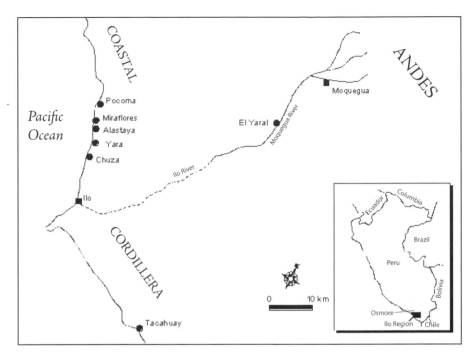

FIG. 1 Map of the Osmore region, Peru, showing sites and features described in text. Dashed line indicates intermittent river flow.

Significantly, prehistoric paleoflood sequences that extend through the Early Holocene and the Late Pleistocene are characterized by a predominance of large-scale deposits, but relatively few deposits approximating those of recent twentieth-century floods (Wells 1990; Keefer et al. 1998, 2003; Fontugne et al. 1999). Similarly, sixteenth- through nineteenth-century historical accounts of ENSO conditions chronicle numerous floods (Caviedes and Waylen 1991; Quinn and Neal 1995), yet there is scant geological evidence of these documented events. Thus, multiple lines of evidence indicate that prehistoric paleoflood sequences underrepresent El Niño–generated floods of recently documented magnitudes, while recording less-frequent events of much larger magnitude than any known from the recent record. We assess biases in paleoflood records by reporting on El Niño conditions, rainfall, runoff, and flood deposits produced during the last two decades of the twentieth century in the Osmore region. There, flash flooding was induced by two El Niño events rated as very strong—one in 1982–83 and a second that was more severe in 1997–98 (Quinn and Neal 1995; Davis 2001). Less-severe intervening El Niño conditions between 1991 and 1995, rated as a strong episode (Davis 2001), generated limited rainfall and runoff in 1992–93. These three episodes of recent flooding lead us to propose that more severe El Niño events typically rework and obliterate the depositional remains of previous smaller ones. We then summarize observations of two large-scale paleoflood deposits (the Chuza and Miraflores units) that evidently survived geologically because they were produced by larger-scale, though less-frequent events.

THE OSMORE REGION

In the Osmore region, there is a Coastal Cordillera averaging 1000 m in elevation, typically separated from the coast by a narrow coastal plain and prominent alluvial fans incised by normally dry channels called *quebradas* (fig. 1). Between this coastal mountain range and the main Andean Cordillera in the study area are the relatively flat plains of the Clemesi Desert and the piedmont Moquegua Valley. The mountain slopes and desert plains are covered with loose, weathered rock and unconsolidated sediments that are readily entrained by runoff from El Niño rainfall events.

With headwaters high in the main Andean cordillera, the Osmore River (also called the Moquegua River near the Andes and the Ilo River near the coast) is ~150 km long (fig. 1). Carrying runoff from altitudes as great as 5000 m, the principal tributaries of this river system converge in the arid sierra below 2000 m in the midvalley area, where the city of Moquegua is located and where seasonal river flow supports irrigation agriculture. Here the drainage is deeply incised, and at an altitude of 1200 m the river enters a deep, narrow canyon that transects the interrange plains occupied by the Clemesi Desert. The river course then cuts through the Coastal Cordillera before disgorging at the port city of Ilo. The lower, coastal valley is also a deep, steep-sided, narrow canyon. Arable land, limited to intermittent, ribbonlike stretches of low fluvial terraces, is only slightly higher in elevation than the active flood plain. Irrigation is supported by small seeps and springs and by several weeks of river runoff. Paralleling the lower river course, steep, incised quebradas descend the coastal range. At an elevation of about 100 m most of these quebradas cut across a coastal aquifer that sources small springs, which also have long sustained small irrigation systems.

In the study area annual rainfall, normally absent along the coast, increases with altitude; mean annual precipitation increases from 5 mm at the coast to 100 mm at the city of Moquegua and 300–400 mm in the high Andes (Satterlee n.d.). Above 3900 m, markedly seasonal precipitation exceeds soil absorption values of ~260 mm a year and supplies runoff to the lower river basin. Before completion of a recent large-scale irrigation project that increased river flow, the river carried sufficient water to reach the coast only for a short period in the austral spring. Under natural conditions, flow then diminishes, watering the lower sierra for several months before disappearing. Under El Niño conditions, the local rainfall regime reverses. Drought frequently prevails at altitudes above 2000 m, while sporadic but intense deluges can occur in the lower watershed and along the desert coast.

Normally, winter fog banks, 800 m or higher, roll in off the ocean and push up the incised Osmore drainage, enveloping the coastal range, the Clemesi Desert plains, and part of the main Andean escarpment at elevations below ~1200 m. At night the stratus layer settles downward, often condenses near the ground, and produces very heavy mist, called *garúa*. Garúa wets the ground but does not generate sufficient moisture to saturate soils and shed runoff. Nonetheless, under favorable circumstances, cloud condensation supports short, seasonal blooms of *lomas* vegetation in the otherwise barren desert. Fog density and lomas distribution are strongly influenced by orographic conditions, particularly along the sloping topography that faces the ocean. Consequently, condensation and plant growth are intermittent in the Clemesi plains and are greatest in the middle and upper slopes of the coastal range and along the basal escarpment of the Andes. El Niño conditions are associated with increased density and intensity of coastal stratus clouds, which under those conditions can push higher and reach Moquegua in the dry sierra.

EL NIÑO EVENTS OF 1982–83, 1992–93, AND 1997–98

During the very strong El Niño events of 1982–83 and 1997–98, warm ocean waters disrupted fishing, and the study area experienced both sunny, windy days and overcast days when fog rested against the lower watershed. In the evenings and continuing through many nights, stratus clouds spilled over the Coastal Cordillera and across the Clemesi plains to the foothills of the Andes, contributing to increased traffic accidents and fatalities on the Pan-American Highway, which runs along the base of the range. Dense garúa also sustained exceptionally abundant and widespread *lomas* plant growth. This in turn provided pasture for herd animals moved in from high altitudes, where drought prevailed. In some areas of the Clemesi plains there was sufficient fog condensation for enterprising farmers to plant and harvest grain, something that did not happen during the weaker 1992–93 El Niño, when there was less deviation from normal conditions.

Stratus clouds and fogs were densest between late afternoon and morning, and the El Niño–induced flash floods in the study area typically resulted from short-lived showers in this time span. Reflecting orographic conditions, rainfall was greatest along the upper and middle slopes of the Coastal Cordillera and the base of the main Andean escarpment. The 1982–83, 1992–93, and 1997–98 El Niños produced several sudden deluges on unvegetated mountain slopes; precipitation from these rainfall events exceeded soil-absorption capacities and led to immediate runoff that entrained loose hillside sediments as rill wash and sheet wash. This entrained material then moved into higher-order drainages, where coarser debris and larger clasts were entrained. Although some flow in the lower reaches of coastal quebradas and in the lower Ilo River valley was initiated by runoff from adjacent hills and gullies, the principal flood surges descending the quebradas and river valley were derived mainly from the higher areas in the Coastal Cordillera. Large surges were often heralded by loud noises, said by eyewitnesses to sound like a train locomotive—presumably the products of violently abrading clasts. Nearing the coast, some quebradas that transect the narrow coastal plain are incised to only relatively small depths, and there floodwaters in the 1982–83 and 1997–98 events locally spilled over banks and inundated vulnerable, low-lying agricultural surfaces with deposits of sand and silt up to 10 or 20 cm thick. Local overtopping of banks also occurred at drainage "choke points," where bedrock outcrops constricted drainages and forced flood surges over channel banks. For the most part, however, deposits from these events, as

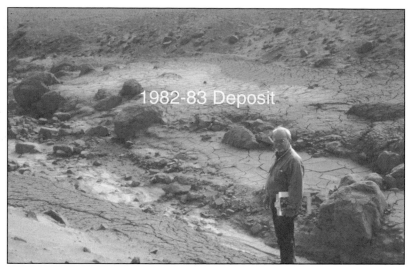

FIG. 2 Deposits from El Niño events of 1982–83 and 1992–93 in coastal quebradas in the Osmore region: a) Flood deposit from the 1982–83 El Niño event in Yara Quebrada. Deposit consists of a sandy silt that is a few cm thick; b) Debris-flow deposits from the 1982–83 and 1992–93 El Niño events in Alastaya Quebrada. Deposit from 1982–83 event has spilled a short distance overbank due to super-elevation at bend in channel. Scale marked in inches and centimeters.

well as essentially all deposits from the weaker 1992–93 El Niño event, were confined to the drainage channels themselves (fig. 2).

Once the flood surges had passed down drainages, relatively clear water typically flowed at low levels for several hours. Raindrops, animals, and people left tracks in the moist sediment, which tended to crack and fissure with desiccation and then harden to a durable, adobelike state within a week. The firm deposits remained well preserved until the next flood, although minor amounts of aeolian sand often accumulated in desiccation cracks exposed for a year or more. Due to annual spring floods, preservation of these deposits was typically poor along the Ilo River, but good in normally dry quebradas.

Under normal, non–El Niño conditions, intermittent accumulation of aeolian sand is taking place throughout the area, particularly near the coast, where daily winds off the sea entrain beach sand supplied by disgorged river sediment. This process breaks down during El Niño events, and even after the resumption of normal conditions, the accumulated aeolian sediment is readily entrained by flash flooding. In the absence of significant intervening sedimentary processes, the deposits from individual flash-flood episodes during the three El Niño events have typically accumulated immediately on top of one another.

During the 1982–83 El Niño event, coastal quebradas and the lower Ilo River valley experienced three distinct episodes of simultaneous flash flooding, separated from each other by a week or more. Precipitation was modest on the Clemesi plains, but increased along the Andean foothills; dense clouds pushed up valley to Moquegua, where one nighttime shower generated small debris flows in several tributary quebradas. These flows lacked sufficient moisture or volume to reach the river channel, which did not flood in this sierra region. It is not clear if the Moquegua rainfall event was synchronous with any of the three coastal events.

During the 1992–93 El Niño there were scant showers and no flooding in Moquegua, the main Andean cordillera, or the Clemesi plains. One evening there was a trace of rain around the port of Ilo, which had seen its last shower during a moderate 1977 El Niño event. In 1992–93, precipitation was somewhat more frequent and intense along the upper slopes of the Coastal Cordillera, and on one occasion there was sufficient nighttime rainfall and runoff to generate small debris flows that reached the mouths of a few coastal quebradas.

During the 1997–98 El Niño event, rainfall and runoff were somewhat greater than in 1982–83. Several weeks apart, the coastal quebradas and the lower Osmore Valley experienced two notable storms with ensuing flash floods that generally equaled, or in some cases exceeded, the severity of the episodes during the 1982–83 El Niño event. In the Clemesi Desert and western Andean foothills, rainfall, runoff, and flooding were sufficient to cut the Pan-American Highway and its trunk lines at multiple points at least once. There were several episodes of rainfall in the Moquegua sierra region. The Moquegua River flooded, with peak discharge of 450 m³/s and considerable sediment transport (Magilligan and Goldstein 2001). Because there are

FIG. 3 Flood and debris-flow deposits in Alastaya Quebrada from the 1982–83 and 1992–93 El Niño events eroded by channel cut during the 1997–98 El Niño event.

no reports of uniform showers stretching from the sea to the sierra, it is not clear whether the Moquegua rainfall episodes were concurrent with those on the coast.

During the 1982–83 El Niño event, one of us (M. E. M.) witnessed the incident of quebrada flooding in the Moquegua region and examined coastal drainages a few days after the last of the three El Niño–induced floods descended the Coastal Cordillera. With tan sediments up to 20 cm thick, deposits of the last flood episode were distinctive and easily recognized. However, this sediment generally overlay, reworked, and obfuscated deposits of the second flash flood that had done the same, in turn, to deposits of the first. Using surface features, such as rain spots and sun cracks that indicated that one deposit had dried and consolidated before the next, it was often possible to distinguish two of the three 1982–83 flood deposits, but rarely all three, owing to poor preservation of the first. In addition, there was little evidence to suggest that the three deposits were products of a single strong El Niño event rather than three separate events.

The 1992–93 flash flood in the Coastal Cordillera was much smaller and deposited finer, lighter sediments largely within the confines of the earlier and larger 1982–83 deposits (fig. 2b). The latter were only slightly reworked by the 1992–93 event, and other than minor sand infilling of desiccation cracks, there was little depositional evidence to indicate that the 1992–93 deposit was produced by a separate El Niño event rather than by yet another individual flood episode associated with the 1982–83 event.

The 1997–98 El Niño event produced two sets of superimposed flood and debris-flow deposits. The later of the two was typically easier to identify, although the earlier of the two was typically as large or larger. The 1997–98 flood episodes tended to rework, erode, or cut channels through deposits from the two previous El Niño events (fig. 3). Although the floods of 1982–83 and 1997–98 were of generally similar magnitude, the latter was evidently somewhat more severe in the many channels where little evidence of the former survives.

These observations of recent El Niño–generated deposits in the Osmore region carry three implications for paleoflood studies. First, El Niño–induced rainfall and runoff have not been uniform within the river basin or the larger region around it. During the very strong El Niño events, the sierra has generally experienced one episode of flash flooding, whereas two or three have occurred at elevations below ~1000 m. Similarly, flooding at lower elevations occurred during a less severe El Niño that did not produce runoff in the sierra. The orographic barrier presented by the Coastal Cordillera almost certainly contributes to this variability, but precipitation and river discharge records also indicate that El Nino–induced rainfall and runoff differ significantly from one coastal valley to another along the central Andean coast (Waylen and Caviedes 1986; Goldberg et al. 1987). Second, presumptions of one-to-one correlation equating a single flood deposit with a single El Niño episode are not supported by recent very strong events that have generated multiple flash floods in the Coastal Cordillera. Multiple storms and floods are not unprecedented in historical records of El Niño activity, and their probability may increase with the severity and duration of ENSO conditions.

Third, earlier El Niño–related flood and debris-flow deposits are reworked, eroded, and obfuscated by subsequent El Niño floods of equal or greater magnitude. In a manner similar to glacial advances, bigger events obscure the evidence of prior smaller ones. In many localities deposits from the 1982–83 event, and especially from the smaller 1992–93 event, were largely or even completely reworked or eroded by the 1997–98 event (fig. 3). Indeed, had the 1997–98 event been stronger, probably even fewer of the earlier deposits would have survived. Thus, recent flash flooding in the Osmore region indicates that long-term geological survival of El Niño–induced flood and debris-flow deposits is biased against those of the magnitude produced by the majority of twentieth-century events.

ORIGINS OF LARGE-SCALE FLOOD DEPOSITS— THE CHUZA AND MIRAFLORES UNITS

Although the 1997–98 El Niño event may have been the most severe in the twentieth century, its sedimentary deposits are small relative to the majority of deposits that make up the Osmore paleoflood sequence (Moseley et al. 1992a; Satterlee n.d.; Keefer et al. 1998, 2003; Satterlee et al. 2001). We hypothesize that larger-scale debris-flow and flood deposits are products of highly variable combinations of water and sediment, because massive flooding is produced both by large quantities of runoff transporting moderate sediment loads and by moderate runoff entraining large quantities of sediment. Relative to the latter, if an El Niño were to occur today, then the ensuing debris flows would probably be relatively large because of the large quantities of mass-wasting sediment generated throughout the Osmore region by the magnitude 8.4

FIG. 4 Exposure of Chuza and Miraflores paleoflood deposits in north bank of Chuza Quebrada. In exposure, top layer is thin deposit from 1982–83 El Niño event. Below that is wall and soil from Spanish colonial-period agricultural terrace with stump and root of olive tree. Below terrace material is Chuza deposit, here 54–61 cm thick, and below that is the Miraflores deposit, here up to 2.14 m thick. Small scale marked in inches and centimeters; longer-scale tape is extended to length of 1.00 m.

earthquake that occurred near the coast of Peru (16.26° S, 73.64° W) on June 23, 2001 (Keefer and Moseley 2004). A recent analogue for such synergistic consequences of tectonic events and ensuing ENSO events is provided by the May 1970 magnitude 7.9 Santa River, Peru, earthquake that generated an estimated 100 to 200 million m^3 of landslide debris (Plafker et al. 1971). Reposing on the lower, normally rainless watershed there, this loose material was subsequently entrained by runoff from the strong El Niño event of 1972–73 and contributed to massive flooding as well as massive coastal deposition and progradation (Moseley and Richardson 1992; Moseley et al. 1992b).

Similar depositional processes are inferred for some components of the Osmore paleoflood sequence (Moseley et al. 1992a; Satterlee n.d.; Satterlee et al. 2001; Keefer et al. 2003), including the Chuza unit, which represents the most recent, large-magnitude flood episode in the region. In comparison to the thinner, finer sediments of recent El Niños that often overlie the unit, Chuza debris flow and flood deposits are thick and coarse, contain abundant, locally derived angular clasts, and overtop channel banks in many localities (fig. 4). In addition, the unit contains well-preserved plant remains, including leaves and wood from olive trees, suggesting that the ratio of water to solid material was relatively low. In many localities, the unit directly overlies volcanic ash from the AD 1600 eruption of Huaynaputina (16.6° S, 70.9° W). We postulate that the Chuza sediments were produced by runoff from the historical El Niño of AD 1607–08 (Quinn and Neal 1995), which may have been unusually severe (Dunbar et al. 1994; Sanford et al. 1985; Meggers 1994; Schimmelmann et al. 1998; Keefer et al. 2003) and which followed an AD 1604 magnitude 8.4 earthquake centered in the Osmore region (18.0°S, 71.5°W; Silgado 1978).

According to modeling of the production of mass-wasting sediment by earthquakes worldwide (Keefer 1994), the 1604 earthquake may have generated on the order of 5 billion m^3 of sediment, much of it consisting of loose, coarse, angular material of the type found entrained in the Chuza deposits. Evidently then, tectonically induced mass wasting was the principal contributor to the Chuza unit's massive debris flows. Certain other units of the Osmore sequence also appear to be products of tectonic events followed by El Niño events (Keefer et al. 2003). This may be a fairly frequent formation process given that Andean historical (Silgado 1978) and instrumental (Espinosa et al. 1985) records indicate that more than thirty earthquakes with magnitudes greater than 7.0 may occur each century, and modeling indicates that each may produce 40 million to 5 billion m^3 of landslide material (Keefer 1994).

A different class of large-scale debris flow and flood deposits was evidently produced by very high ratios of water to sediment in the Miraflores unit, which immediately underlies the AD 1600 Huaynaputina ash deposit. Though typically thicker and even more laterally extensive than the Chuza unit deposits (fig. 4), Miraflores deposits are also typically finer and contain floral impressions that survive only as casts and molds, suggesting high water content. The Miraflores deposits are generally massive, but indications of bedding at some localities in the Ilo Valley indicate that two or three separate flood surges may have taken place during the Miraflores event. During this event, pervasive sheet wash and debris flows inundated numerous Pre-Hispanic settlements and most local agricultural systems on a catastrophic scale that remains unrepeated. Material for radiocarbon dating of this unit has been recovered from one site at the head of the Ilo Valley, where charcoal was dated to 590 ± 35 radiocarbon years before the present (Satterlee et al. 2001).

This radiocarbon determination has calibrated intercepts of AD 1330, 1343, and 1395 with one-standard-deviation error ranges of 1312–1359 and 1388–1403 (as calibrated using the Internet-based Calib 4.2, based on datasets described by Stuiver et al. 1998a, 1998b, and available at http://www.calib.org). Magilligan and Goldstein (2001) report radiocarbon age determinations from four additional samples from the Miraflores unit collected near the city of Moquegua. These age determinations are 620±40, 730±60, 620±70, and 750±60 radiocarbon years before the present, with calibrated one-standard-deviation ranges of AD 1300–1400, 1255–1295, 1290–1410, and 1235–1290, respectively, with a weighted mean date of AD 1300. Such a range of ages for the Miraflores unit is consistent with abundant stratigraphic and archaeological evidence (Satterlee n.d.; Satterlee et al. 2001). The Osmore flood sequence includes several earlier large-scale units with sediments that are similar in important respects to those of the Miraflores Unit, including some consisting of single, massive debris flow events (Keefer et al. 2003). Therefore, we postulate that there have indeed been rare but recurrent "mega-Niños," as implicated in pioneering studies (Caviedes 2001), and that they have generated flash floods that greatly exceed the magnitude of any recent events along the Andean coast.

DISCUSSION AND CONCLUSIONS

Recent episodes of El Niño–induced flash flooding in the Osmore region of southern Peru demonstrate that later, larger events typically mask or destroy the geological evidence of earlier, smaller ones. This can explain why deposits of scale comparable to those of very strong twentieth-century El Niño events are underrepresented in the paleoflood record, whereas large-scale deposits make up most of that record. Although such paleoflood deposits may constitute an incomplete record of past El Niño activity, they provide extraordinarily significant proxies of recurrent processes that are exceptionally calamitous for humans. We have summarized examples of two catastrophic events involving different ratios of sediment and water.

The collateral volcanic, earthquake, and El Niño disasters that produced the Chuza unit in the early seventeenth century resulted in more than a thousand fatalities (Silgado 1978; de Silva et al. 2000). Prehistoric fatalities are difficult to calculate for the Miraflores unit, but in its wake 77 percent of the preflood settlements in the affected area were abandoned (Reycraft 2000), and we also infer that population declined by an equivalent amount. Furthermore, demographic, social, and political conditions never returned to their former state. Unlike the Chuza event, this catastrophe was presumably pan-Andean if not of global reach and approximates the "mega-Niño" severity of erosion and deposition postulated by the first Andean paleoflood study more than two decades ago (Diaz and Markgraf 1992). Independent confirmation of the original identification has been a long time coming due to a lack of support for Andean paleoflood studies, since they fall into the disciplinary cracks between climatology, geology, and anthropology. More is the pity, for these important proxies point to rare but recurrent catastrophic processes that people and their national planners are neither aware of nor immune to.

ACKNOWLEDGMENTS

We thank Southern Peru Copper Corporation for consistent long-term infrastructure support facilitating more than two decades of El Niño observations that draw heavily upon acknowledgeable observations of numerous fishermen, farmers, and technical personnel in the study area. Dennis Satterlee, Jorge Tapia, and Adán Umire have been very important contributors of observational field data. The Miraflores unit was defined with support from the Heinz Charitable Foundation, and the U.S. Geological Survey has supported David Keefer's paleoflood investigations.

REFERENCES CITED

Bawden, Garth
1996 *The Moche*. Blackwell, Cambridge, MA.

Caviedes, César N.
1984 El Niño 1982–83. *Geographical Review* 74: 267–290.

2001 *El Niño in History: Storming through the Ages*. University of Florida Press, Gainesville.

Caviedes, César N., and Peter R. Waylen
1991 Chapters for a Climatic History of South America. In *Beiträge zur regionalen und angewandten Klimatologie* (Wilfred Endlicher and Hermann Gossmann, eds.): 149–180. Freiburger geographische Hefte 32. Institut für Physische Geographie, University of Freiburg, Germany.

Craig, Alan K., and Izumi Shimada
1986 El Niño Flood Deposits at Batán Grande, Northern Peru. *Geoarchaeology* 1: 29–38.

Davis, Mike
2001 *Late Victorian Holocausts— El Niño Famines and the Making of the Third World*. Verso, London.

de Silva, Shanaka, Jorge Alzueta, and Guido Salas
2000 The Socioeconomic Consequences of the A.D. 1600 Eruption of Huaynaputina, Southern Peru. In *Volcanic Hazards and Disasters in Human Antiquity* (Floyd W. McCoy and Grant Heiken, eds.): 15–24. Geological Society of America Special Paper 345. Geological Society of America, Boulder, CO.

Diaz, Henry F., and Vera Markgraf (eds.)
1992 *El Niño: Historical and Paleoclimatic Aspects of the Southern Oscillation*. Cambridge University Press, Cambridge, England.

Donnan, Christopher B.
1990 An Assessment of the Validity of the Naymlap Dynasty. In *The Northern Dynasties: Kingship and Statecraft in Chimor* (Michael E. Moseley and Alana Cordy-Collins, eds.): 243–273. Dumbarton Oaks Research Library and Collection, Washington, D.C.

Dunbar, Robert B., Gerard M. Wellington, Mitchell W. Colgan, and Peter W. Glynn
1994 Eastern Pacific Sea Surface Temperature since 1600 A.D.: The ^{18}O Record of Climate Variability in Galápagos Corals. *Paleoceanography* 9: 291–315.

Espinosa, A. F., Lucia A. Casaverde, J. A. Michael, Jorge Alva-Hurtado, and Julio Vargas-Neumann
1985 *Earthquake Catalog of Peru*. U.S. Geological Survey Open-File Report 85–286.

Fontugne, Michel, Pierre Usselmann, Danièle Lavallée, Michèle Julien, and Christine Hatté
1999 El Niño Variability in the Coastal Desert of Southern Peru during the Mid-Holocene. *Quaternary Research* 52: 171–179.

Goldberg, Richard A., Gilberto M. Tisnado, and R. A. Scofield
1987 Characteristics of Extreme Rainfall Events in Northwestern Peru during the 1982–1983 El Niño Period. *Journal of Geophysical Research* 92 (C13): 14,225–14,241.

Grodzicki, Jerzy
1994 Nasca: Los síntomas geológicos del fenómeno El Niño y sus aspectos arqueológicos. *Estudios y Memorias* 12. University of Warsaw Centre for Latin American Studies, Warsaw.

Keefer, David K.
1994 The Importance of Earthquake-Induced Landslides to Long-Term Slope Erosion and Slope-Failure Hazards in Seismically Active Regions. *Geomorphology* 10: 265–284.

Keefer, David K., Susan D. deFrance, Michael E. Moseley, James B. Richardson III, Dennis R. Satterlee, and Amy Day-Lewis
1998 Early Maritime Economy and El Niño Events at Quebrada Tacahuay, Peru. *Science* 281: 1833–1835.

Keefer, David K., and Michael E. Moseley
2004 Southern Peru Desert Shattered by the Great 2001 Earthquake: Implications for Paleoseismic and Paleo–El Niño–Southern Oscillation Records. *Proceedings of the National Academy of Sciences of the United States of America* 101: 10,878–10,883.

Keefer, David K., Michael E. Moseley, and Susan D. deFrance
2003 A 38,000 Year Record of Floods and Debris Flows in the Ilo Region of Southern Peru and Its Relation to El Niño Events and Great Earthquakes. *Palaeogeography, Palaeoclimatology, Palaeoecology* 194: 41–77.

Magilligan, Francis J., and Paul S. Goldstein
2001 El Niño Floods and Culture Change: A Late Holocene Flood History for the Río Moquegua, Southern Peru. *Geology* 29: 431–434.

McClelland, Donna
1990 A Maritime Passage from Moche to Chimu. In *The Northern Dynasties: Kingship and Statecraft in Chimor* (Michael E. Moseley and Alana Cordy-Collins, eds.): 75–106. Dumbarton Oaks Research Library and Collection, Washington, D.C.

McGlone, Matt S., A. Peter Kershaw, and Vera Markgraf
1992 El Niño/Southern Oscillation Climatic Variability in Australasian and South American Paleoenvironmental Records. In *El Niño: Historical and Paleoclimatic Aspects of the Southern Oscillation* (Henry F. Diaz and Vera Markgraf, eds.): 435–462. Cambridge University Press, Cambridge, England.

Meggers, Betty J.
1994 Archaeological Evidence for the Impact of Mega-Niño Events on Amazonia during the Past Two Millennia. *Climatic Change* 28: 321–328.

Moseley, Michael E.
1987 Punctuated Equilibrium: Searching the Ancient Record for El Niño. *Quarterly Review of Archaeology* Fall: 7–11.

1990 Structure and History in the Dynastic Lore of Chimor. In *The Northern Dynasties: Kingship and Statecraft in Chimor* (Michael E. Moseley and Alana Cordy-Collins, eds.): 1–41. Dumbarton Oaks Research Library and Collection, Washington, D.C.

Moseley, Michael E., Robert A. Feldman, and Charles R. Ortloff
1981 Living with Crises: Human Perception of Process and Time. In *Biotic Crises in Ecological and Evolutionary Time* (Matthew Nitecki, ed.): 231–267. Academic Press, New York.

Moseley, Michael E., and James B. Richardson III
1992 Doomed by Natural Disaster. *Archaeology* 45 (6): 44–45.

Moseley, Michael E., Jorge Tapia, Dennis R. Satterlee, and James B. Richardson III
1992a Flood Events, El Niño Events, and Tectonic Events. In *Paleo-ENSO Records International Symposium, Extended Abstracts* (Luc Ortlieb and José Macharé, eds.): 207–212. ORSTOM and Instituto Geofísico del Perú, Lima.

Moseley, Michael E., David Wagner, and James B. Richardson III
1992b Space Shuttle Imagery of Recent Catastrophic Change along the Arid Andean Coast. In *Paleoshorelines and Prehistory: An Investigation of Method* (Lucy Lewis Johnson and Melanie Stright, eds.): 215–235. CRC Press, Boca Raton.

Nials, Fred L., Eric R. Deeds, Michael E. Moseley, Shelia G. Pozorski, Thomas Pozorski, and Robert A. Feldman
1979 El Niño: The Catastrophic Flooding of Coastal Peru. *Field Museum of Natural History Bulletin* 50 (7): 4–14 and 50 (8): 4–10.

Plafker, George, George E. Ericksen, and Jaime Fernández Concha
1971 Geological Aspects of the May 31, 1970, Perú Earthquake. *Bulletin of the Seismological Society of America* 61: 543–578.

Pozorski, Thomas
1987 Changing Priorities within the Chimu State: The Role of Irrigation Agriculture. In *The Origins and Development of the Andean State* (Jonathan Haas, Shelia Pozorski, and Thomas Pozorski, eds.): 111–120. Cambridge University Press, Cambridge, England.

Quinn, William H., and Victor T. Neal
1995 The Historical Record of El Niño Events. In *Climate since A.D. 1500*. Rev. ed. (Raymond S. Bradley and Philip D. Jones, eds.): 623–648. Routledge, London.

Reycraft, Richard M.
2000 Long-Term Human Response to El Niño in South Coastal Peru. In *Environmental Disaster and the Archaeology of Human Response* (Garth Bawden and Richard M. Reycraft, eds.): 99–120. Anthropological Papers 7. University of New Mexico, Maxwell Museum of Anthropology, Albuquerque.

Sanford, Robert L., Jr., Juan Saldarriaga, Kathleen E. Clark, Christopher Uhl, and Rafael Herrera
1985 Amazon Rain-Forest Fires. *Science* 227: 53–55.

Satterlee, Dennis R.
n.d. The Impact of a Fourteenth-Century El Niño Flood on an Indigenous Population near Ilo, Peru. Ph.D. dissertation, Department of Anthropology, University of Florida, Gainesville.

Satterlee, Dennis R., Michael E. Moseley, David K. Keefer, and Jorge E. Tapia A.
2001 The Miraflores El Niño Disaster: Convergent Catastrophes and Prehistoric Agrarian Change in Southern Peru. *Andean Past* 6: 95–116.

Schimmelmann, Arndt, Meixun Zhao, Colin C. Harvey, and Carina B. Lange
1998 A Large California Flood and Correlative Global Climatic Events 400 Years Ago. *Quaternary Research* 49: 51–61.

Shimada, Izumi
1990 Cultural Continuities and Discontinuities on the Northern North Coast of Peru, Middle– Late Horizons. In *The Northern Dynasties: Kingship and Statecraft in Chimor* (Michael E. Moseley and Alana Cordy-Collins, eds.): 297–392. Dumbarton Oaks Research Library and Collection, Washington, D.C.

Silgado, Enrique F.
1978 Historia de los sismos mas notables ocurridos en el Peru (1513–1974). Boletin 3, serie C del Sector Energía y Minas, Instituto de Geología y Minera, Geodinámica e Ingeniería Geológica, Lima.

Stuiver, Minze, Paula J. Reimer, Edouard Bard, J. Warren Beck, G. S. Burr, Konrad A. Hughen, Bernd Kromer, Gerry McCormac, Johannes van der Plicht, and Marco Spurk
1998a INTCAL98 Radiocarbon Age Calibration 24,000–0. *Radiocarbon* 40: 1,041–1,083.

Stuiver, Minze, Paula J. Reimer, and Thomas F. Braziunas
1998b High-precision Radiocarbon Age Calibration for Terrestrial and Marine Samples. *Radiocarbon* 40: 1,127–1,151.

Waylen, Peter R., and César N. Caviedes
1986 El Niño and Annual Floods on the North Peruvian Littoral. *Journal of Hydrology* 89: 141–156.

Wells, Lisa E.
1987 An Alluvial Record of El Niño Events from Northern Coastal Peru. *Journal of Geophysical Research* 92 (C13): 14,463–14,470.

1990 Holocene History of the El Niño Phenomenon as Recorded in Flood Sediments of Northern Coastal Peru. *Geology* 18: 1,134–1,137.

n.d. Holocene Fluvial and Shoreline History as a Function of Human and Geologic Factors in Arid Northern Peru. Ph.D. dissertation, Department of Geology, Stanford University, Stanford, CA.

MARCHING TO DISASTER:
THE CATASTROPHIC CONVERGENCE OF
INCA IMPERIAL POLICY, SAND FLIES, AND
EL NIÑO IN THE 1524 ANDEAN EPIDEMIC

James B. Kiracofe
John S. Marr

The invasion of the Inca realm by Pizarro and his followers after 1530 caused a sudden and dramatic redirection of Andean cultural evolution. The rapid Spanish conquest followed a period of dynastic destabilization and subsequent civil war between the contending claimants to imperial authority after the death of the last legitimately installed Inca emperor, Huayna Capac, and many of his relatives in a widespread epidemic circa 1524. We have argued at length elsewhere against previously accepted diagnoses of smallpox while advancing our contention that bartonellosis was the real cause of this outbreak (Kiracofe and Marr n.d.). Recent research (Zhou et al. 2002, n.d.) demonstrates the relationship between El Niño events and the cyclical occurrence of bartonellosis outbreaks in the Andean region: El Niño events dramatically increase the intensity of bartonellosis epidemics and permit the spread of the disease into new areas. Here, we briefly describe the disease, explain its mode of transmission, show how the military campaigns of Huayna Capac along the Manabí coast of Ecuador set the stage for a widespread disaster, and review evidence for a convergence of Huayna Capac's fatal campaigns with an El Niño event.

BARTONELLOSIS

Bartonellosis is a potentially fatal, exclusively New World, arthropod-borne bacterial disease, long endemic in the northeastern Andean region. Its acute phase is characterized by irregular fever, headache, muscle aches and pains, and a severe hemolytic anemia that may lead to death if untreated. This acute phase is known as Oroya fever, named after an 1871 epidemic that killed roughly 8,000 railroad workers along the Lima–La Oroya railroad line in the Rimac Valley of Peru.

A chronic phase may follow in weeks or months and is referred to as *verruga peruana* (or Peruvian wart), so named by a highly vascularized, wartlike nodular (and occasionally pedunculated) skin eruption that may cover the

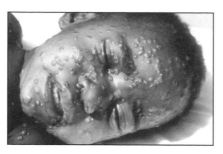

FIG. 1. (Above) Smallpox (courtesy of the World Health Organization). (Left) Bartonellosis (from Odriozola 1944

entire body, but is usually most prominent on the extremities. Prior to 1913, the disease was known as Carrión's disease, an eponym derived from the Peruvian medical student, Daniel Carrión, who in 1885 gave his life, through self-experimentation, discovering the linkage between Oroya fever and *verruga peruana*. Transmitted from person to person by the bite of sand flies, the disease is caused by a bacterium, *Bartonella bacilliformis*, named in honor of the Peruvian bacteriologist, Alberto Barton.

In historic times epidemics of bartonellosis occurred in public-works projects and armies on the march. The Incas mobilized tens of thousands of rotational laborers from various regions of the empire, concentrating them on vast construction projects. Based on the symptoms reported in the early accounts, including fever and a severe skin rash, many writers believed that Huayna Capac succumbed to smallpox. However, these symptoms correspond more closely with clinically documented symptoms of bartonellosis, a disease then unknown to the Spanish (figs. 1 and 2).

In his *Antigualla peruana*, Jiménez de la Espada (1892) mentions an epidemic at the time of Huayna Capac's death that he calls smallpox. In his *Relaciones geográficas de Indias* (Jiménez de la Espada 1965 [1881]; cited by Polo 1913, authors' translation), he also copied the following regarding Tomebamba: "where Huayna Capac was for ten years as it was his major residence rather than any other, and in this time befell a very great illness and pestilence in which innumerable people died of a measles that opened everyone to an incurable leprosy of which died Lord Huayna Capac."

FIG. 2 Pre-Columbian ceramic figurine (Alexander 1995)

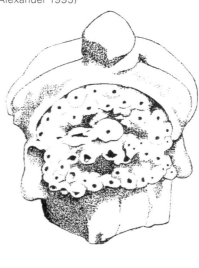

Pre-Columbian Evidence of Bartonellosis

A Pre-Columbian ceramic found near San Isidro, Ecuador shows *verruga peruana* in a severe manifestation, such as that suffered by Pizarro's men when they passed through the same area (Alexander 1995; Cieza de León 1998 [1553]: 469).

Other Pre-Columbian evidence of bartonellosis is found in the mummified remains of a ritually sacrificed young man, demonstrating that armies on the march were an important triggering mechanism for epidemic bartonellosis in the Pre-Columbian world, as they have been in modern times (Allison et al. 1974: 298).

Epidemiology of Bartonellosis

Bartonellosis is endemic to certain mountain valleys throughout the northwestern Andean region, but recent research has expanded the geographical distribution of its vector (various *Lutzomyia* species), as well as the distribution of human cases of the disease. The pathogenic organism resides in both symptomatic and asymptomatic (carrier) human reservoirs, where its potential infectivity may extend for years. Recently, the possibility of a rodent reservoir has been postulated, because infected *Lutzomyia* species were recovered in areas with sparse human populations (Cooper et al. 1997).

The interval between the acute and chronic phases may be variable. As Ellis et al. (1999: 348) notes, "the cutaneous phase of the disease does not always follow the anemic phase." In addition, the anemia caused during the fulminant hemolytic onset may linger and remain a debilitating factor even during the dermatological presentation (Salazar 1858). Between the acute and chronic manifestations there may be an asymptomatic intermediate period (Maguiña and Gotuzzo 2000: 5). The disease is transmitted indirectly from person to person by phlebotomine sand flies (Ellis et al. 1999: 344; Ihler 1996: 2).

The incident at the La Oroya railway bridge was the single most deadly outbreak recorded in modern times (García-Cáceres and García 1991: 61). It is "said to have killed more than 7000 of the 10,000 workers building the railroad between Lima and La Oroya" (Ihler 1996: 2). This would mean a general mortality rate in excess of 70%. This exceeds the typical mortality rates of smallpox, stated to have been 20–40% or more, in the standard text, and the stated case mortality rate for plague of 50–60% (Chin 2000: 456, 381). "Untreated, Oroya fever comes close to having the highest death rate of all infectious diseases, 40–85%, since essentially 100% of the red cells can be parasitized, and because of secondary bacterial infection" (Ihler 1996: 2). Thus, even in isolated areas, with an unapparent asymptomatic human reservoir, bartonellosis can quickly erupt in epidemic fashion with extremely high mortality rates. This happens because infected sand flies can effectively spread the disease, rapidly infecting thousands of susceptible nonimmune individuals entering the area, such as armies or construction workers.

Human-reservoir populations in endemic foci may develop immunity to the disease, so that many members of a community may act as carriers and have minimal, subclinical, or no disease symptoms (Amano et al. 1997: 178).

FIG. 3 Ollantaytambo, water canal (photograph by James B. Kiracofe)

New outbreaks with subsequent development of additional endemic foci may result when asymptomatic infected populations travel through or settle in adjacent nonimmune areas where sand flies are present. Alternatively, nonimmune members from outside communities may pass through endemic areas, become infected and then pass into other areas previously free of this disease. Thus, in any area with human and sand-fly populations, bartonellosis may be introduced by infected insects, by asymptomatic carriers, or by overtly infected humans (Alexander 1995: 357). During a 1990 epidemic in a Peruvian Andean village where bartonellosis had not been endemic, "the case fatality rate for patients who did not receive antibiotics was 88%. . . . The overall case fatality rate was 50%" (Gray et al. 1990: 217, 219).

Significantly, the local water source, running in unprotected open canals through the village, was contaminated with "coliforms"—bacteria indicative of fecal contamination. This finding also illustrates another co-factor associated with high bartonellosis fatality rates. Both food and waterborne bacterial pathogens (particularly salmonellosis) are synergistic co-factors associated with mortality as high as 90% (Gray et al. 1990: 220).

Ollantaytambo in the Urubamba Valley near Cusco is perhaps the only example of Inca urban planning to survive more or less as built in a continuously inhabited community. The ceremonial center cut into the mountainside adjacent to the town was unfinished when the Spanish conquistadors arrived in the 1530s. Under the Spaniards, Mediterranean architectural elements were superimposed on existing dwellings, but the distinctive trapezoidal grid plan of the streets has remained in place together with the hydraulic system that provides water throughout the town through open canals that run down the edges of the stone-paved streets (Protzen 1993; Wright et al. 1999). The water from these canals in Ollantaytambo was still used routinely for household

FIG. 4 Stone walls of dwellings at Machu Pichu (photograph by James B. Kiracofe)

FIG. 5 Adobe and stone dwelling at Pisac (photograph by James B. Kiracofe)

purposes in 1994 (fig. 3). The opportunity for the introduction of salmonella from free-ranging animal populations as well as human contamination of the waters is obvious.

Between April and May of 1998, an epidemic of bartonellosis occurred in the Urubamba region, during which patients reported for treatment in Urubamba, Calca, and Ollantaytambo (Ellis et al. 1999: 345). This area of Peru was not previously thought to be endemic for bartonellosis. However, results of the study "may indicate that Cusco is an endemic area of transmission that was previously unrecognized, and that adults in this area are at least partially immune to [bartonellosis]." Furthermore, the researchers noted, "the only sand fly species encountered in the Urubamba area was *Lutzomyia peruensis*, known to feed on dogs and humans. This is the first report of sequence-confirmed *B. bacilliformis* from this species of *Lutzomyia*, though a report from 1942 suggested that *L. peruensis* may be associated with transmission of bartonellosis." The authors go on to note that "favored harborage and breeding areas for *Lutzomyia* include adobe walls inside homes, and stone walls outdoors. Many of the homes in the area had these constructions, and we collected *Lutzomyia* by aspiration from adobe and stone walls in case and control households" (Ellis et al. 1999: 348).

Stone and adobe were the favored construction materials in the area when the Incas began building Ollantaytambo (figs. 4 and 5). The typical architecture of both elite and commoner domestic residences as well as military barracks incorporated these materials and techniques. Thus, Inca structures were ideal habitats for breeding large sand-fly populations, which favor such small, micro-environmental niches in natural and man-made cracks in rock and adobe.[1]

1 For more information on sand flies, see http://cipa.snv.jussieu.fr/.

BARTONELLOSIS AND THE 1531 SPANISH
EXPEDITION ON ECUADOR'S MANABÍ COAST

Documentary, archaeological, and paleopathological evidence show that bartonellosis existed in Pre-Columbian times, in endemic foci, and in the very regions through which Pizarro passed with his troops in 1531–32. Pizarro's force was struck by an unusual disease described as a type of smallpox. We suggest that the infection was acute and chronic bartonellosis.

In comparing a map of endemic foci with a map of Pizarro's path in 1531–32 (Amano et al. 1997: 175; Hemming 1970: 13), it is clear that Pizarro traversed an area long endemic for bartonellosis (fig. 6). His nonimmune Spanish soldiers paid a terrible price for their intrusion along the Manabí coast:

> In Coaque were found many mattresses of ceyna these being from a tree they grow. It so happened that some of the Spaniards who rested on these woke up paralyzed, so that their arms could not be bent or their legs straightened without considerable effort; this occurred to certain of them and it was understood that this was the onset of a sickness that caused warts, so bad and grievous that it caused many to be much fatigued. They toiled in great pain as if they had buboes, until they broke out in warts that covered the whole body, some as large as eggs. On being burst these produced matter and blood, so that it was necessary to cut them off and place upon the sores strong potions to remove the root. Others were as small as measles that covered the whole of the men's bodies. Few escaped the disease, although some suffered more than others. Others will say that this sickness was caused by some fish that they ate in Porto Viejo, that the Indians in their wickedness gave to the Spanish. (Pizarro 1978 [1571], cited in Alexander 1995: 355–356)

Stiffness of the joints associated with pain and fatigue followed by the eruption of warts and lesions corresponds precisely with the classic signs and symptoms of the progress of bartonellosis from its acute, Oroya-fever phase through the eruptive phase marked by the lesions of *verruga peruana* (Kosek 2000: 866). Of the same incident at the Bay of Coaque, Garcilaso de la Vega (1966 [1609]) reported:

> The excrescences broke out all over their bodies, but principally on their faces. They thought at first that these were warts. But as time passed, they grew larger and began to ripen like figs, of which they had both the size and shape; they hung and swung from a stem, secreted blood and body fluids, and nothing more frightful to see or more painful, because they were very sensitive to touch. The wretched men afflicted with the disease were horrible to look at, as they were covered with these purplish-blue fruits hanging from their foreheads, their eyebrows, their nostrils, their beards and even from their ears nor did they know how to treat them. Indeed, some died of them while others survived. (Cited in Schultz 1968: 508.)

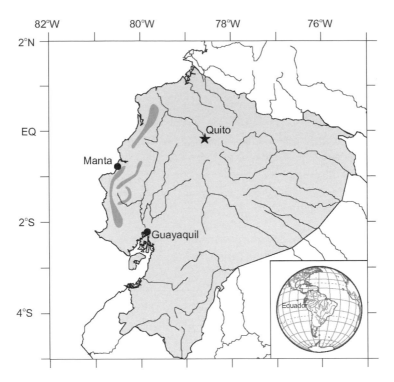

FIG. 6 Map showing areas of Ecuador endemic for bartonellosis today (darker gray)(after Amano et al. 1997)

Myron Schultz did not believe that the illness the Spaniards suffered at Coaque was bartonellosis because the epidemic at Coaque occurred on the coast at less than 500 feet above sea level, and according to Schultz, "it would be the only one that has ever occurred, either in ancient or modern times outside the range of 2,500 to 8,000 feet" (Schultz 1968: 508). However, subsequent published studies of bartonellosis cases and *Lutzomyia* populations show the disease has much wider distribution (Ollague and Guevara de Veliz 1977; Alexander 1995; Amano et al. 1997; Ellis et al. 1999; Kosek 2000; Zhou et al. 2002).

In 1547, sixteen years after the incident at the Bay of Coaque, following the Spanish conquest when Francisco Pizarro rebelled against the Spanish Crown, an army was sent from Panama to quell the revolt. They landed at Manta, on the Manabí coast of Ecuador. On the way from Manta to Porto Viejo, many became ill, as was graphically reported by Juan Cristóbal Calvete de Estrella:

> warts as big or bigger than nuts on their noses, eyebrows or beards,
> of a pestilential humor between red and black. They may last four
> or five months and until they begin to dry up do not cease to cause
> pain. Finally they begin to heal and those that have suffered them
> remain clean and healthy. It is thought by the men of these lands
> that these warts and other diseases which occur below the equator
> are due to certain constellations, the influence of which is stronger
> here than elsewhere. (Calvete de Estrella 1964 [1565–67], cited in
> Alexander 1995: 356)

Citing the same incident and Spanish source, Cook (1998: 108) points out that "[m]edical historian Lastres wondered if it might have been 'smallpox.' But, given the context of the illness infecting mostly the Europeans, a more likely diagnosis is *verruga peruana*, a form of Carrion's disease."

Like others before them, Pizarro and Calvete de Estrella discovered that the Manabí coast of Ecuador is a dangerous land. Even today, "most of the suspected and confirmed cases of bartonellosis in Ecuador have been reported from Manabí, which is an arid, coastal province whose name derives from the Quechua words *mana pi*, meaning "no water" (Alexander 1995: 355).

The Conquest of the Manabí Coast, Huayna Capac's Last Campaign

During the last years of his reign, Huayna Capac sought to complete the conquest and pacification of coastal Ecuador.[2] He assembled an army of fifty thousand soldiers from the highlands north of Cusco and marched them from Puna past Manta and Porto Viejo along the Manabí coast to Coaque on the equator and back, probably rotating them at regular intervals back to their homes in the Inca heartland. As the Spanish expedition led by Pizarro through this very same area discovered less than a decade later, and at great cost, this area was already hyperendemic for bartonellosis.[3] Furthermore, Huayna Capac employed the standard Inca "remedy" for dealing with recalcitrant populations, moving many of the recently conquered tribes from the Manabí coast to other provinces and replacing them with colonists from peaceful areas more closely allied with the Inca state.

Following this campaign along the Manabí coast, Huayna Capac and much of his army became ill and died. But this was not before they learned of an epidemic that had already swept through Cusco, killing many of the Inca ruler's relatives along with countless others. A regularly rotated army of 50,000 highland soldiers and an unknown number of relocated coastal tribes from the endemic Manabí coast may well have carried in their blood the seeds of a great disaster.[4]

To sum up, the Jipijapa-Pajan area is now endemic for bartonellosis and apparently was so even in Pre-Inca time. There is solid archaeological evidence of an Inca presence in this area, because the most direct route from Puna or Guayaquil to Manta passes through Pajan and Jipijapa. The early Spanish sources say that Huayna Capac operated in this area with his army. We have concluded from this evidence that this area is the most likely source of the epidemic of 1524.

2 See "Carta de la provincia de Quito 1750: Pedro Vicente Maldonado," in Gómez E. (1999). This map shows a road from Guayaquil to Xipixapa (Jipijapa), calling it "Camino desierto." Given the geography, the descriptions of his movements in the sources, and his objectives, it seems likely that Huayna Capac would have passed this way. See also Rostworowski (1999: 72 and the map on p. 20); McEwan and Silva (1992, especially pp. 74–75); Garcilaso de la Vega (1966 [1609]: 546, 549, 551, 568, also the map in Garcilaso de la Vega 1961 [1609]: 126, showing the areas of Huayna Capac's activities described in Garcilaso's account); and Sarmiento de Gamboa (1942 [1572]: 130–131).

3 See Alexander (1995: 352) for a discussion of endemic bartonellosis in Jipijapa and Pajan, including a historical perspective.

4 Even after full recovery from bartonellosis, the survivor may become an asymptomatic carrier, as may also happen in the case of survivors of typhoid fever, and staphylococcal infections, among others. These "carriers" can act as "seeds."

The Death of Huayna Capac

Huayna Capac's death is chronicled by Juan de Betanzos (d. 1571), who was married to the former wife of Inca Atahualpa, also the niece of Inca Huayna Capac. "Betanzos questioned his wife's relatives at length. . . . Juan de Betanzos reported that after the conquest of the province of Yaguarcoche, Inca Huayna Capac returned to Quito and was there six years. At the end of that time, he came down with an illness which deprived him of his senses and understanding, and gave him a saran [cutaneous disease] and lepra [leprosy] that made him very debilitated" (Cook 1998: 76, citing Betanzos 1987 [1551]: 200).

Images of advanced cases of leprosy[5] show a similarity to the Pre-Columbian ceramics depicting *verruga peruana* and to Pedro Pizarro's description of the disease that afflicted his compatriots during their early expedition. We do not believe that the illness that killed Huayna Capac or other of his subjects in the 1520s epidemic or that struck Pizarro's men was Hansen's disease (leprosy), but there are evident similarities in appearance, at certain phases, between leprosy and *verruga peruana* that might help to explain Betanzos's use of the term.

Bartonellosis is typically described as a two-stage disease. The first, acute stage—characterized by a severe progressive hemolytic anemia accompanied by high fever—is usually the deadly phase. The second, chronic stage—characterized by a verrucous skin eruption—is usually thought to be a sign that the patient has survived the crisis and will recover, though this is not always true.

Thomas Salazar's bachelor's thesis, "Historia de las verrugas," was published in the *Gaceta médica de Lima* in 1858. It included the first medical photograph ever published in Peru of a patient. Don Aniceto de la Cruz, a patient under Salazar's care, entered the hospital 21 June 1857 suffering from verrugas, which he had first noticed a year earlier (fig. 7). In spite of continuous treatment the verrugas multiplied and increased in size while his general condition steadily deteriorated until his death on 3 December 1857. Don Aniceto "descended into an endemic area from the uplands of central Peru, and died a week after this photograph was made" (García-Cáceres and García 1991: s61), having apparently suffered from the debilitating effects and complications from bartonellosis that probably included secondary infections.

The case of don Aniceto may help us understand the real meaning of Betanzos's account of the death of Huayna Capac. Elsewhere we hear that Huayna Capac suffered from fevers. Certainly these descriptions are consistent with the case of don Aniceto. Since the photograph (fig. 7) clearly shows the unmistakable evidence of verrucous bartonellosis and since don Aniceto died one week after this photograph was taken, there is reason to believe that Huayna Capac may have died under similar circumstances wherein skin lesions were already apparent. Like don Aniceto, Huayna Capac may have died of acute bartonellosis after suffering an eruption of verrugas. The eruption of verrugas may occur as soon as fifteen days after the onset of the fever (personal communication, Jesús González, M.D.). Although some have argued that verrugas are not concomitant with the acute anemic phase, general debility consequent

5 For examples see http://wwwihm.nlm.nih.gov/.

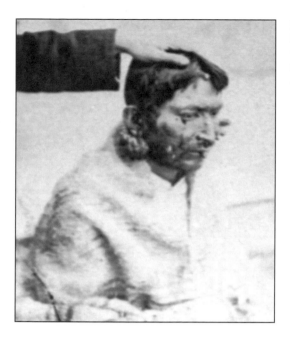

FIG. 7 Aniceto de la Cruz
by Thomas Salazar
(1858; photograph of
original courtesy of
Jorge Lossio, Lima)

to profound anemia (and increased susceptibility to enteric pathogens such as
salmonella) may have contributed to Huayna Capac's death during the later
eruptive phase.

However, even if Betanzos was not referring to verrucous symptoms of
bartonellosis in a manifestation of the disease that would have paralleled the
case of don Aniceto, a pronounced skin rash can be present at the onset of
the acute stage, which might offer an alternative explanation consistent with
Betanzos's description of a "leprous" disease.[6]

Another phrase in Betanzos's account bears careful scrutiny in light of
what we now know about bartonellosis. Betanzos wrote "he came down
with an illness which deprived him of his senses and understanding" (Cook
1998: 76). However, in a slightly different and more complete translation of
Betanzos's account, Betanzos reported the following description of the Inca
Huayna Capac's death:

> At the end of those six years in Quito, he fell ill and the illness took
> his reason and understanding and gave him a skin irritation like
> leprosy that greatly weakened him. When the nobles saw him so
> far gone they came to him; it seemed to them that he had come a
> little to his senses and they asked him to name a lord since he was
> at the end of his days. To them he replied that he named as lord
> his son Ninancuyochi, who was barely a month old and was in the
> province of the Cañares. Seeing that he had named such a baby, they
> understood that he was not in his right mind and they left him and

6 See also the description of the rash given by Daniel Carrión (in Schultz 1968). This is consistent
with the cases studied by Trelles et al. (1969), who reported death thirteen to thirty-four days
after the onset of symptoms. The petichial skin rash developed on the fourth day after onset,
and after the onset of fever.

went out. They sent for the baby Ninancuyochi, whom he had named as lord. The next day they returned and entered and asked him again whom he named and left as the ruler. He answered that he named as lord Atahualpa his son, not remembering that the day before he had named the above-mentioned baby. The nobles went immediately to the lodgings of Atahualpa, whom they told was now lord, and they gave him their respects as such. He told them that he had no wish to be the ruler even though his father had named him. The next day the nobles returned to Huayna Capac and in view of the fact that Atahualpa did not wish to be ruler, without telling him anything of what happened the day before, they asked him to name a lord and he told them it would be Huascar his son. . . . Huascar retired for a period of fasting after he heard the news in Cuzco. Here we will leave him and return to Huayna Capac, who was in his final days. After having named Huascar as ruler in the way we have described, he died in four days. After he died, the messengers who had gone for the baby who had been named as ruler by Huayna Capac returned. The baby had died the same day they arrived of the same leprous disease as his father. (Betanzos 1996 [1551]: 183–184)

Echoing don Aniceto's case, Huayna Capac suffered a skin irritation like leprosy that greatly weakened him. The illness also took his reason and understanding, but then he came a little to his senses, which suggests a period of relative lucidity following delirium. He did not remember what had happened from one day to the next or what he had said. Then, four days after his last reported communication, he died.

Trelles et al. (1969) gave detailed descriptions of nine clinical cases of bartonellosis treated in hospitals in Lima between 1945 and 1955 and involving lesions within the brain (encephalitis). Some of the patients experienced delirium and temporal-spatial disorientation. Case eight experienced loss of vision, nausea, and vomiting, followed by loss of consciousness. These symptoms were preceded by a brief period of delirium and "talking to himself." Nevertheless, the next day the patient was lucid. Subsequently his fever increased dramatically and he died five days later of bronchopneumonia (Trelles et al. 1969: 265–266). Case nine entered the hospital in a comatose condition, but four days later, when he was able to begin the interrogatory process with the doctors, he evidently suffered from amnesia, stating that he had no recollection of events during the previous period of his illness, which had begun two days before he entered the hospital (Trelles et al. 1969: 269). Accounts of Huayna Capac's death parallel the findings of the neurological forms of Carrión's disease (bartonellosis). It seems likely, therefore, that Huayna Capac suffered from a cerebral form of bartonellosis similar to those cases described in Trelles et al. (1969).

Analysis of duration of incubation to onset and death provides further evidence in favor of bartonellosis as opposed to smallpox. In eight of the nine cases[7] of neurological bartonellosis cited by Trelles et al. (1969), death followed onset of symptoms in thirteen to thirty-four days.

7 The ninth was a case of twenty-five years' duration of a different category.

Further Analysis of Huayna Capac's Death

Careful analysis of the duration of the illness after the onset of symptoms is important because sixteenth-century documents state that Huayna Capac first felt ill at Tumibamba (Tomebamba), and that he returned from there to Quito, where he died. According to Cook,

> Cabello Balboa reports that after Huayna Capac completed conquests in the north, he paused in Tumibamba. He continued to the coast and reached the island of Puná, where he may have first been notified of the arrival of Viracocha—the European. At Puná, Huayna Capac also "received very sad news from Cusco, where they informed him that there was an incurable and general pestilence that had taken his brother Auqui Topa Inca and his uncle Apo Illaquita, whom he had left as governors when he left Cusco, and his sister Mama Toca, and other principal lords of his lineage." Profoundly saddened by the news, Huayna Capac continued toward Tumibamba. . . . From Tumibamba, "feeling indisposed and in ill health, he continued to Quito with the greatest and best part of the army, and arriving there his malady worsened, and turned into deadly fevers." Cabello Balboa gives the date of the ruler's death as "according to our count, in the year 1525." (Cook 1998: 80–81, following Cabello Valboa 1951 [1586]: 393–394)

It is impossible to know the exact travel time of Huayna Capac's journey, but John Hyslop's study of the Inca road system suggests 60 km per day as the extreme distance for carriers in the eighteenth century, 50 km for guides on his field trip, and 40 km per day as the speed of Pizarro's army (Hyslop 1984: 297). There are 350 km between Tumibamba and Quito. At 40 km per day, the travel time would be approximately ten days, but at the highest rate of speed reported, the duration would be no less than six days.

Huayna Capac came "a little to his senses" following a period of delirium. We are not told how long he had been in the city following his return from Tumibamba. On the day that he returned to lucidity, and on each of the two following days, the nobles questioned him about his successor. He died after another four days. Taken together, this makes seven days. Added to an estimated ten days of travel time, not including any time that may have elapsed not accounted for in the sources, this makes at least seventeen days, during which time his illness progressed from indisposition to death. As Cabello de Balboa (1951 [1586]: 393–394 in Cook's 1998: 80–81 translation) reports, upon arriving in Quito, "his malady worsened, and turned into deadly fevers," so from this one might assume that there was some additional, but unknown, period of time not accounted for in this estimate of seventeen days.

These symptoms and the sequence of their appearance are consistent with the clinically reported onset of the Oroya-fever phase of bartonellosis. Smallpox has a more rapid onset and progress to crisis, with death usually occurring between the fifth and seventh day after the onset of symptoms, and occasionally as late as the second week (Chin 2000: 455–457). Bartonellosis is thus a more likely diagnosis for Huayna Capac's fatal disease.

TANTALIZING TEMPTATIONS, MISSING LINKS, AND THE GREAT NEW WORLD "PANDEMIC"

The weight of the evidence supports a diagnosis of bartonellosis, but contradicts Schultz, who wrote: "The descriptions of a rash together with the high fever and high mortality of the native Andeans are suggestive of an exanthematous disease. Furthermore, the great rapidity of its spread and the intriguing fact that Huayna Capac fell ill after he had received a messenger from Cusco, the epidemic center, also suggest that this was a disease spread by contact rather than one spread by an arthropod (sic. insect) vector" (Schultz 1968: 506). Skin rash, high fever, and high mortality are all characteristic of acute bartonellosis, as is its rapid spread by the sand-fly vector, as seen in the Lima–La Oroya railroad bridge epidemic.

If the messenger were the carrier of smallpox that came from Cusco, "the epidemic center," as Schultz seems to suggest, there is no mention of his suffering symptoms of smallpox, which would have been advanced and note-worthy as well as debilitating. If the messenger who arrived to infect Huayna Capac was only the last in a series of multiple relay runners from Cusco, it would be even more difficult to explain how a continuous, unbroken chain of transmission was maintained over the thousand-odd miles separating the emperor from Cusco.

Nor is there any mention anywhere in the sources of others around Huayna Capac at that time suffering from the contagious symptoms of smallpox, as would have been noteworthy and necessary for Huayna Capac to have contracted smallpox while in the area. On the other hand, inhabitants of the endemic areas through which Huayna Capac would have recently passed would have been immunologically adapted to bartonellosis over time, and need not have been suffering or presenting visible symptoms to have been reservoirs from whom sand-fly vectors could have transmitted bartonellosis (Amano et al. 1997). Huayna Capac and the others could have remained asymptomatic for up to 210 days after becoming infected, although the typical incubation period has recently been placed at sixty-one days (Maguiña and Gotuzzo 2000: 5).

THREE FINAL POINTS

Mortality Rates and Contamination of Public Water

There are three important points we would like to emphasize before closing the argument. The first is that mortality rates for bartonellosis are dramatically increased when other systemic bacterial infections are present. Among these is *Salmonella typhimurium* (*Salmonella enterica*, serovar Typhimurium), which includes over 2,000 serotypes widely spread in wild and domestic animal populations throughout the world (Maguiña and Gotuzzo 2000). Moreover, "case fatality ratios (CFRs) of untreated Oroya fever exceed 40% but may reach more than 90% when super infection with Salmonella species occurs" (Ellis et al. 1999: 344).

Public water supply systems using open canals may become contaminated by human and animal feces, which may contain salmonella and other pathogenic enteric organisms. Gray et al. (1990) describe the 1987 case of the Peruvian Andean village, and similar water systems can still be seen at Pisac and numerous other Inca sites including Ollantaytambo. If large Inca population centers such as Cusco relied on open-canal public water delivery systems that were susceptible to contamination, this may have led to endemic viral, bacterial, and protozoan gastrointestinal diseases. Even today this happens in many parts of Latin America, including the Andean region.

In rural farming areas drawing water directly from natural springs or rivers, similar contamination may have occurred, as it does today in rural areas of Latin America and elsewhere. While the rural and urbanized Inca population may have routinely experienced episodic bouts of self-limited gastroenteritis, including salmonellosis, if epidemic bartonellosis were introduced, mortality rates would be expected to rise from 30% to 90% from septisemic salmonella superinfections in areas nonimmune to bartonellosis.

Rapid Ecological Change, Reservoir Populations, and Vector Nidalities

Second, we call attention to the magnitude and rapidity of the ecological changes brought about during the reign of Huayna Capac, from approximately AD 1498 to 1524. Indeed, many of the vast public works for which the Inca Empire has become so famous were still works in progress when the Spaniards arrived. In other words, the construction of large, private, elite estates and sophisticated urban systems, and the consequent concentration of population, were still underway. These were young cities and villas, at a time when the empire was still expanding rapidly into new territories, absorbing and relocating new ethnic groups who may have brought with them new diseases to the nonimmune Inca heartland from the remote endemic periphery of the empire. Huayna Capac put into the field armies of 40,000 to 50,000 soldiers, soldiers who were rotated regularly. The Inca state mobilized tens of thousands of rotational laborers to work on public projects. Even larger numbers worked on royal estates. Huayna Capac himself mobilized over 150,000 laborers "from throughout the empire" to work on his estate alone (Niles 1999: 296).

The concentration of a human workforce of this magnitude, composed of rotational laborers working in a relatively confined area for a sustained period of time stretching over several years, certainly calls into question how the workers were fed and housed and how even rudimentary sanitation was managed, if it existed at all. The fact that many if not most of the workers were employed in hydraulic works raises additional questions about exposure to coliforms that, as previously mentioned, dramatically increase mortality rates during bartonellosis epidemics.

Beyond this is the more intriguing question of nonhuman reservoirs recently raised by Ihler, among others: "The source of the bacteria which colonize sand flies is not known. Because of the close phylogenetic relationship to the former *Grahamella* species, which are widely prevalent in small

FIG. 8 Terraces at Pisac (photograph by James B. Kiracofe)

woodland animals, it seems possible that a mammalian reservoir for *B. bacil-liformis* might exist. It seems also possible to me that *B. bacilliformis* might grow on some plant or fruit debris with which the sand flies associate, eat, or obtain fluids" (Ihler 1996: 2).

Another recent study of epidemic bartonellosis in Ecuador closely examined the possibility of a nonhuman reservoir (Cooper et al. 1997). Cooper and his colleagues found dead rodents present only in the case houses and suggest that the outbreak may have followed increased rodent invasion of houses during food shortages caused by drought conditions. Cooper et al. (1997: 544) point out that "[a]ll case houses were located in isolated areas at the margin of forest and the presence of dead rodents was reported only in case houses. We suggest that human bartonellosis is a zoonosis with a natural rodent reservoir and that migrant humans infected in this way may become a temporary reservoir host in populated areas."

There are several reasons why this important finding about wild rodents may have a direct bearing on the epidemic of 1524. First, because one of the major goals of Inca state and imperial policy was to increase the ability to produce food; they did so by mobilizing immense human workforces implementing vast terracing programs on royal estates and elsewhere. Not only did the relatively sudden creation of terrace walls built of coursed stone present millions of square feet of vertical surface as a new habitat for sand flies, creating their "favored [outdoor] harborage and breeding areas," but also the terraces themselves greatly increased food production, not only for the human population, but no doubt for the expanding wild-rodent populations as well (fig. 8). Second, because along with the construction of the famous Inca Road, there was also systematic construction of special infrastructure, built at regular intervals all along the road, for maintaining and distributing food supplies, principally grains (LeVine 1992). Not only would this have provided needed food for the human population, but also it would have again provided the means for expanding the wild rodent populations. Similarly, rapidly expanding human urban populations, such as in Cusco, may also have contributed to a corresponding increase in wild-rodent populations within the urban precincts.

If human bartonellosis is a zoonosis with a natural rodent reservoir, as Cooper et al. (1997) suggest on the basis of their recent investigation, then the relatively sudden expansion of the rodent reservoir in response to greater availability of food from intensified human agriculture may have played an all-important role in the epidemic of 1524. Furthermore, if, as seen in the case Cooper et al. (1997) studied, hydroclimatological conditions, such as drought, tend to intensify rodent invasion of houses, leading in turn to epidemic outbreaks or intensified epidemic outbreaks, then this, too, must be considered in the analysis of the 1524 event.

Clearly the reign of Huayna Capac saw unprecedented ecological changes resulting from construction and military campaigns, the former resulting in the concentration of enormous human workforces and the expansion of urban populations, while the latter moved large bodies of soldiers great distances into new territories and back. This was not a time of stasis, but a time of large-scale ecological change brought about deliberately by human activity organized and manipulated on a vast scale by imperial command at the state level. These changes created new conditions and new opportunities, not only for the human population, but also, perhaps, for the florescence of microbial pathogens.

Relationship between El Niño and Intensified Bartonellosis Epidemics

Our third major point concerns recent, groundbreaking research by Jiayu Zhou and colleagues demonstrating a linkage between epidemic outbreaks of bartonellosis and El Niño weather patterns (Zhou et al. 2002, n.d.). When seasonal air temperatures rise above normal during periods of El Niño weather phenomena, the vector sand-fly population increases correspondingly, resulting in a dramatically increased incidence of bartonellosis outbreaks. In the 1997–98 event that Zhou and colleagues studied, "the monthly disease case number was almost doubled and the disease transmission lasted much longer." But, perhaps even more importantly for this investigation of the epidemic of 1524–25, their research showed that "during 1997/98 El Niño event, the disease epidemics expanded to the southern part of the country, where bartonellosis had not been recorded earlier." We believe that the outbreak of 1524–25 may have been exacerbated by a convergence of factors including an El Niño event. Quinn et al. (1987) listed 1525–26 as both a moderate regional El Niño year and a moderate ENSO (El Niño–Southern Oscillation; see Maasch, this volume) event, based on documentary data. Hocquenghem and Ortlieb (1990) strongly contested the designation of 1525–26 and 1531–32 as El Niño years, but Quinn (1993: 17–18) retained those years on the list and added Nile River discharge data in support of his El Niño chronology (Quinn 1993: 27). Caviedes (2001: 43–44) reviewed the evidence and supports Quinn's view.

Until we identify mummified remains of a human known to have died in the 1524–25 epidemic, it will be impossible to demonstrate absolutely that bartonellosis was the cause of this disaster. Nevertheless, as we have shown here, the application of current medical science to the existing archaeological and documentary evidence makes bartonellosis a more likely diagnosis than smallpox for the death of Huayna Capac and his subjects, particularly when considered in light of recent discoveries of the influence of El Niño weather patterns in spreading epidemics of this disease.

REFERENCES CITED

Alexander, Bruce
1995 A Review of Bartonellosis in Ecuador and Colombia. *American Journal of Tropical Medicine and Hygiene* 52 (4): 354–359.

Allison, Marvin J., Alejandro Pezzia, Enrique Gerszten, and Daniel Mendoza
1974 A Case of Carrion's Disease Associated with Human Sacrifice from the Huari Culture of Southern Peru. *American Journal of Physical Anthropology* 41: 295–300.

Amano, Yasuji, José Rumbea, Jürgen Knobloch, James Olson, and Michael Kron
1997 Bartonellosis in Ecuador: Serosurvey and Current Status of Cutaneous Verrucous Disease. *American Journal of Tropical Medicine and Hygiene* 57 (2): 174–179.

Betanzos, Juan de
1987 [1551] *Suma y narración de los Incas* (María del Carmen Martín Rubio, ed.). Atlas, Madrid.

1996 [1551] *Narrative of the Incas* (Roland Hamilton and Dana Buchanan, eds. and trans.). University of Texas Press, Austin.

Cabello Valboa, Miguel
1951 [1586] *Miscelánea antártica: Una historia del Perú antiguo.* Universidad Nacional Mayor de San Marcos, Facultad de Letras, Instituto de Etnología, Lima.

Calvete de Estrella, Juan Cristóbal
1964 [1565–67] *Rebelión de Pizarro en el Perú y vida de don Pedro Gasca.* In *Crónicas del Perú* (Juan Pérez Tudela Bueso, ed.). Biblioteca de Autores Españoles 167. Atlas, Madrid.

Caviedes, César N.
2001 *El Niño in History: Storming through the Ages.* University Press of Florida, Gainesville.

Chin, James (ed.)
2000 *Control of Communicable Diseases Manual.* 17th ed. American Public Health Association, Washington, D.C.

Cieza de León, Pedro de
1998 [1553] *The Discovery and Conquest of Peru* (Alexandra Parma Cook and Noble David Cook, eds. and trans.). Duke University Press, Durham, NC.

Cook, Noble David
1998 *Born to Die: Disease and New World Conquest, 1492–1650.* Cambridge University Press. Cambridge, England.

Cooper, Philip, Ronald Guderian, P. Orellana, Carlos Sandoval, H. Olalla, M. Valdez, Manuel Calvopiña, Angel Guevara, and George Griffin
1997 An Outbreak of Bartonellosis in Zamora Chinchipe Province in Ecuador. *Transactions of the Royal Society of Tropical Medicine and Hygiene* 91: 544–546.

Curie, Elizabeth J.
1995 *Prehistory of the Southern Manabí Coast, Ecuador: López Viejo.* BAR International Series 618. Tempus Reparatum, Oxford.

Ellis, Barbara A., Lisa D. Rotz, John A. D. Leake, Frine Samalvides, José Bernable, Gladys Ventura, Carlos Padilla, Pablo Villaseca, Lorenza Beati, Russell Regnery, James E. Childs, James G. Olson, and Carlos P. Carrillo
1999 An Outbreak of Acute Bartonellosis (Oroya Fever) in the Urubamba Region of Peru, 1998. *American Journal of Tropical Medicine and Hygiene* 61 (2): 344–349.

García-Caceres, Uriel, and Fernando U. Garcia
1991 Bartonellosis: An Immunodepressive Disease and the Life of Daniel Alcides Carrion. *American Journal of Clinical Pathology* 95 (4, suppl. 1): s58–s66.

Garcilaso de la Vega, El Inca
1961 [1609] *The Incas: Royal Commentaries of the Inca, Garcilaso de la Vega 1539–1616* (Maria Jolas, trans. from the French edition; L. Alain Gheerbrant, ed.). Orion, New York.

1966 [1609] *Royal Commentaries of the Incas and General History of Peru, Part One* (Harold V. Livermore, trans.). University of Texas Press, Austin.

Gómez E., Nelson
1999 *Nuevo Atlas del Ecuador.* Ediguias C., Quito, Ecuador.

Gray, Gregory C., Alberto Angulo Johnson, Scott A. Thornton, William Alexander Smith, Jürgen Knobloch, Patrick W. Kelley, Ludovico Obregon Escudero, Maria Arones Huayda, and F. Stephen Wignall
1990 An Epidemic of Oroya Fever in the Peruvian Andes. *American Journal of Tropical Medicine and Hygiene* 42 (3): 215–221.

Hemming, John
1970 *The Conquest of the Incas.* Harcourt Brace Jovanovich, New York.

Hocquenghem, Anne-Marie, and Luc Ortlieb
1990 Pizarre n'est pas arrivé au Pérou durant une année El Niño. *Bulletin de l'Institut Français d'Études Andines* 19: 327–334.

Hyslop, John
1984 *The Inca Road System.* Academic Press, Orlando, FL.

Ihler, Garret M.
1996 *Bartonella bacilliformis*: Dangerous Pathogen Slowly Emerging from Deep Background. *FEMS Microbiology Letters* 144: 1–11.

Jiménez de la Espada, Marcos
1892 *Una antigualla peruana.* M. Gines Hernández, Madrid.

1965 [1881] *Relaciones geográficas de Indias: Perú* (José Urbano Martínez Carreras, ed.). Biblioteca de Autores Españoles 183–185. Atlas, Madrid.

Kiracofe, James B., and John S. Marr
n.d. The Great Andean Epidemic of 1524–4: Smallpox or Bartonellosis? A paper presented at a symposium entitled "Disease and Disaster in Pre-Columbian and Colonial America," at the Inter-American Institute for Advanced Studies in Cultural History Conference, Washington D.C., 2002.

Kosek, Margaret
2000 Natural History of Infection with *Bartonella bacilliformis* in a Nonendemic Population. *Journal of Infectious Disease* 182: 865–872.

LeVine, Terry Y. (ed.)
1992 *Inca Storage Systems.* University of Oklahoma Press, Norman.

McEwan, Colin, and María Isabel Silva I.
1992 ¿Que fueron hacer los Incas en la Costa Central del Ecuador? In *5000 Años de Ocupación, Parque Nacional Machalilla* (Presley Norton and Marco Vinicio García, eds.): 71–102. Centro Cultural Artes: Ediciones Abya Yala, Quito, Ecuador.

Maguiña, Ciro, and Eduardo Gotuzzo E.
2000 Bartonellosis New and Old. *Emerging and Re-emerging Diseases in Latin America, Infectious Disease Clinics of North America* 14 (1): 1–22.

Niles, Susan A.
1999 *The Shape of Inca History, Narrative and Architecture in an Andean Empire.* University of Iowa Press, Iowa City.

Odriozola, Ernesto
1944 La enfermadad de Carrión;
o Verruga Peruana. *Revista de la
Sanidad de Policía.* Lima.

**Ollague Loayza, Wenceslao,
and A. Guevara de Veliz**
1977 Verruga Peruana en el Ecuador.
*Medicina Cutánea Ibero-Latino-
Americana* 4: 235–240.

Pizarro, Pedro
1978 [1571] *Relación del descubrimiento y
conquista del Perú.* Pontificia Universidad
Católica del Perú, Fondo Editorial, Lima.

Polo, José Toribio
1913 Apuntes sobre las epidemias del
Perú. *Revista histórica* 5: 50–109.

Protzen, Jean-Pierre
1993 *Inca Architecture and Construction at
Ollantaytambo.* Oxford University Press,
New York.

Quinn, William H.
1993 The Large-Scale ENSO Event,
The El Niño and Other Important
Regional Features. *Bulletin de l'Institut
Français d'Études Andines* 22: 13–34.

**Quinn, William H., Victor T. Neal,
and Santiago Antúnez de Mayolo**
1987 El Niño Occurrences over the
Past Four and a Half Centuries.
Journal of Geophysical Research 92 (C13):
14,449–14,463.

Rostworowski de Diez Canseco, María
1999 *History of the Inca Realm* (Harry B.
Iceland, trans). Cambridge University
Press, Cambridge, England.

Salazar, Thomas
1858 Historia de las verrugas. *Gaceta
médica de Lima* 2 (2) (March): 161–164;
175–178.

Sarmiento de Gamboa, Pedro
1942 [1572] *Historia indica* (Roberto
Levillier, ed.). Espasa-Calpe Argentina,
Buenos Aires.

Schultz, Myron G.
1968 A History of Bartonellosis.
*American Journal of Tropical Medicine
and Hygiene* 17 (4): 503–515.

**Trelles, Julio Oscar, Luis
Palomino, and Luis Trelles**
1969 Formas neurológicas de la
enfermedad de Carrión. *Revista de neuro-
psiquiatría* 32 (4): 245–306.

**Wright, Kenneth R., Alfredo Valencia
Zegarra, and William L. Lorah**
1999 Ancient Machu Pichu Drainage
Engineering. *Journal of Irrigation and
Drainage Engineering* November:
360–369.

**Zhou, Jiayu, William K.-M. Lau,
Larry W. Laughlin, Penny M.
Masuoka, Richard C. Andre,
and Judith Chamberlin**
n.d. The Effect of Regional Climate
Variability on Outbreak of Epidemics
of Bartonellosis in Peru. Preprints of
the 3rd Symposium on Environmental
Applications: Facilitating the Use
of Environmental Information.
American Meteorological Society
82nd Annual Meeting, 3–17
January, Orlando, Florida, 2002.

**Zhou, Jiayu, William K.-M. Lau,
Penny M. Masuoka, Richard C. Andre,
Judith Chamberlin, P. Lawyer, and
Larry W. Laughlin**
2002 El Niño Helps Spread
Bartonellosis Epidemics in Peru. Eos,
Transactions, *American Geophysical
Union* 83 (14): 157, 160–161.

CENTRAL AMERICA
AND MESOAMERICA

ARMAGEDDON TO THE GARDEN OF EDEN: EXPLOSIVE VOLCANIC ERUPTIONS AND SOCIETAL RESILIENCE IN ANCIENT MIDDLE AMERICA

Payson Sheets

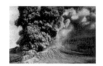

Technology is a blessing to be sure, but every blessing has its price.
The price of increased complexity is increased vulnerability.
Garret Hardin (1993: 101)

Large explosive volcanic eruptions can be disastrous or even apocalyptic for people, as well as for flora and fauna, and they may well have seemed like Armageddon (Rev 16:16) to many people in preindustrial societies. On the other hand, periods of quiescence can generate very fertile, volcanically derived soils, providing for a veritable Garden of Eden. Comparative analyses of eruptions and their effects on ancient Middle American societies have shown that peoples varied dramatically in their vulnerabilities to explosive eruptions. Although I have considered only a very small sample, simpler egalitarian societies apparently were more resistant to sudden massive stress than were more complex societies. The greater resilience of egalitarian societies apparently involved lower population densities, smaller social units, less reliance on the "built" environment, greater access to refuge areas, less hostility among neighbors, and greater reliance on a wide range of wild and domesticated foods. But this comparative study has found that other factors, beyond complexity, can render societies highly vulnerable to even small sudden stresses, and it is suggested here that the concept of "scaled vulnerability" can help explain patterns and variations in societies' susceptibility to collapse in the face of explosive volcanism.

The frequent eruptions of Arenal volcano in Costa Rica (fig. 1) provide cases where sudden ashfalls necessitated emigration by egalitarian sedentary societies, and those simpler societies are striking for how resilient they were to sudden massive stresses. In contrast, the eruption of Baru volcano in nearby Panama, which affected the Barriles chiefdoms, shows the importance of

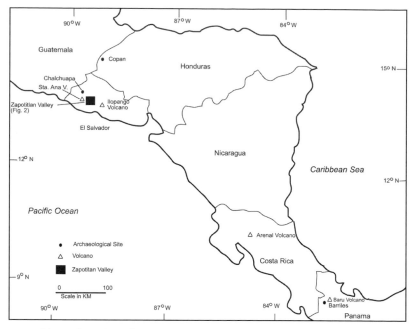

FIG. 1 Map of ancient Central America with archaeological sites and volcanoes indicated.

political factors in understanding human-environmental interactions. Baru's relatively small eruption had great repercussions, evidently because of the institutionalized hostility among chiefdoms.

In Mesoamerica volcanoes often erupted explosively, from the small eruption of Loma Caldera that buried the Cerén site in El Salvador (fig. 1) to the massive eruptions of Ilopango and Popocatepetl that had widespread and long-lasting effects. In some cases full cultural recovery was achieved relatively rapidly, and in other cases not at all.

Volcanic eruptions can have disastrous effects on people, their societies, and their environments, and there is much written on the topic. However, disasters can have positive or creative effects, as people learn from them and change their adaptations, their "built" environments, their loci of settlement, and their belief systems to deal with uncertainty and fear. Scholars from the technologically sophisticated, secular Western world can easily underestimate the importance of religious belief and ritual for indigenous people as they deal with sudden massive stresses. Disasters, or hazards as disasters waiting in the wings, certainly can figure heavily in human interactions with the supernatural world. This chapter explores the interplay of natural, cultural, and supernatural elements in response to explosive volcanic eruptions.

Burton, Kates, and White (1978) provide a useful framework for comparing human reactions to sudden massive stresses. The minimal adjustment is Loss Absorption, which occurs after the first threshold of Awareness is crossed. People simply accept the losses and get on with their lives. When the stress is greater and the Direct Action threshold is crossed, then Loss Reduction is the result. Still greater stress crosses the Intolerance threshold, and people must take Radical Action. An example of Radical Action would be a forced migration to an area far from the disaster and different from the homeland, thus necessitating significant changes in the society, culture, and/or adaptation.

This chapter examines the relationships among a wide range of factors including explosive volcanism, different organizations of human societies, environments, flora and fauna affected by eruptions, and how people deal with volcanic hazards using natural-cultural-supernatural means. The pertinent variables thus cover a great range, including the pre-eruption environment, climate, society, culture, demography, the nature of the eruption, surrounding societies, the political landscape, and natural as well as cultural recoveries. It is not reasonable to expect compelling conclusions about robust patterns among so many variables with only thirty-six eruptions considered. The cases vary greatly in how thoroughly they document natural and cultural variables before and after each eruption. Precision in dating eruptions also varies. Thus, this study should be considered exploratory. Another thing to consider is that an explosive eruption will have varying effects on nature and people depending on distance from the source, wind direction, and other factors. Often there are zones of total devastation surrounded by zones of lesser impact that are, in turn, surrounded by a zone that has experienced some beneficial effects.

This chapter uses research results from as wide a range of disciplines as possible. There is a surprising number of disciplines doing disaster research—a total of thirty, according to Alexander (1997), ranging from the social sciences to the physical sciences and engineering. Alexander (1995) tabulated the research funding for disaster studies, and found that 95% went to the physical and technological sciences, and only 5% went to the social sciences. Although disaster research was initiated by Gilbert White (e.g., 1945), a cultural geographer who started working in the first half of the twentieth century, it is now dominated by the physical sciences and engineering.

The principal variables included in this study are societal complexity, adaptation, architecture, economics, politics, ideology, demography, environment, and the nature of eruptions. Most of the disciplines Alexander (1997) discusses deal with the immediacy of an eruption; in contrast, archaeology studies phenomena over centuries or millennia.

THE CASES

The following cases of explosive eruptions affecting Pre-Columbian societies in Middle America, extending from Panama to Mexico, are organized from simpler to more complex societies, because organizational complexity apparently was an important factor in resilience. Other key factors, such as the political landscape, are highlighted when they are relevant. Because the more complex societies developed in the northwestern end of the area covered here, the cases begin in Lower Central America and then move to Mesoamerica.

FIG. 2 Six of the ten white volcanic-ash layers from the major eruptions of Arenal volcano. The dark layer under the lowest white-ash layer is the clay-laden soil prior to Arenal beginning to erupt. The dark layers on top of the white-ash layers are soils that sustained vegetation and some cultivation. The small lens of whitish ash at the man's chest level is the remains of the 1968 eruption.

Arenal, Costa Rica

The Arenal project found a virtually continuous record of human habitation in northwest Costa Rica for the past ten millennia (Sheets 1994). That record began with PaleoIndian occupation (est. 10,000–7000 BC),[1] as evidenced by the Clovis-style projectile point recovered from the south shore of Lake Arenal, which proves that PaleoIndians had adapted to moist tropical rainforest environments.

The very low PaleoIndian population densities were followed by higher densities during the Archaic Period (est. 7000–3000 BC). Settled, sedentary villages with at least a small fraction of the diet from maize emerged near the end of this Archaic Period, perhaps as early as the fourth millennium BC (calibrated ^{14}C date TX-5275 from Tronadora Vieja site G-163 is 3780 [3765] 3539 BC). Arford (n.d.) reports the earliest maize pollen at about 1900 BC in the lake sediment cores from Lake Cote, only 3 km north of Lake Arenal's north shore. That date is consistent with the archaeological record of the Lake Arenal area (Sheets 1994).

The succeeding Tronadora phase may begin as early as the mid-fourth millennium BC, but we conservatively place its beginning at 2000 BC. The villages were small, with circular houses around 5 m in diameter that presumably were roofed with thatch; the dead were buried beside the houses. Houses were provisioned with metates and manos for food grinding, ceramics, cooking stones, and a basic core-flake lithic technology, all of which showed remarkable perseverance up to the Spanish Conquest, a span of three and a

1 Dates and ranges for all periods and sites are expressed as calendar years BC/AD based on calibrated radiocarbon dates (Stuiver and Reimer 1986; Stuiver and Becker 1993).

FIG. 3 Arenal volcano erupting. Part of the eruptive column collapsed into a pyroclastic flow, seen racing down the slope into tropical rainforest. The rainforest was vaporized.

half millennia. Melson (1994) dated Arenal's first big explosive eruption to about 1800 BC, but that may have been Cerro Chato's final eruption (Soto et al. 1996), and the Arenal and Chato eruptive sequence before 1000 BC may be more complex than we thought. Soto et al. found evidence of eight eruptions from about 5000 to 1000 BC, for an average of one every 550 years, but we have little or no direct documentation of sites interdigitated with most of those tephras.

Although the source of the ca. 1800 BC tephra is in doubt, what is more pertinent for this study is that ten large explosive eruptions affected people in the Arenal area over the course of the last four millennia (fig. 2), for an average periodicity of four hundred years, a span of time well within human societies' abilities to maintain knowledge within an oral-history tradition. The ancient inhabitants of the Arenal area may have been less surprised by explosive eruptions than were the recent residents when it erupted in 1968 and killed almost a hundred people (fig. 3).

The paleolimnological and tephrostratigraphic work at Lake Cote has detected a major climatic change at about 500 BC (Arford n.d.). Arford found that the climate from 2000 to 500 (or 400) BC was dryer than it is at present, indicated by the relatively high concentrations of grass pollen and charcoal in sediments from that time span. He notes the decline of grass pollen and charcoal in post-400 BC layers and interprets that as a shift to wetter conditions. That climatic change correlates with the cultural change from the Tronadora to the Arenal phase (Sheets 1994), which is evident in ceramics and in many other aspects of culture. In addition, cemeteries were located at a moderate to considerable distance from the villages and village size increased, as did regional population density. Ceremonial feasting after the dead were interred began in a major way.

In fact, at least in this area of Costa Rica, postinterment ceremonialism peaked during the Arenal phase. Because all these changes are consonant with cultural changes that were underway throughout lower Central America, it is not clear whether any of these changes had anything to do with the climatic change. Alternatively, the climatic change detected at Lake Cote could have been more widespread than is currently understood. At most, the increase in site size and density could have been facilitated by the increase in moisture, but that is only speculation at this point. However, the eastern end of the research area is now so humid that maize agriculture is not feasible, because soils are saturated eleven months of the year and seeds rot rather than germinating, discouraging even the most determined cultivator. While the increase in moisture may have discouraged seed-based agriculture, it must have increased the biomass and probably the diversity of wild flora and fauna.

Recounting each eruption and looking at each case for possible cultural effects is not necessary here. That information is available elsewhere (Melson 1994; Sheets 1994, 1999). Because we have yet to find an Arenal eruption that coincided with, and may therefore have led to, significant culture changes, we can deal with them in the composite here. Population densities from the Archaic to the Spanish Conquest varied significantly, reaching a peak during the Arenal phase (500 BC–AD 600), but populations never reached the densities of Mesoamerica or the Andes during the same long time span. Societal complexity remained low, with egalitarianism the rule from the PaleoIndian period up to the Spanish conquest. The only time local societies probably were pushing that boundary, and thus perhaps could be considered "trans-egalitarian," was during the Arenal and Silencio phases. During both of those phases slight variations were noted in the upper and lower areas of cemeteries. The upper areas of cemeteries in both phases received slightly more elaborate treatment of the dead during burial and with subsequent feasting, than did the lower areas. Skeletal preservation is insufficient, especially during the Arenal phase, to explore the possibility that gender was a factor in this differential treatment.

Barriles, Panama

Compared to the Arenal-area societies, the Barriles societies (Linares and Ranere 1980), which I evaluate as ranked societies or chiefdoms, were more complex, having clearly crossed the threshold from egalitarian (fig. 4). But they were

FIG. 4 Life-sized sculpture of a commoner in the Barriles society, nude, holding an elite or chief on his shoulders. The top individual wears a peaked cap and a pendant as symbols of high status and holds an axe and a severed human head in his hands.

FIG. 5 Barriles site (BU-24) test pit, with the thin tephra layer from the Baru volcano eruption, indicated by the T on the wall. The soil developed from that has largely incorporated the tephra, and a few more centuries of weathering and soil formation will eliminate it as a visible layer. The thick fertile soil that sustained the Barriles society is visible below the T and the tephra from which it was derived is visible at the bottom of the pit.

not markedly more complex than at Arenal, especially when compared to the state-level societies of Mesoamerica, and both area societies lived in similar environments. The population density along river courses was greater at Barriles than at Arenal, but certainly did not reach the densities of Mesoamerica. Thus, if we were conjecturing solely in terms of complexity, we would expect the Barriles chiefdoms to be moderately more sensitive to sudden explosive volcanism than were Arenal societies. The Baru volcano erupted, probably in the seventh century AD, and buried the nearby Barriles chiefdoms in the upper reaches of the Chiriqui Viejo River under a relatively thin blanket of volcanic ash (fig. 5). The ash depth, probably in the 10–20 cm range, is comparable to the ash blankets that affected Arenal-area societies at the west end of the lake. Those societies recovered completely; we are unable to detect any culture change that coincides with any eruption and thus may have been forced by that eruption.

Based upon the above factors of adaptation, population density and distribution, climate, complexity of society, and magnitude of the eruption, we would predict that the Baru volcano eruption would have caused some short-term dislocations, probably crossing the Direct Action threshold but not beyond. Thus, the same societies would have reestablished their settlements and culture within a few decades. But that expectation is strikingly different from what actually happened. For reasons that initially were unclear, the Barriles chiefdoms in the upper reaches of the Chiriqui Viejo River resorted to Radical Action, permanently abandoned their settlements, and moved over the divide and down to the Caribbean coast (Linares and Ranere 1980: 244–245). They had to change their architecture and fundamentally change their subsistence adaptation to the much wetter environment, and they never returned to the Pacific drainage.

Politics rendered the Barriles chiefdoms more vulnerable than did ecology, eruption, adaptation, or demography (Sheets 1999). The Barriles chiefdoms engaged in chronic warfare, often capturing victims from adjacent polities and decapitating them. Although they shared the same culture, presumably spoke the same language, and shared the same adaptation, peoples' allegiances were intra-chiefdom, and the political landscape was one of hostility. Therefore, when a relatively small volcanic eruption deposited 10–20 cm of volcanic ash in their terrain, no refuge area was available to them, and they had to emigrate and change much of their culture. Thus, the factor of interpolity hostility generated a high vulnerability, even though all factors except politics would indicate a relatively low composite vulnerability. Politics alone evidently increased Barriles chiefdom vulnerability. Vulnerabilities are therefore a composite of complexities and other key factors such as institutionalized hostility.

El Salvador, Focusing on the Zapotitan Valley

The first estimates of the dating of the Coatepeque eruption (Williams and Meyer-Abich 1955) to perhaps 10,000 years ago indicated that it might have affected human populations (Sheets 1984). However, recent volcanological research has dated its big eruptions from about 77,000 to 57,000 years ago (Rose et al. 1999; Pullinger n.d.).

People have occupied what is now El Salvador for at least the past 10,000 years (Sheets 1983a, 1983b, 1984), and they often have interacted with volcanoes (Sheets 1980) (fig. 6). By the first centuries AD, the "Miraflores" cultural

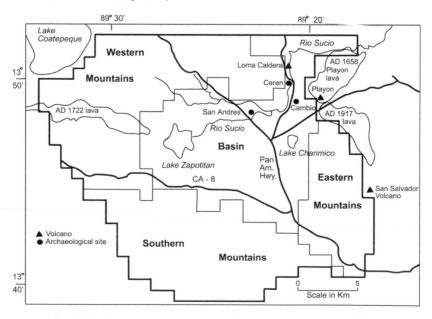

FIG. 6 Map of the Zapotitan Valley, including sites, volcanoes, and lava flows

sphere had developed, extending from El Salvador well into the Guatemalan highlands (Dull et al. 2001). This precocious cultural florescence included large chiefdoms or small states with hierarchical settlement and economic systems in the intermontane valleys (Sharer 1974). Monumental construction was made of adobe brick or rammed earth, while domestic vernacular architecture was the highly earthquake-resistant wattle-and-daub (Sheets 1992). Stelae commemorating rulership were carved, and included hieroglyphics and calendrics (Sharer 1974). This cultural florescence was forever truncated by the huge eruption of Ilopango volcano, initially dated to calibrated one sigma AD 260±114 (Sheets 1983a), but recently redated to AD 408 (429) 536 calibrated two-sigma range (Dull et al. 2001). The plinian and phreatomagmatic eruption was one of the greatest in Central America during the Holocene, and at least 1000 km² were rendered uninhabitable for peo-

FIG. 7 Payson Sheets cleaning a profile of tephra from the Ilopango eruption overlying the pre-eruption soil (labeled "Soil" on the photo) and a small archaeological site. The volcanic ash is 6.7 m deep, and soil formation after almost two millennia of weathering is not as thorough as with the pre-eruption soil. This locality is 11 km from the source.

ple as well as fauna and flora (fig. 7). The estimated minimal population density of 30/km² allows for an estimated death toll of some 30,000 people. The surrounding 10,000 km² received tephra depths greater than .5 m, creating stresses greater than traditional agriculturalists could handle. This disaster crossed the Intolerance threshold, necessitating Radical Action in the form of emigration by an estimated 300,000 people, though we have no idea how many asphyxiated or how many survived to emigrate. Many could have avoided asphyxiation by breathing through fine-weave cotton cloth. Areas farther from the source received thinner tephra blankets, resulting in less disruption. Areas receiving only a few centimeters could have benefited from the mulch effect as well as increasing soil porosity and suffocation of some insect pests.

Lothrop (1927) was the first to suggest that the reoccupation of the area devastated by Ilopango (Porter 1955) was from the north, by Maya-affiliated people, and that identification has stood the test of time. Thus, cultural recovery by descendants of the original inhabitants of the Miraflores cultural sphere never occurred. One of the earliest settlements yet found representing the reoccupation of the devastated area is Cerén, where polychrome ceramics and architecture show close relationships with the Classic Maya site of Copán and environs (Beaudry-Corbett 2002; Sheets 2000, 2002; Webster et al. 1997). We estimate Cerén was established about a century after the Ilopango eruption, and it functioned for about another century. Early- to mid-seventh century, the Loma Caldera volcanic vent opened nearby and buried the village under 5 m of tephra (Miller 2002, n.d.). I doubt we will ever know the year in which it erupted, but we do know the month and time of day, an irony of archaeological dating. The eruption occurred in August, the middle of the rainy season, based on maturation of annuals and sensitive perennials. As evidenced by multiple artifact patterns, the eruption occurred in the early evening, after dinner was served but before the dishes were washed, that is, around 6:00 to 7:00 pm, if Cerénians were following eating patterns similar to those of traditional Central American households today. The eruption devastated only a few km² of the valley.

Recovery from the Loma Caldera eruption is best documented at the Cambio site (fig. 6), 2 km south of Cerén (Chandler 1983). All of the artifact categories are identical before and after that eruption, with only one minor exception. After the eruption the people reoccupying the area had a new kind of polychrome pottery called Arambala, but all other types and wares continued unchanged (Beaudry-Corbett 2002).

Life went on in the Zapotitan Valley for about two centuries without a significant explosive eruption that we have detected, but toward the end of the Classic Period San Salvador volcano erupted a fine-grained wet pasty tephra that severely affected the eastern half of the valley. Hart (1983) named it the "San Andres Talpetate Tuff" and argued that it was phreatomagmatic, being erupted through a lake that occupied the large crater called "Boqueron." It devastated some 300 km², thus placing it midway between Loma Caldera and Ilopango in magnitude. Estimated population sizes range from 21,000 to 54,000 people, who would have had to migrate to arable areas. Cambio again best documents the recovery from the Boqueron eruption (Chandler 1983), a recovery that was complete by the end of the Classic Period at about AD 900. The moderately complex large chiefdoms or small states were sufficiently resilient to withstand these two eruptions—small and medium in scale, respectively—recovering thoroughly within a few decades.

Extensive documentation is available for the numerous historic eruptions in the Zapotitan Valley (fig. 6), beginning with the Playon eruption in 1658, and including large lava flows in 1722 and 1917, and the growth of the Izalco volcano from 1770 until 1966. Browning (1971) provides a fascinating account of native Pipiles having to emigrate from the area devastated by Playon, and their decades of struggle before they finally gained access to some land for subsistence. These eruptions further emphasize how volcanically active the Zapotitan Valley landscape has been, but they are beyond the Pre-Columbian focus of this chapter.

Mexico: Tuxtla Mountains

The Tuxtla Mountains, on Mexico's southern gulf coast (fig. 8), were volcanically active in the Holocene (Reinhardt n.d.) and have often affected populations. The Tuxtlas average four explosive eruptions per thousand years, a periodicity comparable to the Zapotitan Valley, and about twice that of Arenal. Santley (1994) has investigated the archaeological site of Matacapan. The earliest eruption that must have affected people dates to 3000–2000 BC, but few direct data are available. The second eruption that affected people dates to about 1250 BC, and is known as the CMB tephra. The affected society was egalitarian and sedentary. I measure the tephra depth from Reinhardt's (n.d.: 97) profiles at about 60 cm, and Santley (1994) notes that the area was abandoned for nine centuries. The CMB tephra may have had a seriously deleterious effect on local settlement, but until regional research is done we will not know how large an area was affected. The radiocarbon dates on these Tuxtla Mountain eruptions were calibrated using the computer program CALIB (Stuiver and Reimer 1986).

The third eruption deposited the CN tephra in about AD 150, which I measure from Reinhardt's section (n.d.: 86) at about 40 cm. Based upon the dating, one would expect societies after the decline of Olmec civilization but before the Classic-Period florescence, and thus most settlements probably were egalitarian farming villages with some ranked societies, perhaps midway between Arenal and Barriles in complexity. Santley reports that the area was abandoned but recovery was relatively rapid. The next two eruptions (CP and LN tephras) occurred in such rapid succession that no soil development occurred on the earlier, so for purposes here they are considered together. I measure their combined thickness at about 20–30 cm on Reinhardt's (n.d.: 100) section, and Santley notes that people continued to live at the site. Apparently this tephra depth was within the domain of Loss Acceptance, that is, within the abilities of agriculturalists to continue to farm their lands. The societies affected may have been slightly more complex than the earlier ones, perhaps ranked societies.

The final Classic-Period eruption occurred about AD 600, and I measure the LC tephra at about 45 cm in thickness on Reinhardt's (n.d.: 97) section. Santley (1994) found a ridged maize field below the tephra. Societal complexity was at a peak at the time, but then began a steady decline. I suspect the consistent, direct correlation between tephra depth and societal impact in these Tuxtla cases is more than coincidence, because the greater depths caused greater dislocations. However, Santley and Arnold (n.d.) do not agree with my interpretation here: they argue that the more complex societies can harness more energy and labor than less complex societies, and are thus able to better withstand volcanically induced stresses.

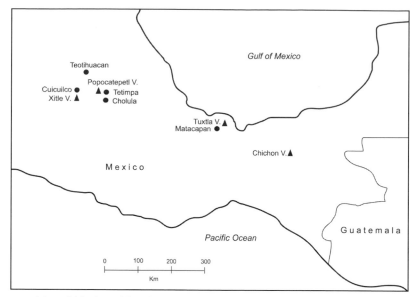

FIG. 8 Map of Mexico with volcanoes and archaeological sites indicated

El Chichón

El Chichón volcano (fig. 8), some 200 km ESE of the Tuxtlas, has erupted often during the past few thousand years (Espindola et al. 2000), and each time must have affected human populations. Espindola et al. (2000) document eruptions at 5750 BC, 1750 BC, 1150 BC, 550 BC, 50 BC, AD 50, AD 350, AD 450, AD 700, AD 1050, and AD 1400. All dates are calibrated using Stuiver and Reimer (1986). The eruptions of AD 700 and AD 1400 deposited significant plinian pumice falls. And, of course, Chichón erupted a couple of decades ago, in AD 1982, blasting fine ash into the stratosphere over 20 km in height. Espindola et al. (2000) found ceramic artifacts between some of the tephra deposits and reasonably infer that the eruptions must have affected people. Here is an excellent opportunity for interdisciplinary research by archaeologists and geologists to study the eleven Pre-Columbian eruptions, effects on societies, resilience, and recoveries. The archaeology of the area has barely begun.

Central Highland Mexico

The Basin of Mexico and nearby Puebla (fig. 8) were affected by explosive volcanism at various Pre-Columbian times, and Popocatepetl continues to erupt tephra and gasses today. Plunket and Uruñuela (1998) document a fascinating persistence of ritual activity during the past two millennia by people trying to persuade Popocatepetl not to erupt. Plunket and Uruñuela have broached a topic that, in my opinion, is woefully underdeveloped by archaeologists: people turning to the supernatural to help them with their anxieties. Miller and Taube (1993: 119–121) summarize the deliberate symbolism of pyramids as constructed sacred mountains in Mesoamerica, and the Maya referring to pyramids as mountains with their word *witz*. One can easily see how the doorway into the temple symbolizes the mouth of the volcano, and the copal incense smoke billowing forth symbolizes the gasses emitted. Miller and Taube (1993: 120–121) note how the twin pyramids with temples in the Aztec capital of Tenochtitlan symbolized the twin volcanoes of Popocatepetl and Iztaccihuatl, and how intricately rituals, feasting, and sacrifice were integrated into volcano and deity worship. It appears that Popocatepetl has been a major volcanic presence in northern Mesoamerica farther back than the Postclassic Period.

Sedentary human communities settled along the resource-rich shores of Lake Texcoco in the Basin of Mexico as early as the sixth millennium BC (Niederberger 1979). The descendants of those communities in the southern part of the basin were devastated by a series of tephra deposits at about 3000 BC, especially pyroclastic flows. These date to about the same time as Siebe et al. (1996) have documented a large eruption from Popocatepetl, so the tephra that devastated those egalitarian communities could be the north-western component of that same eruption. Adams (1991: 38) was convinced by Niederberger's interpretations, and he noted that it took about five centuries for people to move back into the area.

Siebe et al. (1996) document three large eruptions of Popocatepetl in the past five millennia: the first is the above-mentioned one, dated to between 3195 and 2830 BC, the second to between 800 and 215 BC, and the third to between AD 675 and 1095. All three large eruptions began with small ashfalls and ashflows, but soared to massive plinian eruptions with extensive ash and pumice falls and pyroclastic flows, and ended with extensive mudflows.

Siebe (2000: 61) revised the dating of the second large "cataclysmic" eruption to 250–50 BC and evaluated the magnitude at a Volcanic Explosivity Index or VEI=6. He notes the devastation was particularly extensive toward the east of Popocatapetl, including the site of Tetimpa and throughout much of the Puebla Valley (fig. 8), due to airfall tephra and pyroclastic flows. He notes that the devastation would also include the northwestern slopes that extend down into the Basin of Mexico, particularly the Amecameca-Chalco regions.

Plunket and Uruñuela (n.d.) summarize the most recent archaeological research pertaining to the second of these big eruptions. Their excavations (1998) at the site of Tetimpa have been well documented, and paint a fasci-nating picture of farmers visiting the site only as long as necessary to keep their fields functioning, and then retiring to what they must have perceived as a safer location. The farmers maintained volcano-effigy shrines in their households to use supernatural intervention to decrease risk. Plunket and Uruñuela (n.d.) suggest that the populations displaced after the big eruption could have been very useful to the elite of Cholula (fig. 9), to be used in great public-works projects such as building the massive central pyramid. The idea that a volcanic disaster can serve as a slave-delivery system is a new concept in archaeo-volcanology.

Plunket and Uruñuela (n.d.) turn their attention to the archaeology of the northwestern flank of Popocatepetl, the area Siebe (2000) documented as devastated. As they note, this area has long been known as the most fertile, moist, and arable of the Basin of Mexico, and Sanders et al. (1979) had docu-mented a long occupation leading to quite dense populations up to about 100

FIG. 9 Cholula site, in Puebla, Mexico. Talud-tablero architecture was buried by a lahar from Popocatepetl volcano.

BC, while the more arid northern part of the basin remained very sparsely populated. Sanders et al. stated that the dramatic population decline in this southern area included at least ten large regional centers along with some hundred other settlements. The reason or reasons why the most arable and densely populated area of the Basin of Mexico was suddenly depopulated has puzzled archaeologists for many decades. As Sanders et al. (1979: 107) state, "we know of no other situation in the historical or archaeological record in which so large a sedentary regional population was involved in such a drastic relocation." Plunket and Uruñuela use Sanders's population figures to estimate that about 70,000 people were affected by this eruption in the basin, and for the first time one can see a possible explanation for both the sudden diminution in the population of the southern basin and the sudden rise in Teotihuacan's population, as both were forced by the eruption. The sudden arrival of tens of thousands of refugees would have strained the newly emerging political, economic, and adaptive systems. Those systems survived the test, and Teotihuacan emerged as Mesoamerica's first urban-based expansionistic empire. Thus, I think Plunket and Uruñuela have gone far to solve two of the Basin of Mexico's longest-lasting Pre-Columbian puzzles.

The most controversial eruption in Mesoamerica surely is that of Xitle (fig. 8), approximately two millennia ago, in the southern Basin of Mexico. It has aroused controversy in its dating, its direct effects on societies, and its possible indirect effects on the rise of Teotihuacan, Mesoamerica's first megacity and aggressively expansionistic civilization. As Cordova et al. (1994) have noted, some scholars dated the eruption to as early as 400 BC, while other investigators dated it as late as AD 400. They side with the later date. If they are correct, they would reverse the common interpretation that Xitle's

FIG. 10 Cuicuilco, Mexico. The principal pyramid, circular in plan, was partially buried by the lava flow from Xitle volcano.

eruption forced Cuicuilco (fig. 10) residents northward, thus contributing to the growth of Teotihuacan as well as the demise of competition. If they are correct, it is more likely that Teotihuacan outcompeted and absorbed Cuicuilco, and that the demise of Cuicuilco was for cultural reasons, not because of a volcanically induced disaster. More recent research (Urrutia-Fucugauchi 1996), however, places the eruption squarely in the center of that time range, at about the time of Christ, approximately contemporary with the cataclysmic Popocatepetl eruption. The most recent geological dating of the Xitle eruption is somewhat later, 1670 ±35, or AD 245–315, calibrated using Stuiver and Becker (1993), based on charcoal below the lava (Siebe 2000). Archaeologists do not agree on whether Cuicuilco was a thriving community at the time of the eruption, in decline, or even abandoned.

SUMMARY AND ISSUES

Of the thirty-six cases of explosive eruptions in ancient Middle America considered here, about a third are so minimally documented that they contribute little to our knowledge. The other cases vary dramatically in the magnitude of the eruption, the environment and ecosystem affected, the nature of the society impacted, and recovery. Vulnerabilities and recoveries vary significantly, but perhaps an inkling of a pattern can be detected. The egalitarian and minimally agricultural societies in the Arenal area of Costa Rica were remarkably resilient, in part because they were not deeply invested in the "built environment" of major architecture and intensive agriculture, occupational specialization, political hierarchy, and a redistributive economy. The Barriles societies of neighboring Panama were somewhat more complex

from Daniel Sandweiss and Jeffrey Quilter to participate in their symposium and to contribute to their book. It is a privilege to be a part of both. This manuscript benefited from discussions with symposium participants and from review by the editors and another reviewer. In this chapter I have tried to present the factual material accurately, and to clearly identify my own interpretations and speculations. Errors of omission or commission are mine alone.

REFERENCES CITED

Adams, Richard
1991 *Prehistoric Mesoamerica*. University of Oklahoma Press, Norman, OK.

Alexander, David
1995 A Survey of the Field of Natural Hazards and Disaster Studies. In *Geographical Information Systems in Assessing Natural Hazards* (Alberto Carrara and Fausto Guzetti, eds.): 1–19. Kluwer, Dordrecht.

1997 The Study of Natural Disasters, 1977–1997: Some Reflections on a Changing Field of Knowledge. *Disasters* 21: 284–304.

Arford, Martin
n.d. Late Holocene Environmental History and Tephrostratigraphy in Northwestern Costa Rica: A 4000-year Record from Lago Cote. Master's thesis, Department of Geography, University of Tennessee, Knoxville, 2001.

Beaudry-Corbett, Marilyn
2002 Ceramics and Their Use at Cerén. In *Before the Volcano Erupted: The Ancient Cerén Village in Central America* (Payson Sheets, ed.): 117–138. University of Texas Press, Austin.

Browning, David
1971 *El Salvador: Landscape and Society*. Clarendon Press, Oxford.

Burton, Ian, Robert Kates, and Gilbert White
1978 *The Environment as Hazard*. Oxford University Press, New York.

Chandler, Susan
1983 Excavations at the Cambio Site. In *Archeology and Volcanism in Central America: The Zapotitan Valley of El Salvador* (Payson Sheets, ed.): 98–118. University of Texas Press, Austin.

Cordova, Carlos, Ana Lillian Martin del Pozzo, and Javier López Camacho
1994 Palaeolandforms and Volcanic Impact on the Environment of Prehistoric Cuicuilco, Southern Mexico City. *Journal of Archaeological Science* 21: 585–596.

Dull, Robert, John Southon, and Payson Sheets
2001 Volcanism, Ecology, and Culture: A Reassessment of the Volcán Ilopango TBJ Eruption in the Southern Maya Realm. *Latin American Antiquity* 12: 25–44.

Ember, Carol, and Melvin Ember
1992 Resource Unpredictability, Mistrust, and War. *Journal of Conflict Resolution* 36: 242–262.

Espindola, Juan Manuel, Jose Luis Macias, Robert Tilling, and Michael Sheridan
2000 Volcanic History of El Chichón Volcano (Chiapas, Mexico) during the Holocene, and Its Impact on Human Activity. *Bulletin of Volcanology* 62: 90–104.

Hardin, Garrett
1993 *Living within Limits: Ecology, Economics and Population Taboos*. Oxford University Press, Oxford.

Hart, William
1983 Classic to Postclassic Tephra Layers Exposed in Archaeological Sites, Eastern Zapotitan Valley. In *Archeology and Volcanism in Central America: The Zapotitan Valley of El Salvador* (Payson Sheets, ed.): 44–51. University of Texas Press, Austin.

Kennedy, Donald
2002 Science, Terrorism, and Natural Disasters. *Science* 295: 405.

Linares, Olga, and
Anthony Ranere (eds.)
1980 *Adaptive Radiations in Prehistoric
Panama*. Peabody Museum monographs
5. Peabody Museum of Archaeology
and Ethnology, Harvard University,
Cambridge, MA.

Lothrop, Samuel
1927 Pottery Types and Their Sequence
in El Salvador. *Indian Notes and
Monographs* 1 (4): 165–220.

Melson, William
1994 The Eruption of 1968 and Tephra
Stratigraphy in the Arenal Basin. In
*Archaeology, Volcanism, and Remote
Sensing in the Arenal Region, Costa Rica*
(Payson Sheets and Brian McKee, eds.):
24–47. University of Texas Press, Austin.

Miller, Daniel
2002 Volcanology, Stratigraphy,
and Effects on Structures. In
*Before the Volcano Erupted: The
Ancient Cerén Village in Central
America* (Payson Sheets, ed.): 11–23.
University of Texas Press, Austin.

n.d. Summary of 1992 Geological
Investigations at Joya de Ceren. In
*1992 Investigations at the Cerén Site, El
Salvador: A Preliminary Report* (Payson
Sheets and Karen Kievit, eds.): 5–9. MS
on file, Department of Anthropology,
University of Colorado, Boulder, 1992.

Miller, Mary, and Karl Taube
1993 *The Gods and Symbols of Ancient
Mexico and the Maya: An Illustrated
Dictionary of Mesoamerican Religion*.
Thames and Hudson, London.

Molino, Jean-François,
and Daniel Sabatier
2001 Tree Diversity in Tropical Rain
Forests: A Validation of the Intermediate
Disturbance Hypothesis. *Science* 294:
1702–1704.

Niederberger, Christine
1979 Early Sedentary Economy in the
Basin of Mexico. *Science* 203: 131–142.

Plunket, Patricia, and
Gabriela Uruñuela
1998 Appeasing the Volcano Gods:
Ancient Altars Attest a 2000-year-old
Veneration of Mexico's Smoldering
Popocatepetl. *Archaeology* 51 (4): 36–43.

n.d. To Leave or Not to Leave: Human
Responses to Popocatepetl's Eruptions in
the Tetimpa Region of Puebla, Mexico.
Paper presented at the 67th Annual
Meeting of the Society for American
Archaeology, Denver, 2002.

Porter, Muriel
1955 Material Preclásico de San
Salvador. *Comunicaciones del
Instituto Tropical de Investigaciones
Científicas de la Universidad de
El Salvador* 4 (3–4): 105–114.

Pullinger, Carlos
n.d. Evolution of the Santa Ana Volcanic
Complex, El Salvador. Master's thesis,
Department of Geology, Michigan
Technological University, 1998.

Reinhardt, Bentley
n.d. Volcanology of the Younger
Volcanic Sequence and Volcanic
Hazards Study of the Tuxtla Volcanic
Field, Veracruz, Mexico. Master's
thesis, Department of Geology, Tulane
University, New Orleans, 1991.

Rose, William, Michael Conway,
Carlos Pullinger, Alan Deino,
and William McIntosh
1999 A More Precise Age Framework
for Late Quaternary Silicic Eruptions in
Northern Central America. *Bulletin of
Volcanology* 61: 106–120.

Sanders, William, Jeffrey
Parsons, and Robert Santley
1979 *The Basin of Mexico: Ecological
Processes in the Evolution of a Civilization*.
Academic Press, New York.

Santley, Robert
1994 The Economy of Ancient
Matacapan. *Ancient Mesoamerica* 5:
243–266.

Santley, Robert, and Philip Arnold III
n.d. Prehispanic Settlement Patterns
in the Tuxtla Mountains, Southern
Veracruz, Mexico. MS in possession of
the authors.

Sharer, Robert
1974 The Prehistory of the Southeastern
Maya Periphery. *Current Anthropology*
15: 165–187.

Sheets, Payson
1980 *Archaeological Studies of Disaster: Their Range and Value.* Working Paper 38, Hazards Center, Institute of Behavioral Science, University of Colorado, Boulder, CO.

1983a Introduction. In *Archeology and Volcanism in Central America: The Zapotitan Valley of El Salvador* (Payson Sheets, ed.): 1–13. University of Texas Press, Austin.

1983b Summary and Conclusions. In *Archeology and Volcanism in Central America: The Zapotitan Valley of El Salvador* (Payson Sheets, ed.): 275–293. University of Texas Press, Austin.

1984 The Prehistory of El Salvador: An Interpretive Summary. In *The Archaeology of Lower Central America* (Fred Lange and Doris Stone, eds.): 85–112. University of New Mexico Press, Albuquerque.

1992 *The Ceren Site: A Prehistoric Village Buried by Volcanic Ash in Central America.* Harcourt Brace Jovanovich, Fort Worth, Texas.

1994 Summary and Conclusions. In *Archaeology, Volcanism, and Remote Sensing in the Arenal Region, Costa Rica* (Payson Sheets and Brian McKee, eds.): 312–326. University of Texas Press, Austin.

1999 The Effects of Explosive Volcanism on Ancient Egalitarian, Ranked, and Stratified Societies in Middle America. In *The Angry Earth: Disaster in Anthropological Perspective* (Anthony Oliver-Smith and Susanna Hoffman, eds.): 36–58. Routledge, New York.

2000 Provisioning the Ceren Household: The Vertical Economy, Village Economy, and Household Economy in the Southeastern Maya Periphery. *Ancient Mesoamerica* 11: 217–230.

Sheets, Payson (ed.)
2002 *Before the Volcano Erupted: The Ancient Ceren Village in Central America.* University of Texas Press, Austin.

Siebe, Claus
2000 Age and Archaeological Implications of Xitle Volcano, Southwestern Basin of Mexico-City. *Journal of Volcanology and Geothermal Research* 104: 45–64.

Siebe, Claus, Michael Abrams, Jose Luis Macias, and Johannes Obenholzer
1996 Repeated Volcanic Disasters in Prehispanic Time at Popocatepetl, Central Mexico: Past Key to the Future? *Geology* 24: 399–402.

Soto, Gerardo, Guillermo Alvarado, and Marcello Ghigliotti
1996 El registro eruptivo del Arenal en el lapso 3000–7000 años antes del presente y nuevas deducciones sobre la edad del volcán. *Boletín Observatorio Sismológico y Vulcanológico de Arenal y Miravalles* 9 (17–18). San Jose, Costa Rica.

Stuiver, Minze, and Bernd Becker
1993 High-precision Decadal Calibration of the Radiocarbon Time Scale, AD 1950–500 BP. *Radiocarbon* 35: 35–36.

Stuiver, Minze, and Paula Reimer
1986 A Computer Program for Radiocarbon Age Calibration. *Radiocarbon* 28: 1022–1030.

Urrutia-Fucugauchi, Jaime
1996 Palaeomagnetic Study of the Xitle-Pedregal de San Angel Lava Flow, Southern Basin of Mexico. *Physics of the Earth and Planetary Interiors* 97: 177–196.

Webster, David, Nancy Gonlin, and Payson Sheets
1997 Copan and Ceren: Two Perspectives on Ancient Mesoamerican Households. *Ancient Mesoamerica* 8: 43–61.

White, Gilbert
1945 *Human Adjustments to Flood.* University of Chicago Press, Chicago.

Williams, Howell, and Helmut Meyer-Abich
1955 Volcanism in the Southern Part of El Salvador. *University of California Publications in the Geological Sciences* 32: 1–64.

THE COLLAPSE OF MAYA CIVILIZATION:
ASSESSING THE INTERACTION OF
CULTURE, CLIMATE, AND ENVIRONMENT

Jason Yaeger
David A. Hodell

A s scientists collect more paleoclimatic data from the Maya lowlands, two important facts have emerged: there has been significant regional variation in the paleoclimate of the Maya lowlands, and a long-lasting phase of decreased rainfall began around AD 800 in many areas of the lowlands. Archaeological research has proven that the social and political transformations that we usually call the Classic Maya collapse were also regionally variable in their timing and nature. Despite this variability, most areas of the Maya lowlands experienced changes of great magnitude between AD 750 and 950—changes that included significant demographic decline, shifts in settlement location, new trade patterns, political decentralization, and new political ideologies. The transformations of the ninth and tenth centuries changed the sociopolitical organization of the Maya lowlands in fundamental ways, and scholars have proposed many different models to explain them.

One factor that has received renewed attention is the role of climatic change. In this paper, we evaluate the archaeological evidence for the social and political transformations at the end of the Late Classic and Terminal Classic periods (ca. AD 700–1000) and the paleoclimatic data for that same period. We compare and contrast the data within and between the northern, central, and southern lowlands of the Yucatan Peninsula. The Maya collapse was undoubtedly a complex process structured by many different factors, of which climatic change was important in many regions of the Maya lowlands. Demographic and environmental transformations such as population increase, deforestation, and topsoil erosion also influenced the collapse.

To understand the role that climatic and environmental change might have had, however, we must assess how such changes would have shaped people's options and decision-making in the Classic Period. Thus, we discuss paleoclimatic data in light of environmental effects, and we evaluate the archaeological data for social and political transformations. In so doing, we explore the complex interactions of paleoclimate, environment, and culture that could have led to the collapse, and critically evaluate the difficulties of synthesizing data sets that are based on three different chronological frameworks (radiocarbon, ceramic chronology, and calendrical texts).

Pre-Columbian Maya civilization of the central and southern lowlands reached its apogee of social complexity during the Late Classic Period (AD 600–800). Scores of political centers, the capitals of rival polities, crowded a landscape marked by high population densities and were supported by a mosaic of intensive and extensive agricultural strategies. Inside these centers, members of the ruling elite lived in elaborate palaces and were laid to rest in sumptuous tombs deep in the hearts of funerary pyramids. With their unprecedented wealth and power, the divine kings or *k'uhul ajawob'* of Maya polities sponsored skilled artisans who crafted elaborate objects for personal adornment and ceremonial use and gifted artists who carved hundreds of stone monuments depicting the political and ceremonial activities of the rulers and their elite peers and clients. Most of these monuments include hieroglyphic texts that describe these important events, recount the genealogies of the royal houses, and extol the legitimacy of the rulers as the leaders of their people.

Late in the eighth century AD, however, these polities began to undergo a series of radical changes that would transform Maya society in fundamental ways over the course of the next two centuries. These changes entailed a substantial demographic decline, the abandonment of most large centers, and the apparent disappearance of the ruling elite who are so salient in the archaeological record of the Classic Period. Although they occurred first in the central and southern lowlands, many regions in the northern lowlands underwent broadly parallel changes approximately fifty or seventy-five years later.

These pan-lowland changes are often termed the collapse of Classic Maya civilization or the Maya collapse, phrases that are useful shorthand referents, but that also mask the complexity of the processes involved (see also Cowgill 1988). As new data have proven the existence of strong continuities from the Classic to Postclassic periods (Chase and Rice 1985; Sabloff and Andrews 1986), many scholars feel uncomfortable with the term collapse, given its connotations of disjunctive change and degeneration. Furthermore, it is now clear that different regions and sites went through very different trajectories in the ninth and tenth centuries, and some did not collapse at all (Demarest et al. 2004; Marcus 1995). This variability underscores the futility of trying to impose a unitary or eventlike model of collapse on the Maya lowlands. Recognizing this, we will use the term collapse advisedly to refer in a general way to the complex sets of processes that restructured Maya civilization in the lowlands in the two centuries following the Late Classic Period.

Scholars have proposed many different models to explain the collapse. In 1995, David Hodell and his colleagues (1995) reported the first physical evidence of a drought during the Terminal Classic Period in sediments from Lake Chichancanab in the north-central Yucatan Peninsula. The Chichancanab record indicated that this drought was one of the most severe of the last 7,000 years, leading to speculation that drought may have been a causal factor in the collapse. Since that time, the role of climatic change in the collapse has received renewed attention. Additional paleoclimatic records from the Maya lowlands have revealed important differences in the climatic histories of subregions of the Yucatan Peninsula during the Terminal Classic Period.

In this chapter, we review the archaeological evidence from the end of the Late Classic Period through the Terminal Classic Period (AD 800–1000) and the paleoclimatic data for that same period. We critically evaluate the drought hypothesis by comparing the timing, magnitude, and spatial extent of paleoclimatic and paleoenvironmental change with the social and political transformations in three regions of the Maya lowlands during the eighth through tenth centuries AD. This evidence suggests that the collapse was a complex process structured by many different factors, of which climatic change was important in many regions of the Maya lowlands.

THE CLASSIC MAYA COLLAPSE

The collapse has long been a topic of scholarly and popular debate, eliciting scores of explanations. Early models tended to focus on single causal factors, often events that were catastrophic in nature and timing (Sabloff 1992: 102), such as a widespread peasant revolt (Thompson 1966), earthquakes (MacKie 1961), and epidemics. In contrast, other scholars argued that the collapse was the result of declining crop yields caused by gradual processes of environmental degradation such as the silting up of productive wetlands (Ricketson and Ricketson 1937) and savanna encroachment (Cooke 1931).

The 1970 Advanced Seminar at the School of American Research punctuated a shift in views of the collapse, as is evident from the contributions in the resulting publication, *The Classic Maya Collapse* (Culbert 1973). An expanding body of archaeological data suggested the collapse was best understood not as an abrupt event, but instead as a transformative process that took generations to play out. Consequently, the models postulated in the publication eschewed unicausal, prime-mover theories in favor of complex, multicausal explanations (e.g., Willey and Shimkin 1973). Most gave weight to social, political, and historical factors rather than environmental or climatic explanations, with a few exceptions (e.g., Sanders 1973).

In the four decades since the advanced seminar, studies of the collapse have largely followed the trajectory set forth therein, and most scholars continue to favor multicausal models that emphasize the social and political aspects of the collapse (Demarest et al. 2004). Although environmental changes—especially environmental degradation and soil exhaustion—continue to be invoked, these changes are often framed as the anthropogenic results of population growth and short-sighted, politically motivated managerial choices (e.g., Culbert 1988; cf. Turner 1990). In the last decade, however, scholars have begun to reconsider the role played by climatic change.

Climatic Change and the Collapse

Two factors have encouraged scholars to reassess the role of climatic change, especially decreased rainfall, in the ninth- and tenth-century transformations in the Maya lowlands. Perhaps the most important stimulus has been the accumulation since the 1980s of a great deal of paleoenvironmental data, including paleoclimatic data, discussed below. Maya archaeologists have begun to incorporate this information into their explanations of the collapse.

The other factor relates to the social context in which Maya archaeology is practiced. As Richard Wilk (1985) has noted, the factors that scholars invoke to explain the collapse correlate closely with the concerns that those researchers face in their daily lives. In the last decade, global warming and El Niño have become household phrases, and the United Nations Environment Programme (2002) found that weather-related natural disasters in the 1990s had doubled in frequency compared to the 1970s, costing trillions of dollars. Within this context, it is not surprising that scholars are increasingly interested in understanding the role of climatic change and climatic fluctuations in social and cultural change across the globe (e.g., Brenner et al. 2002; deMenocal 2001; Diamond 2005; Fagan 1999; Weiss and Bradley 2001).

In the Maya area, Richard E. W. Adams (1997), William Folan (Folan et al. 1983a, 2000), Richardson Gill (2000), Joel Gunn (Gunn and Adams 1981; Gunn et al. 2002) and others (Dahlin 1983; Lucero 2002) have argued more specifically that long-lasting pan-lowland droughts were a primary stimulus or trigger for the collapse of the complex Classic sociocultural system. As Brian Fagan (1999: 158) evocatively states, many of these scholars believe that climatic change "delivered the coup de grâce to rulers no longer able to control their own destinies because they had exhausted their environmental options in an endless quest for power and prestige." Unfortunately, some of these accounts are simple recapitulations of earlier mono-causal, catastrophic models of the collapse (e.g., Gill 2000), which we believe obscure more than they explain. The irrefutable evidence of dramatic regional differences in the cultural transformations of the Terminal Classic Period across the Maya lowlands (e.g., Demarest et al. 2004), coupled with increasing evidence of the complexity of the climatic changes that occurred during that same period, demands more complex understandings of the relationship between Maya society and culture, climate, and the natural environment.

The Collapse in the Archaeological Record

All models that attempt to explain the collapse must be judged ultimately against the same set of archaeological data, which several scholars have laid out in some detail (Adams 1973; Sabloff 1973a; Sharer 1977, 1994; Webster 2002). Most agree that the following general empirical trends mark the collapse:

1. Abandonment of palaces and temples and cessation of the construction of monumental architecture;
2. Reduction or cessation of the creation of public sculpture;
3. Substantial decline in the use of hieroglyphic writing on stone media and disappearance of Long Count calendar;

4. Reduction in the production of sumptuary or "luxury" items, like polychrome pottery and carved shell and jade objects;

5. Significant decline in investment in mortuary architecture and grave goods;

6. Significant decline in population and population density.

These six empirical factors must be understood as generalizations, however, and when examined in more detail, the data prove more complex, demonstrating variability at three different scales. First, there is significant variability between the different regions of the lowlands in the timing and nature of the changes that are routinely grouped together under the term collapse. Second, even within regions we find significant differences among sites: some did not decline as rapidly as others, and some, such as Lamanai, continued to be occupied until the Spanish Conquest (Pendergast 1991). Some, such as Chichen Itza in the northern lowlands, even flourished during the Terminal Classic Period (Andrews and Robles 1985; Cobos 2004). Finally, within specific polities, the processes that caused the collapse affected different groups in different ways. Although decline and abandonment are the general trends, settlement survey and household archaeology data from the Petexbatun region, Copán, and upper Belize River valley demonstrate that some families and villages survived for decades if not generations longer than their neighbors (Palka 1997; Webster and Freter 1990; Yaeger 2003). We mention this complexity not to downplay the scope of the changes of the Terminal Classic Period, but to emphasize that the collapse was neither a unitary nor an eventlike phenomenon; it was a complex set of processes spanning generations that played out quite differently at sites across the lowlands (see also Demarest et al. 2004; Marcus 1995).

The three primary data sources we use to understand the social and cultural changes of the Terminal Classic Period are archaeological, epigraphic, and art-historical, the former two being employed most frequently. Hieroglyphic inscriptions provide detailed histories of those sites that possess a large corpus of inscriptions, but synthesizing the epigraphic and the archaeological data is a complex task (Fash and Sharer 1991; Houston 1989; Stuart 1992). The histories of many sites contain lacunae that can be attributed to gaps in the creation of texts, the later destruction of text-bearing monuments, and, in some cases, a relative lack of fieldwork. Furthermore, Classic-Period hieroglyphic texts are restricted almost exclusively to accounts of the activities of the rulers and nobles of Maya polities: raids against neighboring polities, the births of royal heirs, the consecration of new temples, the deaths and burials of powerful nobles, for example (Marcus 1992; Martin and Grube 2000). They do not speak to demographic trends, population movements, or agricultural productivity, and they bear only indirect witness to many other topics pertinent to a full understanding of the collapse, such as the production of luxury goods. Archaeological data can provide information with which to understand these topics, but chronologically coordinating these two kinds of data is problematic. Often the trends that can be documented by the much finer-grained hieroglyphic chronologies are given priority in interpretations, and the other patterns are assumed to follow suit.

Dating the Collapse

Many early studies of the collapse focused on determining when the collapse occurred or when a particular site or region was abandoned. This approach was driven by the culture-historical framework employed by most Maya archaeologists in the first half of the twentieth century and was informed by an implicit understanding of the collapse as a panregional event or a series of collapses of individual sites. It was further encouraged by the nature of early Maya epigraphy. The glyphs relating to the Maya calendar were deciphered decades before other aspects of the writing system, so early scholars could identify the last recorded dates at sites before they could understand the events recorded for those dates. Furthermore, with few exceptions, hieroglyphic texts constituted the only data set for inferring the timing of the abandonment of many lowland sites (e.g., Morley 1946).

As scholars have reconceptualized the collapse in terms of cultural processes and historically contingent transformations instead of a one-time catastrophic event at the site or regional levels, their interests in chronology necessarily have shifted from determining the terminal date of a site's occupation to charting changes in Maya society during the eighth through tenth centuries. This requires a firm chronological control of the timing and pace of different social and political transformations, which at most sites is obtained by combining calendrical information from hieroglyphic inscriptions, ceramic sequences, and radiocarbon dates.

Hieroglyphic texts are almost ubiquitous on Classic-Period Maya monuments, and they usually contain calendrical information that allows epigraphers to assign the events described in the texts to the Gregorian date on which they occurred. Many studies of the collapse have charted regional patterns in the final hieroglyphic dates at sites across the lowlands, using the last inscribed date at a site as a proxy for the date of abandonment or political decentralization of that site (Gill 2000; Haug et al. 2003; Lowe 1985). Although the cessation of creation of stone monuments and hieroglyphic texts is a central symptom of the collapse, it does not reliably correlate with other aspects of the collapse. At some sites, the last dated stone monument predates, in some cases significantly, the center's final abandonment (Harrison 1999; Pendergast 1991); at others, it is followed by politically charged stone sculpture lacking hieroglyphic texts (Martin and Grube 2000). Other studies, however, have looked for patterns in the events described in the inscriptions of the later Late Classic and Terminal Classic periods. They have found evidence for increasing political competition as reflected in the number of people and offices mentioned in the texts, increasing frequency of warfare, and decentralization in the larger polities as reflected in the proliferation of sites whose rulers use their own emblem glyphs and the title *k'uhul ajaw* or divine king (Houston 1993; Marcus 1976, 1992; Martin and Grube 2000; Stuart 1993).

Most of the deposits studied by archaeologists in the Maya lowlands lack associated calendrical texts, however, and the principal tool archaeologists use in dating these deposits is the associated ceramic vessels, following the groundbreaking work of Robert E. Smith (1965) at Uaxactun. Smith defined ceramic complexes by the co-occurrence of distinct vessel forms and decorative modes, and then dated the sequence of complexes using associations between diagnostic vessel types and hieroglyphic texts. Even after the development of radiocarbon dating provided archaeologists with another tool for assigning absolute dates to the ceramic sequences, the range of error was often so great that cross-ties to the ceramic sequences of Uaxactun and other sites, together with the local association of ceramic material with hieroglyphic dates, have remained critical for assigning chronological dates to ceramic sequences (e.g., Chase 1994).

Ceramic phases tend to be relatively long, on the order of seventy-five to two hundred years in the Late and Terminal Classic periods, although some scholars have established finer-grained chronologies (e.g., Taschek and Ball 1999). Consequently, the temporal resolution in the archaeological record at most sites is between three and eight generations. This gross chronology seriously compromises efforts to tightly correlate archaeologically visible changes—shifts in demography or household economy, for example—with the political events that we can follow from year to year in the hieroglyphic inscriptions. Further, because of the palimpsest nature of the archaeological record, many archaeological deposits are actually cumulative products of years or even generations of human behavior, and short-term changes can be lost or aggregated and averaged out.

Further clouding the picture, the absolute dates for ceramic phases that correspond with the Terminal Classic Period are often less secure than those for earlier phases. At most sites, the Terminal Classic Period is marked by reduced occupation and fewer monuments. The consequent paucity of associations between Terminal-Classic ceramic vessels and radiocarbon dates or hieroglyphic texts makes it difficult to accumulate enough dates to define the boundaries of the phase with much precision. Consequently, there is a risk of using ceramic cross-ties to date these phases, making them essentially contemporaneous, and masking significant variability between regions. It is especially difficult to ascertain the ending date for the Terminal Classic phase, a significant empirical fact, given that it is likely that most sites continued to be occupied after their rulers ceased to commission texts with hieroglyphic dates. In these cases, the only way to date the end of the Terminal Classic phase independently is through radiocarbon dating, and the lack of later occupation makes it impossible to bracket the radiocarbon dates to more precisely define the last use of the assemblage. This is unfortunate, as this date would serve as a much more accurate proxy for the abandonment of a site than does its last hieroglyphic date.

The third dating technique that Mayanists use frequently is radiocarbon dating. Although radiocarbon dates are often presented as Gregorian calendar dates, they are more accurately thought of as time spans, not exact points in time. The probabilistic nature of radiocarbon decay, the fluctuations in atmospheric radiocarbon frequencies, and the nature of organic preservation all introduce an inherent multidecadal fuzziness in the precision of radiocarbon dates. The error range of samples dated with accelerated mass spectrometry (AMS) techniques is often ±30–50 years and two to three times that for samples dated with standard procedures. Consequently, dating events or contexts in the archaeological record, even when it involves combining dates from multiple independent samples, is subject to a decadal or generational precision that is much fuzzier than the temporal precision that pertains for events recorded in hieroglyphic texts, which can often be dated to the day they occurred.

Further complicating the precision of a date is the need to calibrate a radiocarbon date to correct for fluctuations in the frequency of radiocarbon isotopes in the atmosphere. Calibration curves have improved markedly in the past twenty years (Stuiver et al. 1998), but portions of the curve are inherently problematic because the rate of change in the amount of radiocarbon in the atmosphere was such that a given radiocarbon date can have multiple intercepts. Unfortunately, these "flat" zones on the calibration curve include three periods during the eighth through tenth centuries (AD 680–760, 790–880, and 900–950). Samples dating to these time periods are likely to return probability curves with long spans of time in which multiple dates are equally likely. For example, a sample with a radiocarbon date of 1195±45 ^{14}C BP yields seven equally probable intercepts between AD 780 and 857, a one-sigma range of AD 774–889, and a probability curve that is essentially flat from AD 785–885—the century of greatest interest to those trying to understand the Maya collapse.

Finally, because of the hot, humid conditions in much of the Maya lowlands, plant remains are rarely preserved unless carbonized, charcoal is rarely preserved, and most preserved carbonized material derives from long-lived deciduous trees. Because it is usually impossible to determine how much time passed between the point at which a segment of a tree became heartwood and thus removed from active carbon-exchange and the point at which people cut down the tree and used it, an additional level of uncertainty—decadal in many cases—is introduced into many carbon dates, save those that derive from preserved seeds, endocarps, tubers, twigs, and other short-lived or annual vegetative elements.

Despite these limitations, significant strides have been made in refining the regional chronological frameworks used by Maya archaeologists in the decades since the 1970 symposium on the collapse. Although the advent of AMS radiocarbon dating was one factor in these advances, basic empirical data collection was much more important. The establishment of refined ceramic sequences at many sites and the related ability to determine the temporal relationships between different regional ceramic traditions have allowed archaeologists to recognize two important facts regarding the timing of the collapse.

First, the collapse of many Maya polities in the central and southern lowlands in the ninth century was roughly contemporaneous with the growth in size and political power of sites in the northern lowlands (fig. 1). Many have argued that the decline of the southern polities and rise of the northern polities

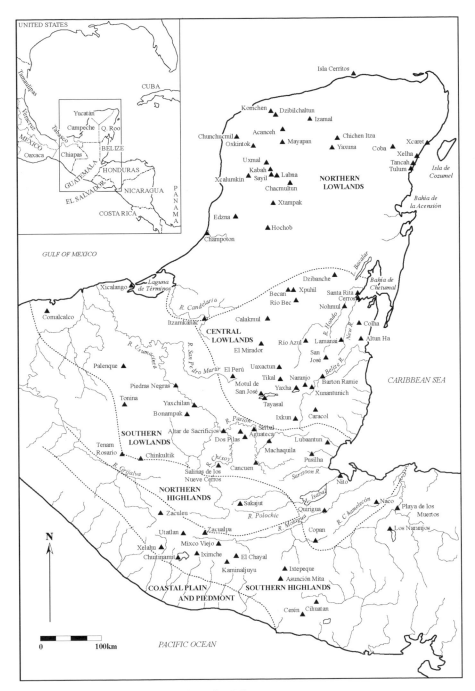

FIG. 1 The Maya area (after Sharer 1994: fig. 1.1)

were related (Erasmus 1968; Lowe 1985). Some have argued that changing political economies and regional exchange networks refocused trade along circum-Peninsular routes that were controlled by Putun, Itza, and other Mexicanized groups and bypassed inland polities (Ball 1977; Braswell 2003; Freidel 1986; Sabloff 1973b; Sabloff and Rathje 1975); others have suggested that the disruptions in the south led to significant immigration to northern cities, increasing their size and economic might (Carmean et al. 2004). Although many northern polities underwent a decline and abandonment in the tenth century, data from the north, discussed below, demonstrate that the collapse did not affect that region uniformly.

The second important development arose from a chronological reevaluation of Chichen Itza, the largest site and clear regional power center in the Early Postclassic northern lowlands. Scholars had argued that Chichen Itza and its associated Sotuta ceramic complex largely postdated the decline of other northern sites like Coba and Uxmal, the latter associated with the Cehpech ceramic complex (e.g., Smith 1971). New chronological data indicate that there is considerable temporal overlap between the Sotuta and Cehpech complexes, and that the Chichen Itza polity was contemporaneous with Coba and the Puuc polities like Uxmal (Anderson 1998; Cobos 2004; Kepecs 1998; Ringle et al. 1998). Many researchers now conclude that Chichen Itza played a central role in the decline and abandonment of other northern sites during the tenth century:

> The rise and fall of the Itzá state in the northern Maya lowlands of the Yucatan peninsula, Mexico, during the Terminal Classic and Early Postclassic times, was probably the single most important process in late Maya history (Andrews 1990: 258).[1]

This new understanding of the Chichen Itza chronology has led many Maya archaeologists, especially those working in the northern lowlands, to reconsider the dating of the Terminal Classic Period. In the central and southern lowlands, most scholars use the term Terminal Classic in two senses. Chronologically, it refers to ceramic complexes roughly contemporaneous with Smith's Tepeu III complex at Uaxactun, usually dated to AD 800–ca. 900, although sometimes adjusted to begin in AD 830, corresponding to the Long Count date 10.0.0.0.0 (see Rice and Forsyth 2004 for a detailed discussion of Terminal-Classic ceramic chronologies). Culturally, scholars use the term to refer to the period of social and political transformations that comprise the collapse. Because data from Uxmal and other sites indicated continued tenth-century occupation, many archaeologists working in the northern lowlands used a somewhat expanded range of AD 800/830–1000 (Andrews and Andrews 1980; Smith 1971), but they lacked many firm absolute dates for Terminal-Classic ceramic complexes (Andrews and Sabloff 1986).

1 William Ringle et al. (1998: 192) suggest that Chichen Itza was abandoned earlier, between AD 950 and 1000. Linda Schele and Peter Mathews (1998: 98) argue that the site's public architecture dates to the period of AD 800–948.

More recently, however, scholars have argued that the dates of the Terminal Classic must be revised if the term is to refer to sociopolitical transformations. George J. Bey and colleagues (1997: 238) suggest that the Terminal Classic in the northern lowlands be used to refer to a period that postdates the end of monumental construction at Uxmal and other Puuc sites, AD 925–1100. Charles Suhler and colleagues (1998) broaden the period to AD 730–1100 so that it encompasses the rise and fall of the Puuc centers and the later apogee of Chichen Itza. In contrast, Kelli Carmean and colleagues (2004) prefer a narrower date range of AD 770–950 for the Puuc Hills region of the northern lowlands, one that encompasses the florescence and decline of the Puuc polities and is more in line with the dates used for that period in the southern and central lowlands.

For our discussion, we define the Terminal Classic as AD 800–1000. This date range is widely used in the paleoclimatic literature and, to a lesser extent, among archaeologists. These dates correspond with the strongest evidence for drier climatic conditions, and they encompass the abandonment of most large polities in the central and southern lowlands and in the Puuc region of the northern lowlands. We would point out, though, that using the same dates for the Terminal Classic period for all parts of the lowlands can generate confusion and obscure important differences at the local and regional scales, as processes and events that can be distinguished temporally become grouped under the broader rubric of Terminal Classic.

CLIMATE AND ENVIRONMENT

Modern Hydrologic Setting

The spatial and seasonal distribution of rainfall is highly variable across the Yucatan Peninsula today and is affected by climatic variability of both Pacific and Atlantic origin (e.g., ENSO [El Niño/Southern Oscillation] and NAO [North Atlantic Oscillation], respectively). The northwest coast is the driest area with an annual rainfall of 450 mm/yr near Progreso, Yucatán, Mexico, but rainfall increases steadily to the south, reaching 1600 mm yr at Flores, Peten, Guatemala (fig. 2). This represents an increase of 1150 mm over a distance of 500 km, a gradient that is significantly greater than that of the drought-prone Sahel (1000 mm over 750 km).

Precipitation is highly seasonal, and most rain falls during a distinct rainy season from May to October (fig. 3), interrupted by the *canícula* or "little dry season" in July and August, when conditions are typically somewhat drier and less cloudy (Magaña et al. 1999). The rainy season coincides with the Northern Hemisphere summer, when the Intertropical Convergence Zone (ITCZ) and North Atlantic subtropical high-pressure system (also known as the Azores–Bermuda high) move northward (Hastenrath 1966, 1967, 1976, 1984, 1991). Tropical storms and hurricanes during this period can contribute greatly to rainfall averages for a single year (Gray 1987, 1993). The dry season occupies the Northern Hemisphere's winter months of November through April. During this time, precipitation is suppressed as the ITCZ swings

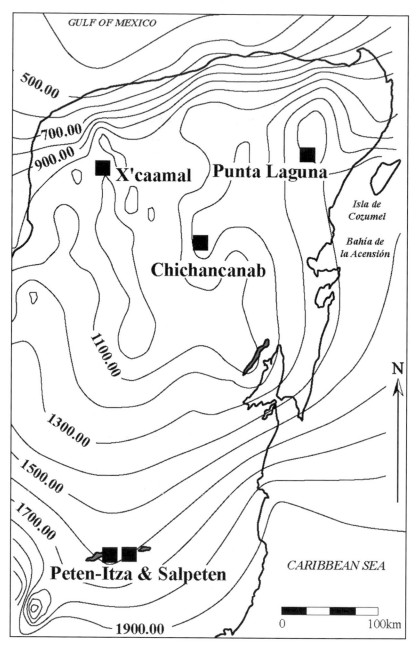

FIG. 2 The Yucatan Peninsula, showing location of lakes used for paleoclimate study and isohyets of annual precipitation (100 mm contour interval)

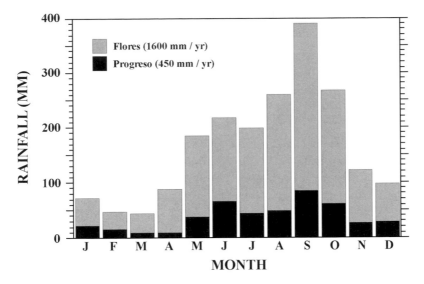

FIG. 3 Average monthly rainfall (in mm) at Progreso, Mexico, on the northwest coast of the Yucatan Peninsula, and Flores, Guatemala, on the shores of Lake Peten-Itza

south of the equator and the North Atlantic subtropical high-pressure zone moves south and dominates in the Intra-Americas Sea (Gray 1993).

Interannual variability in rainfall is controlled by a mechanism similar to that of the annual cycle, involving competition between the North Atlantic high-pressure system and the eastern Pacific ITCZ (Giannini et al. 2000, 2001a, 2001c). The strength of the subtropical North Atlantic high is governed mainly by the NAO. A positive phase is associated with a stronger-than-usual subtropical high-pressure center, stronger trade winds, cooler SST (sea-surface temperature) in the tropical North Atlantic, and decreased rainfall in the Caribbean.

During the lifetime of an ENSO cycle, the Caribbean experiences both dry and wet extremes (Giannini et al. 2000, 2001a; see also Maasch, this volume). A warm El Niño phase is generally associated with drier-than-average conditions during the boreal summer of year (0) and wetter-than-average conditions during the spring of year (+1). However, the dry season that coincides with the mature phase of ENSO is wetter than average in the Yucatan. Following the strong El Niño of 1997–1998, much of Mexico and Central America experienced drought conditions that resulted in massive wildfires.

The relationship between ENSO and Mesoamerican precipitation is complex, however, because of interactions with conditions in the North Atlantic. The interaction of the NAO and ENSO can produce constructive interference that leads to anomalously dry or wet conditions in the Caribbean (Giannini et al. 2001a, 2001b). For example, summers following winters characterized by a positive phase of the NAO and a developing warm ENSO event (0) produce extremely dry conditions in the Caribbean. In contrast, winters characterized by a negative phase of the NAO and a warm ENSO event (+1) should result in anomalously wet conditions in the Caribbean. This is similar to the findings of Enfield and Alfaro (1999), who concluded that oppositely signed SST anomalies in the Pacific and tropical North Atlantic are associated with enhanced rainfall departures over the Caribbean and Central America. For example, when an El Niño event (associated with warm temperatures in the eastern tropical Pacific) is coupled with a cool tropical North Atlantic (positive NAO), the summer rainy season in the Caribbean tends to be anomalously dry.

Climate and the Classic Maya

The interannual variability in precipitation caused by the cycles described above would have had a significant impact on rainfall agriculture in the Yucatan Peninsula. As important as the total amount of rainfall, however, is intra-annual distribution of rain and the onset of the wet and dry seasons. In modern swidden agricultural systems, predicting the onset of the rainy season is especially important in the cycle of burning and planting of fields (Gunn et al. 1995). If the rainy season begins before farmers have burned the forest they felled to make their farming plots, they get an incomplete burn, resulting in a field full of half-burned trees and a loss of nutrients that derive from the burned plant matter. In contrast, if the rainy season begins late, weedy growth will spring up in a field before the planted maize has germinated, requiring extensive labor investment in weeding.

In the Classic Period, nonswidden farming practices were probably quite common, reducing some of the risk entailed by an unexpected onset of the rainy season, but the timing and distribution of rainfall over the course of the growing season would have been critical nonetheless. Shortages or excesses in water availability at certain key times in a crop's life cycle—especially during germination, pollination, and fruit maturation—can affect yields substantially. In the northern lowlands where average annual rainfall is just sufficient for maize production and in areas where the Maya grew crops during the dry season using pot or canal irrigation, raised fields, or other strategies, slight reductions in precipitation could have had a disproportionate impact on agricultural productivity, especially if they occurred at the key times mentioned above. The risk of crop failure due to reduced rainfall could be abated by planting multiple kinds of crops, and multiple varieties of maize that had somewhat different maturation rates. In contrast, in the central and southern lowlands, annual rainfall is several times greater than the amount needed to produce a full crop of maize. Because there is no linear relationship between rainfall and maize yields, the excess rainfall does not increase maize productivity, and thus a significant reduction in annual precipitation could have had little impact on maize yields, especially if the reduced rainfall affected the rainy season.

Both the NAO and ENSO can potentially affect the start and end dates of the rainy season. For example, the ITCZ tends to migrate more slowly to northern latitudes during El Niño years, thereby delaying the onset of the rainy season (Mesoamerica Climate Outlook Forum 1998). In southern Central America, Enfield and Alfaro (1999) found that a warm tropical North Atlantic favors an expansion of the rainy season at both ends (onset and end), while a cool North

Atlantic leads to a contraction of the rainy season. In contrast, Pacific ENSO affects the rainy season end dates but not the onset date.

The Classic Maya required water for other purposes besides agriculture, including drinking, cooking, construction, and bathing (Folan et al. 2000; Lucero 2002), and it is likely that the Maya in many areas of the lowlands would have needed to store water in the rainy season in order to have sufficient supplies throughout the dry season. Water supply is influenced not only by precipitation, but also by access to groundwater. In the northern lowlands, rainfall quickly percolates into the porous limestone and surface drainage is negligible. The water table is close to the surface (e.g., ~28 m at Chichen Itza), however, and can be accessed through *cenotes*, the lakes reached by sinkholes that perforate the bedrock. Because of higher elevation caused by faulting along the Sierrita de Ticul, the water table is significantly lower in the Puuc region (Dunning 1992). There, the Maya built cisterns called *chultuns* to store rainwater for domestic use (McAnany 1990; Thompson 1897).

In the central and southern lowlands, surface waters are perched above the regional water table, which is more than a hundred meters below the surface (e.g., ~130 m at Tikal). The Maya of the central and southern lowlands were generally more reliant on surface water supplies for drinking water than was the case in the northern lowlands. These water supplies included lakes, rivers and streams, and springs, as well as reservoirs both in large cities (Lucero 2002; Scarborough 1996) and in hinterland residential areas (Weiss-Krejci and Sabbas 2002).

Climate, Environment, and the Maya

Under the direction of Edward S. Deevey from 1972 to 1988, the Central Peten Historical Ecology Project (CPHEP) advanced our understanding of the relationship between the prehistoric Maya and their environment by combining archaeological data on population density with paleolimno-logical information from several lake basins in the central lowlands (Rice 1996). The project demonstrated that a thick deposit of inorganic colluvium called Maya Clay underlies many Peten lakes. As the name implies, the researchers concluded that this sediment deposit was anthropogenic, the end result of growing Maya population densities, which caused deforesta-tion, topsoil erosion, increased colluviation, and phosphorus sequestration in lake sediments. The CPHEP data led to a conceptual model, applied widely throughout the Maya lowlands, that the Maya were primary agents of envi-ronmental transformation and "stressors" of the natural ecosystem (Deevey et al. 1979), and that human-induced environmental degradation contributed significantly to the collapse.

A fundamental assumption of CPHEP was that climatic change was negligible during the period of Maya occupation, as explicitly expressed by Deevey and colleagues (1980: 420): "[c]limatic changes during and since Maya time were unimportant, and the major environmental perturbations arose from human settlement and technology." Recent paleoclimatic evidence derived from analysis of sediment cores from the Yucatan Peninsula and circum-Caribbean region demonstrates that Holocene climate, and specifically rainfall, has not been constant throughout the period of Maya occupation (Hodell et al. 1991, 1995, 2001, 2005a; Curtis et al. 1996, 1998, 1999; Haug et al. 2000, 2003). Furthermore, the data show important differences in the climatic histories of subregions of the Yucatan Peninsula during the Terminal Classic Period, as we demonstrate below (see also Messenger 2002; Shaw 2003).

Charting Paleoclimatic Changes

In the Maya lowlands, instrumental records of climatic change are limited to the last century (e.g., 100-yr-long record of rainfall at Mérida), although colonial records and Maya chronicles such as the Books of Chilam Balam extend back to the fifteenth century (Folan and Hyde 1985; Craine and Reindorp 1979). Several approaches have been used to infer paleoclimatic change in Mesoamerica for earlier periods. Initially, climate-modeling studies were undertaken to retrodict climate in the Maya region on the basis of various forcing mechanisms, such as solar insolation and volcanic activity, and their correlations to global temperature trends (Gunn and Adams 1981; Messenger 1990; Sanchez and Kutzbach 1974; for review see Gunn et al. 2002). Other studies have relied on more detailed paleoclimatic records from other regions (mostly from the high-latitude Northern Hemisphere), employing models of teleconnective climatic linkages between regions to predict how climate would have changed in Maya lowlands (Folan et al. 1983a; Gill 2000; Gunn et al. 1995; cf Messenger 2002).

A more direct approach to climate reconstruction uses sediment cores taken from closed-basin lakes in the Maya lowlands (Curtis et al. 1996, 1998; Hodell et al. 1995, 2001, 2005a; Rosenmeier et al. 2002a, 2002b). Mark Brenner and colleagues (2002, 2003) have reviewed the use of sediment cores to reconstruct climate in the Maya lowlands elsewhere, but a brief summary of the use of oxygen isotope ratios and mineral concentrations as geochemical proxies for inferring changes in the ratio of evaporation to precipitation (E/P) is called for here.

In closed basin lakes, the volume, concentration of dissolved solutes, and $^{18}O/^{16}O$ ratio of the lake water are controlled by a balance between water lost by evaporation relative to water gained by precipitation and runoff:

$$dVolume_{lake} = precipitation\ (P) + runoff\ (R) + groundwater\ (G) - evaporation\ (E)$$

The amount of runoff and groundwater input is generally related to precipitation, so the hydrologic budget of a lake is essentially dependent only on precipitation and evaporation. In some cases, however, runoff and groundwater input can be affected by human- or naturally induced vegetation changes. For example, deforestation reduces evapo-transpiration and soil moisture storage, thereby increasing surface and groundwater flow to the lake basins in a way that mimics increased rainfall (Rosenmeier et al. 2002b).

When water evaporates, water with the lighter isotope of oxygen ($H_2^{16}O$) evaporates at a faster rate than the heavier form ($H_2^{18}O$), thereby increasing the $^{18}O/^{16}O$ ratio of lake water. Furthermore, the dissolved salts in the lake water become more concentrated. During a period of drier climate, a closed lake loses more water to evaporation than it receives from precipitation and, consequently, the lake volume decreases, dissolved solutes become more concentrated, and the $^{18}O/^{16}O$ ratio of lake water increases. The reverse occurs during wet periods, when the lake basin receives more water through precipitation and runoff than it loses to evaporation.

When organisms such as ostracods, gastropods, or bivalves precipitate shells of calcium carbonate ($CaCO_3$), the $^{18}O/^{16}O$ ratio of the carbonate-bound oxygen is related to the $^{18}O/^{16}O$ ratio of the water from which the carbonate precipitated. Temperature also affects the $^{18}O/^{16}O$ ratio of $CaCO_3$, but temperature changes in the Maya region during the late Holocene were small relative to changing E/P. Therefore, by measuring the changes in the $^{18}O/^{16}O$ ratio ($\delta^{18}O$) of shells down the length of a core, one can reconstruct the relative changes in the $^{18}O/^{16}O$ ratio of lake water and, consequently, E/P.

Many lakes in the Maya lowlands have sulfate as their dominant anion (e.g., Lakes Chichancanab, Salpeten, and Peten-Itza), because gypsum ($CaSO_4$) is a common mineral in evaporite deposits in the bedrock. When rainwater or groundwater comes into contact with gypsum, Ca^{2+} and SO_4^{2-} ions are dissolved and are delivered to the lake. These ions build up until the lake water becomes supersaturated. In cases of a sudden reduction in lake volume from increased E/P, supersaturation can be exceeded and these ions precipitate out. The sulfur content of sediments, measured as wgt. %S, can be used as a qualitative proxy of E/P in those lakes, like Lake Chichancanab, that are at or near gypsum saturation.

Although both oxygen isotopes and sulfur concentrations are valuable proxy measures, the combined measurements of both on the same sample is an especially powerful tool for reconstructing changes in E/P. For example, if both $\delta^{18}O$ and wgt. %S increase simultaneously in a sediment profile, then one can eliminate a change in the $\delta^{18}O$ due to the rainfall because this process would have no effect on gypsum saturation.

Dating Climatic Changes

Prior to the advent of the AMS dating technique, most radiocarbon dates in lake sediments from the Maya lowlands were measured on shell or bulk organic material using traditional analytical techniques (gas counting and liquid scintillation). These dates were subject to hard-water lake error because the weathering of limestone in a lake's watershed produces dissolved inorganic carbon that is devoid of ^{14}C and thus dilutes ^{14}C in the lake (Deevey and Stuiver 1964). Consequently, the dates make the samples appear older than their true age. Although there are corrective algorithms, hard-water lake error has historically limited the accuracy of core chronologies on the Yucatan Peninsula. AMS ^{14}C analysis now permits the dating of milligram-sized terrestrial organic material preserved in sediment cores, such as wood, seeds, twigs, and charcoal. This tool has improved tremendously the dating accuracy of lake sediment cores in the Maya region, but several factors still serve as obstacles to correlating paleoclimatic changes observed in different lake cores, and to correlating patterns observed in the cores with the archaeological record.

As is the case with radiocarbon dates from archaeological contexts, there are errors introduced by the probabilistic nature of radioactive decay and other factors. Similarly, preservation of charcoal or other organic matter is an important consideration. Because the age of sediment between dated horizons is usually interpolated by assuming a constant sedimentation rate between points or by fitting higher-order functions to age-depth pairs, the number and position of dated points in a core are critical. In some cores, terrestrial organic matter is sparse or not present in datable quantities during the time periods of greatest interest, leading to interpolated age estimates for sediments far from any secure reference points.

We clearly face challenges in precisely and accurately dating both the paleoclimatic and archaeological records, challenges that are amplified when trying to correlate the two data sets. First, there are very few cases in which horizons or events like volcanic tephra deposits left visible signs in sediment cores and the archaeological record that we could use as shared chronological reference points. Thus, our correlations are almost always subject to the uncertainties inherent in radiocarbon dating, making it difficult to achieve more than a fuzzy temporal correlation between paleoclimatic and archaeological data.

Second, observations in the archaeological record using ceramic information and slices from sediment cores can only provide temporal resolution on a multidecadal scale. Within these slices of time, rapid changes can be hidden or "averaged out," making them indistinguishable from gradual ones. These two difficulties should be kept in mind when attempting to correlate the archaeological and paleoclimatic evidence for the Terminal Classic.

DATA FROM THREE REGIONS

There is indisputable evidence of regional diversity in the social and cultural transformations of the Terminal Classic Period, and the paleoclimatic sequences from sites across the lowlands also present distinct local pictures. These facts preclude a single monolithic summary of the relationship between climatic changes and the collapse. Consequently, we will discuss three different regions that have rich archaeological and paleoclimatic records for the eighth through tenth centuries. We feel that it is important to begin our case studies in the early eighth century, during the heart of the Late Classic Period, and continue into the eleventh century wherever possible. In this way, we span the entire range of time that encompasses the collapse. These regions are the central, the north-central to northeast, and northwest sectors of the lowlands.

The Central Lowlands

The Paleoclimatic Record. One of the more striking geological features of the central lowlands is an east-west chain of lakes known as the Peten Lakes that follows a geological fault in the limestone bedrock (Deevey et al. 1979, 1980). Of all areas of the lowlands, this region has been subjected to the most intensive paleoclimatic research, thanks largely to Deevey's CPHEP project. As described above, CPHEP was more interested in human–environment interactions than climatic change. Consequently, Mark Brenner, David Hodell, and Jason Curtis have been directing more recent coring projects in Lake Peten-Itza and Lake Salpeten to recover paleoclimatic proxy records (fig. 2).

Two facts affect the interpretion of these records. First, because the central lowlands were the most densely populated region in the Classic Period, the oxygen isotope records must be interpreted with caution. Anthropogenic landscape changes such as human deforestation can alter a lake's hydrologic budget by changing surface runoff and groundwater inflow in ways that mimic climatic change (Rosenmeier et al. 2002a, 2002b).

Second, the rate and nature of sediment deposition in Lake Peten-Itza and Lake Salpeten have resulted in proxy records with a multidecadal temporal resolution. Consequently, we complement them with a brief discussion of new paleoclimatic data from the Cariaco Basin of Venezuela, which has an extraordinarily fine temporal resolution.

Lake Peten-Itza. Lake Peten-Itza is the largest lake in the Maya lowlands, with a surface area of 100 km^2 and maximum depth of 165 m. Its large volume buffers its hydrologic budget from human activities such as deforestation, but at the same time renders it relatively insensitive to changes in E/P compared to smaller lakes. As a result, observed changes in the oxygen isotope record were small (<0.5‰) during the past two millennia (fig. 4). The lowest $\delta^{18}O$ values occurred during the Early Classic and Late Classic periods, followed by a steplike increase in the Terminal Classic Period (fig. 5). The Maya Clay in Lake Peten-Itza is represented by relatively high magnetic susceptibility that reflects increased erosion of clastic material from the watershed (Curtis et al. 1998).

The increase in $\delta^{18}O$ in the Terminal Classic Period coincides with a decrease in magnetic susceptibility, which is consistent with a reduction in the quantity of eroded, magnetic minerals in the sediment (fig. 5). At the same time, changes in pollen frequencies in the profile indicate a decline in disturbance taxa and an increase in lowland forest taxa (Islebe et al. 1996). Together these changes reflect the recovery of forests and stabilization of soils as the declining population density associated with the collapse reduced human pressures on the landscape. The $\delta^{18}O$ increase in the Terminal Classic Period may represent either an increase in E/P (i.e., drought) similar to Lakes Chichancanab and Punta Laguna to the north, or a decrease in runoff related to reforestation of the watershed.

Lake Salpeten. Lake Salpeten is located just east of the northern basin of Lake Peten-Itza. It measures only 2.6 km^2 in area, however, and its hydrologic budget is consequently more sensitive to changes in water input and

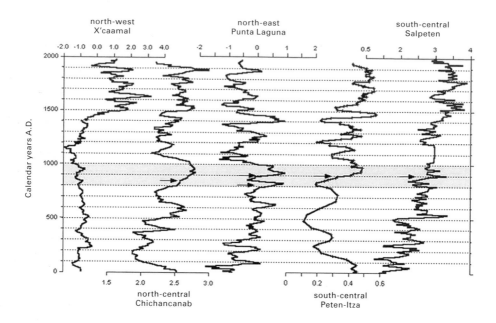

FIG. 4 Oxygen isotopic records from Aguada X'caamal (N 20. 61°, W 89. 72°) and Lakes Chichancanab (N 19. 88°, W 88. 77°), Punta Laguna (N 20. 65°, W 87. 65°), Peten-Itza (N 16. 92°, W 89. 83°), and Salpeten (N 16. 98°, W 89. 67°). Gray area highlights the Terminal Classic Period; arrows indicate $\delta^{18}O$ values that represent increases in E/P.

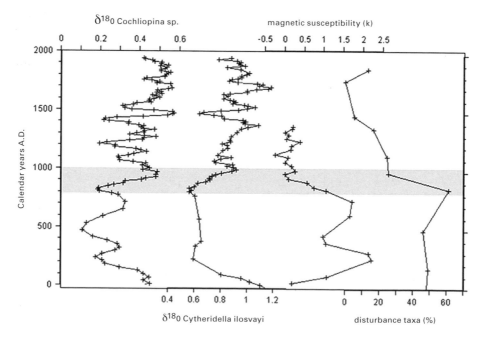

FIG. 5 Oxygen isotope records of the gastropod *Cochliopina sp.* and ostracod *Cytheridella ilosvayi*, magnetic susceptibility, and percent disturbance taxa in pollen assemblages from Lake Peten-Itza (Curtis et al. 1998). Shaded gray area represents the Terminal Classic Period.

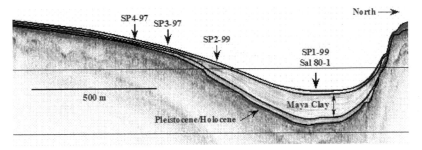

FIG. 6 Single-channel, high-resolution seismic reflection profiles from Lake Salpeten (Brenner et al. 2003). Approximate core locations are indicated by vertical arrows.

evaporation. The lake's depth reaches 32 m along the steep, fault-controlled northern shore, but the southern shore is shallow and shelving (fig. 6). This basin geometry has resulted in intense sediment focusing and the deposition of a thick sediment package in the deep basin. The Maya Clay is the thickest unit, representing erosion of catchment soils as a consequence of human-induced deforestation. The base of the Maya Clay in Salpeten core 80-1 was dated to 3,160±80 ^{14}C yrs BP by AMS-^{14}C dating of terrestrial material, which converts to an age range of 1700–1100 BC, in the Early Pre-Classic Period (Rosenmeier et al. 2002a). The top of the Maya Clay corresponds to the Terminal Classic Period, when forests recovered and soils stabilized as human pressures on regional vegetation were reduced (Anselmetti et al. 2007).

During the last two millennia, the oxygen isotopic record of Lake Salpeten was marked by low values between AD 1–150, followed by a series of step-like increases around AD 150, 550, 900, and 1400 (fig. 4). These events may have been caused by increases in E/P and/or decreased hydrologic input to the lake as a consequence of land-use change (Rosenmeier et al. 2002b). It is often difficult to differentiate the relative roles of climatic change and human landscape modifications in creating the proxy records we study (see also Dunning and Beach 2000; Rice 1993: 44). Nonetheless, paleolimnological records from both Lakes Salpeten and Peten-Itza suggest that climatic and/or human-induced changes on the environment of the central lowlands were profound during the time of Maya occupation.

Cariaco Basin. Although not located in the Maya lowlands, paleoclimatic records from cores retrieved in the anoxic Cariaco Basin off northern Venezuela are highly relevant for reconstructing precipitation changes in the Maya lowlands. As discussed above, rainfall in both regions is related to the seasonal migration of the ITCZ, and southward displacement of the ITCZ during summer should result in lower rainfall in both northern Venezuela and the Maya lowlands. Gerald Haug and his colleagues (2003) used the concentration of titanium in annually laminated sediments to infer changes in precipitation at a temporal resolution that is unmatched by any other terrestrial or marine record in the Neotropics. Ti is a detrital element delivered to the Cariaco Basin by rivers, and therefore its concentration is related to runoff and precipitation.

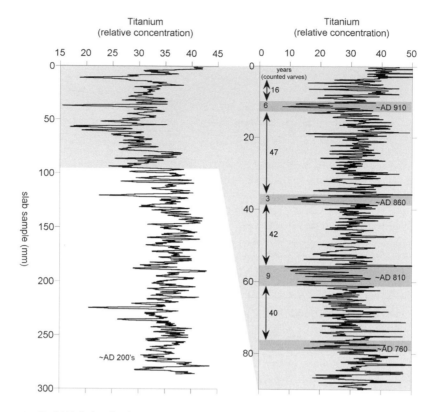

FIG. 7 (Left) Relative titanium concentrations (30-point running mean) from a 300 mm slab sample from Ocean Drilling Program Hole 1002D. (Right) Relative titantium concentrations (3-point running mean) from the upper 90 mm of the slab sample. Redrawn from Haug et al. (2003).

Cariaco Ti concentration was low in the Terminal Classic Period (fig. 7), indicating generally drier conditions, consistent with findings from north-central Yucatan lakes discussed below (Curtis et al. 1996; Hodell et al. 1995, 2001, 2005a). Superimposed upon this generally drier climate, however, were four periods of drought dated to approximately AD 760, 810, 860, and 910. These severe droughts lasted between three and nine years and were spaced forty to forty-seven years apart (fig. 7). If we accept that the teleconnective links between Cariaco and the Maya lowlands would create roughly parallel climatic changes, the extraordinary temporal resolution of the Cariaco record permits a comparison between paleoclimatic events of the Terminal Classic Period and the archaeological evidence of the collapse (Haug et al. 2003). Although the Cariaco chronology is based on varve counting, the absolute dates for the individual drought events "float" in time because the chronology is referenced to an assumed date of AD 930 for a rise in Ti that marks the local onset of Medieval Warm conditions. Nonetheless, the pattern and relative timing of events in the Cariaco Ti signal should be robust. New fine-grained paleoclimate data from Belizean stalagmites indicate peak drought conditions in the central lowlands at AD 754–798, 871, and 893–922, with successive droughts of increasing severity (Webster et al. 2007).

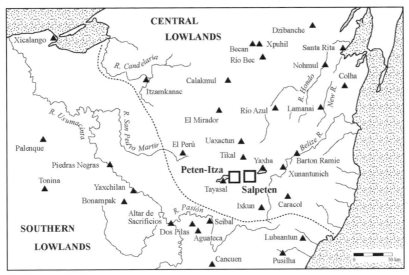

FIG. 8 The central and southern Maya lowlands (after Sharer 1994: fig. 1.1)

The Archaeological Record. Throughout the central and southern lowlands, the last half of the eighth century was a remarkable time. Different polities had distinct historical trajectories, but Tikal, located in the middle of the central lowlands (fig. 8), provides a good example. The kingdom's *k'uhul ajaw*, Yax Nuun Ayiin II, began his rule in AD 768 and ruled at least until AD 794 and perhaps through AD 810 (Valdés and Fahsen 2004: 143); he continued a program of growth and political revitalization that his great grandfather Nuun Bak Chak and grandfather Jasaw Chan K'awiil had initiated in the seventh century (Valdés and Fahsen 2004; Martin and Grube 2000). These powerful rulers built towering funerary monuments and added palace buildings to the royal court (Harrison 1999; Jones 1991; Martin and Grube 2000; Schele and Mathews 1998). The polity's population densities reached new highs (Culbert et al. 1990; Ford 1986; Puleston 1983), presumably creating an unprecedented pool of labor tribute and staple surplus. Consequently, the divine kings who occupied the apex of Tikal's political economy reached a zenith of wealth and power, as reflected in their building projects and the craft goods and public art they commissioned.

Across the region, rulers at more sites than ever before erected new carved monuments during this period, a trend that culminated around AD 790 (Martin and Grube 2000: 226; Morley 1937–38). Although sometimes interpreted as a sign of great prosperity and cultural florescence, this proliferation of monuments is probably better understood as a sign of political decentralization and intensified competition between the rulers of the largest polities and their allies and vassals. Nowhere is this clearer than in the Petexbatun region, where competition turned to chronic warfare that wracked the entire region in the late eighth century (Demarest 2004; Demarest et al. 1997). After the fall of the ruler of Dos Pilas, K'awiil Chan K'inich, in AD 761, the region's polities, no longer subject to Dos Pilas, entered a period of endemic conflict and competition. There is little evidence of environmental degradation in the region (Dunning et al. 1997), and the frequency of indicators of dietary stress, malnutrition, and disease change little from the Early Classic through the Terminal Classic (Wright 1997). The end product of this fierce competition was the destruction and abandonment of most cities by AD 800 (Demarest 1997), and the last hieroglyphic monument in this area dates to AD 807 (Martin and Grube 2000). Although it is especially noteworthy in the Petexbatun, epigraphic evidence of increasing warfare during the later eighth century is found in many regions of the southern and central lowlands in polities like Naranjo, Yaxchilan, Piedras Negras, and Tonina. Warfare continued to be common across the region into the ninth century.

The next fifty years mark perhaps the period of greatest political change during the collapse in the central and southern lowlands. Many important sites—Palenque, Yaxchilan, Piedras Negras, Aguateca, Dos Pilas, Quirigua—have yielded no firmly dated monuments that postdate the k'atun-ending celebration of AD 810 (Martin and Grube 2000), suggesting the end of centralized political organization based on divine kingship in those polities, a system that had been pervasive in the Classic-Period Maya lowlands (Sharer and Golden 2004). At Tikal, the frequency of monument dedication decreases significantly in the ninth century, and there is no evidence of a long-reigned ruler after Yax Nuun Ayiin II. There is a sixty-year hiatus in monument dedication at Tikal beginning in AD 810, although inscriptions at other sites mention Tikal rulers during this time. This hiatus spans the completion of the important tenth bak'tun in AD 830, which went uncelebrated at Tikal, although the ruler of Zacpeten, a former Tikal subordinate, erected a monument to record the date (Martin and Grube 2000). Taken together, these facts suggest an extended period of political turmoil and reduced control of labor and resources by Tikal's rulers.

This fragmentation of larger regional polities was a widespread pattern (Marcus 1998). As Calakmul's rural population declined significantly in the Terminal Classic (Braswell et al. 2004), the rulers of once-subordinate sites such as Oxpemul, Nadzcaan, and La Muñeca began commissioning their own stelae, dating to the ninth century (Marcus 1976; Martin and Grube 2000). Copán had begun to undergo increasing political decentralization after AD 763, as reflected by the proliferation of carved monuments and hieroglyphic texts in nonroyal compounds and other trends in the art and architecture of the site's epicenter (Fash et al. 2004); the site's last monument dates to AD 822 (Stuart 1993). By AD 830, the royal lines of kings who had ruled the Late Classic polities had disappeared from most sites of the central and southern lowlands (Martin and Grube 2000: 227).

This decentralization was not universal, however. At Caracol, K'inich Joy K'awiil and his successors presided over a renaissance beginning around AD 798, which followed 118 years of political crisis (Martin and Grube 2000). They reconquered rebellious subordinates and forged alliances with neighboring Ucanal, restoring Caracol to its role as a regional political power for fifty years or perhaps a century, until its palaces and ceremonial spaces were abandoned, apparently in a sudden fashion, sometime around AD 895 (Chase and Chase 2004).

Seibal, adjacent to the Petexbatun region, presents an even more remarkable case. Following the region's devastating wars of the later eighth century, a new royal dynasty was established in AD 830 and presided over a remarkable florescence. The city grew to its maximum population (Tourtellot 1988), and its rulers commissioned seventeen stelae between AD 830 and 889 and several undated monuments that are probably later (Martin and Grube 2000: 227).

Recent research does not support earlier reconstructions that Seibal was conquered by foreigners (Tourtellot and González 2004), but the presence of certain ceramic types, hairstyles, costume elements, writing conventions, and architectural styles at Seibal does indicate close ties between the polity's elite and groups who were culturally affiliated with societies in the Gulf Coast region, presumably Itza and/or Putun Maya (Sabloff 1973b). These contacts likely linked Seibal, strategically located on the Pasión River, into the expanding circum-Caribbean canoe-trade networks. Several of the polities that maintained some degree of centralization during the ninth century, including Caracol, show similar connections (Chase 1985; Chase et al. 1991; Sabloff 1973b), and groups affiliated with northern polities apparently immigrated into the Peten Lakes region in the ninth and tenth century (Rice and Rice 2004).

The general trends of political fragmentation and decentralization in the early ninth century correspond roughly with marked population declines in much of the southern and central lowlands (Culbert 1988). Tikal is typical of many large sites, in that its population after AD 830 was reduced to around fifteen percent of its Late Classic apogee in the city itself and twenty percent in the city's sustaining hinterland (Culbert et al. 1990). Unfortunately, most of the region's Late and Terminal Classic ceramic phases are a century or more in length (Rice and Forsyth 2004), a fact that precludes precisely correlating the political transformations of the early ninth century with trends in Maya demography or economy. At Tikal, the Imix complex dates to AD 700–830 and the Eznab complex to AD 830–930, making it impossible to assess empirically demographic changes between AD 775 and 825, for example.

During the second half of the ninth century, most of the remaining large cities were essentially abandoned. At a few of the old capitals, including Tikal and Calakmul, rulers erected monuments during this time, but most show only a few decades of monument dedication. Tikal's ruler, Jasaw Chan K'awiil II, his name evoking the great *k'uhul ajaw* who rebuilt Tikal's political prestige nearly two centuries earlier, dedicated the site's last carved monument, Stela 11, to celebrate the k'atun-ending date 10.2.0.0.0 (AD 869). The fall of the royal dynasty was followed by the movement of people into abandoned palaces and plazas, where they made makeshift houses in the crumbling city. Near Tikal, the rulers of Ixlu, Xultun, and Jimbal—polities that had once been Tikal's subjects—erected monuments in the four decades beginning in AD 859. In the texts they name themselves *k'uhul ajawob'* and, using the Tikal emblem glyph, indicate either their autonomy or an extension of that title to noble vassals beyond the Tikal sovereign (Valdés and Fahsen 2004). The rulers of Jimbal and Uaxactun continued to exercise the royal prerogative to commission stelae after the last monument at Tikal, celebrating the k'atun ending in AD 889, but they were also much reduced in power.

Seibal, with its apparently strong ties to trade routes along the Gulf Coast, continued to thrive during the late ninth century (Tourtellot and González 2004), as did Quirigua, where the old site center was reoccupied by what appears to be a population from the Caribbean coast (Sharer 1988). Whatever their bases, these polities were relatively short-lived, apparently abandoned by the early tenth century. A few other larger sites showed continued occupation into the tenth century. Rulers at Calakmul and Tonina commissioned the region's last Long Count dates to celebrate the k'atun ending on 10.4.0.0.0 (AD 909), and a few undated monuments at Seibal and Calakmul are likely somewhat later (Folan et al. 1995; Martin and Grube 2000). By the early tenth century, however, over a century of political competition and warfare, political fragmentation, and depopulation had radically changed the region, reducing the area's once-great cities into crumbling ruins and transforming its once densely populated countryside into dispersed hamlets and farmsteads.

Discussion. The growing evidence of temporal correlations—albeit imprecise—between changes in the environment of the central lowlands and the tumultuous political and social changes of the eighth through tenth centuries must be incorporated into our understandings of the collapse. Although we must repeat that the evidence from Peten lake cores is ambiguous and we cannot single out climatic change as responsible for the proxy data there, the data do indicate that the natural environment in the central Peten changed dramatically over the course of the Late Classic Period. Toward the end of this period, Maya populations in this region were probably confronted by declining crop yields due to soil erosion, perhaps exacerbated by decreasing rainfall coupled with maximum population densities. It is important to note, however, that multiple lines of evidence suggest that the Petexbatun region did not suffer similar environmental degradation during the eighth or ninth centuries (Dunning et al. 1997; Emery et al. 2000).

The Cariaco record suggests that the Terminal Classic Period was marked by somewhat reduced rainfall, punctuated by drought events of three to nine years in length roughly every fifty years, that began in the Late Classic Period (AD 760) and continued into the Terminal Classic Period (ca. AD 810, 860, and 910). Although scholars have cited these droughts as explanations for regional demographic trends (Haug et al. 2003, following Gill 2000), they seem to correspond better to documented political changes: increasing warfare and political destabilization in some areas after AD 760; the absence of monuments in many large cities after AD 810 and increasing political decentralization of many regional states in the early ninth century; and the cessation of dated hieroglyphic monuments at many sites over the second half of the ninth century, culminating with the last Long Count dates in AD 909.

That said, the impact of any climatic changes cannot be meaningfully isolated from environmental degradation, nor from the peak eighth- and early ninth-century population densities, nor the unprecedented demands on farmers' surplus and labor by polity rulers, nor from the cultural constructs by which the Maya themselves interpreted and expected these changes (Freidel and Shaw 2000; Puleston 1979; Rice 2004). Fine-grained archaeological and paleoenvironmental data now available for many regions in the southern and central lowlands show considerable variation in these factors (Demarest et al. 2004), which likely accounts for much of the variability in the sociopolitical changes that occurred in different polities across the southern and central lowlands during the eighth and ninth centuries. In some regions, droughts and drier conditions do not appear to have played a significant role. For example, one of the first areas to be abandoned was the Petexbatun, despite the region's high annual precipitation rates, many large rivers, and the absence of significant environmental degradation. Caracol provides a counter example. Despite the polity's high population densities and a scarcity of water that continues to be a limiting factor at the site today, K'inich Joy K'awiil oversaw a renaissance that does not appear to have been significantly affected by the droughts hypothesized for ca. AD 810 and 860.

Beyond the central and southern lowlands, however, the ninth century is a period of Maya geopolitical recentering, as most of the large polities in the south declined and powerful polities like Uxmal, Coba, and Chichen Itza in the northern lowlands grew. These two phenomena are almost certainly related, and some have suggested that the disruptions in the south led to significant immigration to northern cities. Convincing empirical evidence of this remains lacking to date, although strontium isotope studies now offer the possibility of evaluating this hypothesis (Hodell et al. 2003; Price et al. 2000).

North-central and Northeastern Lowlands

The Paleoclimatic Record. *Lake Chichancanab.* The strongest physical evidence for a drought during the Terminal Classic Period comes from Lake Chichancanab in the north-central part of the Yucatan Peninsula (fig. 2; Hodell et al. 1995, 2001, 2005a). Chichancanab means "little sea" in Yucatec Maya, which accurately describes its relatively high-salinity (4000 mg l^{-1}), sulfate-rich (2545 mg l^{-1}) waters. Lake Chichancanab's water is saturated for both gypsum ($CaSO_4$) and celestite ($SrSO_4$), and therefore the S content of the sediment provides a qualitative proxy of E/P. Because sediment containing gypsum is denser than shell-bearing organic matter, sediment density can be used as an indicator of gypsum precipitation. One advantage of using sediment density over weight-percent sulfur (wgt. %S) as a gypsum proxy is that density can be measured nondestructively at a high spatial resolution of every 0.5 cm by gamma ray attenuation while the sediment is still in polycarbonate core tubes. Measurement of wgt. %S requires that the core be extruded or split so that discrete samples can be taken for geochemical analysis.

In 1993, David Hodell, Jason Curtis, and Mark Brenner (1995) retrieved a 4.9 m core in 6.9 m of water from Chichancanab's central basin. This core provided the first physical evidence for a protracted drought during the Terminal Classic Period, among the most severe in the last 7000 years. The $\delta^{18}O$ of gastropod and ostracod shells and the wgt. %S record showed a pronounced increase at a depth of ~65 cm in the core. Hodell and colleagues (1995) dated this event by AMS-^{14}C analysis of a seed that was extracted from the portion of the core that showed peak signals for $\delta^{18}O$ and wgt. %S. The

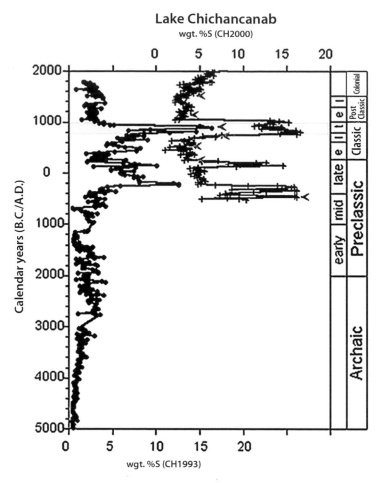

FIG. 9 Comparison of wgt. %S records in the 1993 (left; Hodell et al. 1995) and 2000 (right; Hodell et al. 2001) cores from Lake Chichancanab. Arrows represent the position of AMS ^{14}C dates on terrestrial organic matter.

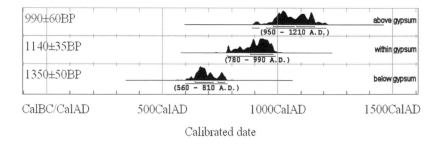

FIG. 10 Probability distributions of calibrated ages obtained using OxCal v. 3. 5 (Ramsey 1995, 1998) and the INTCal98 data set (Stuiver et al. 1998) on AMS ^{14}C dates of terrestrial organic matter from below, within, and above the gypsum layer in the Lake Chichancanab cores.

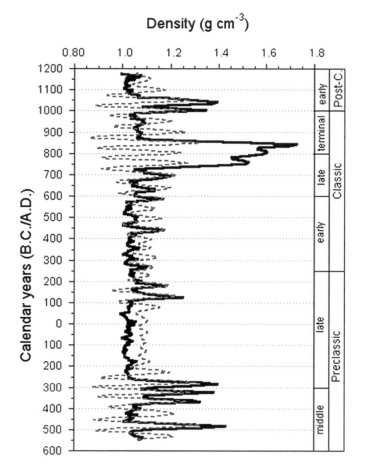

Density (g cm^{-3})

FIG. 11 Sediment bulk density record from Lake Chichancanab (solid line) shown relative to a bandpass filter (dashed line) of the signal filtered at ~50 yr. Density reflects the concentration of gypsum (CaSO$_4$) in the sediment, which is related to the E/P ratio of the lake. Note the prominent peak in density (gypsum) between ca. AD 750 and 850.

seed's age was 1140±35 ^{14}C yrs BP, which yielded a calibrated age range of AD 780–990 (95.4% probability) or AD 880–980 (68.2% probability). Although the age estimate spans a century, it strongly suggests that a major drought occurred during the Terminal Classic Period in the north-central lowlands.

In 2000, Hodell and colleagues returned to Lake Chichancanab with the aim of retrieving new cores from deeper water that might possess higher sedimentation rates than the 1993 cores (~0.5 mm yr^{-1}). They were successful in obtaining higher-resolution cores with an average sedimentation rate of 0.7 mm yr^{-1} from ~11 m of water. The wgt. %S record from the 2000 cores was very similar to the 1993 cores and contained a pronounced gypsum layer in the Terminal Classic Period (fig. 9). This gypsum lens formed when the lake volume dropped because of drought, the lake water became supersaturated for CaSO$_4$, and gypsum precipitated. Two AMS-^{14}C dates on terrestrial organic matter from just below the gypsum layer yielded identical dates of 1350±50 ^{14}C yrs BP (fig. 10), which converted to a calibrated age range of AD 560–810 (95.4% probability). This

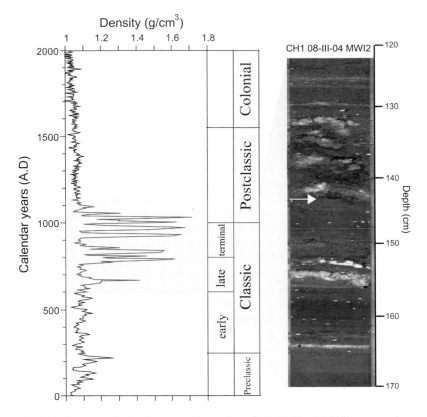

FIG. 12 Sediment bulk density record from Core CH1 08-III-04 MWI2 versus time (Hodell et al. 2005a). (right) Image of Core CH1 08-III-04 MWI2 between 120 and 170 cm showing gypsum layers (light tan sediment) interbedded with organic rich strata (dark sediment) containing shell material (white specks). The interval between 120 and 170 cm represents the period from ca. AD 600 to 1150. The white arrow marks the position of a radiocarbon sample that dates to 1130 ± 35 ^{14}C yrs (AD 780–1000).

range, encompassing the entire Late Classic Period, illustrates the limitations of temporal comparison between paleoclimatic data and Maya cultural periods. Nonetheless, the date indicates that the gypsum was first deposited after AD 560–810. This date, when combined with the AD 780–990 age of the seed from the 1993 core, supports reconstructions of drier conditions during the Terminal Classic Period. Sediment just above the gypsum layer was dated to 990±60 ^{14}C yrs BP (fig. 10), which translates to a calibrated age range of AD 950–1210 (92.7% probability), dating to the Early Postclassic Period.

Spectral analysis of the density signal from the 2000 Chichancanab core demonstrated that drought has been a recurrent phenomenon on the Yucatan Peninsula with periods of dry conditions of about 208 and 50 years (fig. 11; Hodell et al. 2001, 2005a). The 208-year period is believed to be related to solar forcing, and the 50-year period is consistent with the spacing of drought events inferred from the Cariaco Basin during the Terminal Classic (Haug et al. 2003).

In March 2004, a series of sediment cores were retrieved in Lake Chichancanab along a water-depth transect ranging from 4.3 m to ~14.7 m (relative to 2004 water levels), near the deepest point in the lake (Hodell et al. 2005a). Results demonstrated that the Terminal Classic gypsum layer is condensed in shallow water and expanded in deeper-water sections where it consists of numerous interbedded gypsum and organic-rich strata (fig. 12). This indicates a series of dry events separated by intervening periods of relatively moister conditions. A radiocarbon date on wood was obtained within the interval of gypsum deposition in deep-water core CH1 08-III-04-MWI-2. The age was 1130±35 ^{14}C yrs BP, resulting in a calibrated age range of AD 780–1000 (95.4% probability). This age is indistinguishable from the ^{14}C date obtained on a seed from the gypsum horizon in the shallow-water core taken in 1993 (1140±35 ^{14}C yrs BP), confirming that gypsum precipitation was occurring during the Terminal Classic Period. Hodell et al. (2005a) were able to correlate density records among cores and derive a fine-grained chronology for the deep-water reference section. Absolute dates on paleoclimatic events remain uncertain because of the inherent errors associated with radiocarbon dating, but the pattern and frequency of climatic variations are robust. The investigators found evidence for two phases of dry climate (ca. AD 770 to 870 and 920 to 1100) separated by an intervening 50-year period of relatively moister conditions (ca. AD 870 to 920). Within each of the dry phases, the climate was marked by alternating dry and moist conditions with an interval of drought recurrence of about fifty years.

In summary, Lake Chichancanab provides one of the most sensitive paleoclimatic records of the Maya lowlands because of its substantial water loss to evaporation, precipitation of gypsum (CaSO4), and relatively low-density occupation of its watershed. A low density of occupation, combined with the dry scrub vegetation surrounding the lake, diminished any significant human impact on lake hydrology, and therefore we believe that the observed changes in sediment geochemistry reflect paleoclimatic change.

Lake Punta Laguna. Lake Punta Laguna is located 145 km northeast of Lake Chichancanab and 20 km from the Maya site of Coba. It is in the wettest part of the northern lowlands, where annual rainfall is ~1400 mm/yr. The core recovered from Punta Laguna in 1993 provided a high-resolution paleoclimatic record because of the high sedimentation rates of 0.1–0.2 cm/yr (Curtis et al. 1996). Oxygen isotopes indicate that the Classic Period was generally drier than the Preclassic or Postclassic (fig. 13). Two increases in δ^{18}O during the Terminal Classic Period coincided with the two density peaks in the Chichancanab cores, within the limits of chronological uncertainty (fig. 13). In the case of the δ^{18}O record, the time of rapid δ^{18}O increase (first derivative of the signal), rather than the peak δ^{18}O value, indicates drier climate. The ages of the two δ^{18}O increases in the Terminal Classic Period are centered on about AD 800 and 900, based on fitting six dates in the core with a second-order polynomial equation (fig. 14). Both Punta Laguna and Chichancanab cores indicate a relaxation of drought conditions by ca. AD 1100 in the Early Postclassic (Hodell et al. 2007), as does the Ti record from the Cariaco Basin (Haug et al. 2003). Drought conditions are not strongly reflected in the proxy records at nearby Lake Coba, an absence perhaps explained as the result of significant anthropogenic transformation of the circumlacustrine landscape (Leyden et al. 1998).

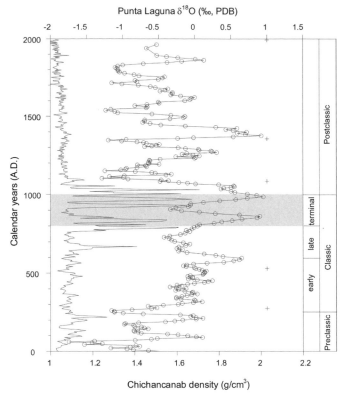

FIG. 13 Sediment density in Lake Chichancanab Core CH1-7-III-04 (left; Hodell et al. 2005a) and δ18O of the ostracod *Cytheridella ilosvayii* in the 1993 Punta Laguna core (right; Curtis et al. 1996). Gray band indicates drier conditions.

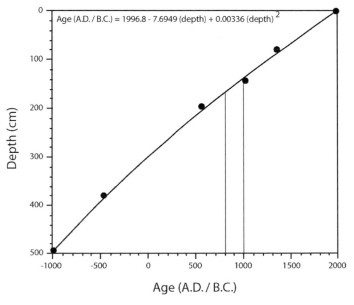

$$\text{Age (A.D. / B.C.)} = 1996.8 - 7.6949 \, (\text{depth}) + 0.00336 \, (\text{depth})^2$$

FIG. 14 Age-depth relationship for the 1993 Punta Laguna core (Curtis et al. 1996). Vertical lines represent the Terminal Classic Period.

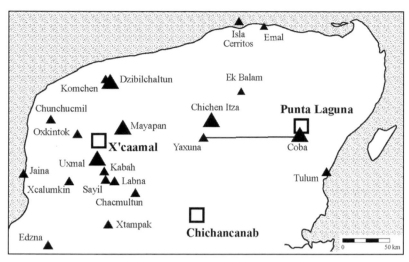

FIG. 15 The northern Maya lowlands (after Sharer 1994: fig. 1.1)

The Archaeological Record. The decades of the 1980s and 1990s witnessed significant archaeological research in the northeastern and north-central lowlands, with long-term research projects at Chichen Itza, Yaxuna, and Ek Balam (fig. 15). Most archaeologists now see long-term interaction between Chichen Itza and Coba—the region's two largest cities—and with other neighboring polities as the central factor in their models of the decline and abandonment of northern sites during the tenth century (Andrews 1990; Andrews and Robles 1985; Ringle et al. 2004; Robles and Andrews 1986; Stanton and Gallareta 2001; Suhler et al. 1998, 2004).

Coba was an urban metropolis located in the moister northeastern part of the Yucatan Peninsula (Folan et al. 1983b). The city grew to its greatest size during the Machukaani phase (AD 600–800), when it reached 70 km^2 and may have had as many as 55,000 inhabitants (Folan 1983a). The core of Coba consists of three large architectural groups built on a massive man-made terrace. A series of causeways or *sacbes* connect this nuclear zone to other architectural groups in the city's urban core (Folan 1983b). The rulers of Coba carved twenty-three stelae in the seventh through ninth centuries. These monuments glorify Coba's *k'uhul ajawob'* in image and text, and the leaders of Coba show clear connections to elite traditions of the central and southern lowlands in the architecture they commissioned and in the layout of their city.

The published data from Coba itself are equivocal as to the relationship between Coba and Chichen Itza. William Folan argues that the city reached its peak during AD 600–800, subsequently experiencing a slight decline in AD 800–900 before undergoing a significant reduction in its political and economic importance around AD 900–1000 (Folan 1983c). Folan's dates for Coba appear to follow traditional ceramic chronologies, and he did not postulate any significant role for Chichen Itza in Coba's decline (Folan 1983c).

A recent revision of the ceramic sequence puts the site's apogee during the period marked by the Oro ceramic complex (Robles 1990), which has significantly later dates (AD 730–1100). Based on this new chronology, Anthony

Andrews and Fernando Robles (1985: 66) assert that Coba continued to grow well past the eleventh century. They envision a scenario in which Chichen Itza gradually surrounded and weakened Coba between AD 1000 and 1200, conquering its territory and establishing trading outposts along the eastern coast of the peninsula. Charles Suhler and colleagues (2004: 457) suggest that the appearance of the foreign Piza subcomplex at Coba may indicate that Chichen Itza conquered Coba sometime around AD 1100.

The other key city in this region was Chichen Itza. Before the Terminal Classic Period, long-distance trade probably along circumpeninsular canoe networks had become important to the elite of Chichen Itza, despite the city's distance from the sea. Before AD 800 the site controlled the trade port of Isla Cerritos (Andrews and Gallareta 1986) and had established political control over a corridor to the coastal salt-making center of Emal (Kepecs 1998; Smith 2001). This focus on salt and canoe trade indicate that the political economy of Chichen Itza may have been organized differently than that of contemporary northern polities.

The same is true of the polity's political organization. Based on hieroglyphic texts, iconography, the layout of elite residences and monumental architecture, and the plan of the causeway system at Chichen Itza, various scholars have argued that Chichen Itza was not ruled by divine kings, but rather by a council of elite leaders, a system called *mul tepal* in early colonial documents (Cobos and Winemiller 2001; Marcus 1993; Schele and Freidel 1990; Wren and Schmidt 1991).[2] The art of Chichen Itza is replete with images of war and conflict (Kurjack and Robertson 1994), suggesting that the power of the polity's leaders lay not only in their mercantile connections, but also in their abilities to control military force.

The site of Yaxuna, located just south of Chichen Itza, has been critical for our understanding of the relationship between Coba and Chichen Itza. Yaxuna has a relatively long occupation history. Founded around 500 BC, the site experienced a burst of growth in the Yaxuna III phase (ca. AD 600–730; Suhler et al. 2004). During this period the polity grew significantly in population, and its rulers sponsored monumental construction in the North Acropolis. The site's ties to Coba were marked materially by a 100 km long causeway built to connect the two sites during this phase (Suhler et al. 1998). Ceramic and architectural patterns indicate interaction between the people of Yaxuna and early Puuc populations to the west, while central and southern lowland diagnostics are absent, signaling a shift from the Early Classic interactions with the Peten (Suhler et al. 1998, 2004; Stanton and Gallareta 2001).

The Yaxuna IVa (AD 730–900) and IVb (AD 900–1250) phases encompass the Terminal Classic Period of greatest interest in this paper. The first phase witnessed intense building activities at the site, including the construction of crude fortification walls near the end of the phase, much like those known from the Petexbatun (Suhler et al. 2004). Numerous buildings were

2 Changes in the organization of the causeways that connected the elite residences at Chichen Itza have led Rafael Cobos and Terance Winemiller (2001) to argue that the mul tepal political organization that existed at Chichen Itza in the ninth century was replaced by more centralized governance and economic structures around ca. AD 900.

destroyed and ritually terminated, suggesting that conflict and warfare were common during this period. Charles Suhler and colleagues (1998: 180; also Suhler et al. 2004) argue that Yaxuna was an outpost by which Coba sought to contain the growing influence of Chichen Itza in the ninth century.

The Yaxuna IVb phase, marked by the presence of Sotuta ceramics, apparently corresponds with the site's conquest by Chichen Itza (Suhler et al. 1998). A Puuc-style elite residential complex was razed and replaced by a large residential structure whose inhabitants used Sotuta ceramics (Stanton and Gallareta 2001: 234), and a structure in the North Acropolis interpreted as a *popol na* council house was burned and torn down (Suhler et al. 2004: 475). The many whole and partial vessels found smashed on the building's floor may be evidence of celebrations or ceremonial activities associated with the ritual destruction of the building following the site's conquest (Stanton and Gallareta 2001: 235; Suhler et al. 2004: 475). Despite these events in the city's center, there is relatively little new construction at Yaxuna during this phase, which led the excavators to conclude that the city was largely depopulated, perhaps converted into a staging point for the Itza conquest of Coba (Suhler et al. 1998: 178).

The site of Ek Balam, a large center located 51 km northeast of Chichen Itza, provides a contrasting case. Intensive investigations at Ek Balam have shown that the site witnessed a peak occupation during the Late Yumcab phase (AD 700–1050; Bey et al. 1998: 115; Ringle et al. 2004). The Late Yumcab is a long ceramic phase, which certainly clouds our understanding of sociopolitical changes during this important period, but the presence of three distinct architectural styles—Pre-Florescent masonry with slab vaults and poor masonry covered by thick plaster, Florescent veneer masonry, and a later lower-quality masonry style—permit finer chronological resolution (Bey et al. 1998: 115). The latter, they argue, is coeval with post-monumental Uxmal (i.e., tenth century or later). This finer-grained chronology allows William Ringle and colleagues (1998: 226) to marshal evidence that the rulers of Ek Balam continued the tradition of divine kingship until sometime around AD 900.

Discussion. Two periods of significantly decreased rainfall are indicated in the proxy records of Lakes Chichancanab and Punta Laguna for the ninth and tenth centuries and again in the twelfth century. The role that these climatic shifts might have played in the events of the northeastern and north-central lowlands is an interesting question. Unless the current ceramic chronologies are radically misdated, the polities of this region generally do not seem to witness a decline or collapse during this time of decreased precipitation, although demographic patterns must be tentative given the paucity of full-coverage settlement surveys. Chichen Itza, Ek Balam, and Coba all prospered during the peak drought conditions of the ninth century (there may have been a reprieve from drought conditions in the middle of the Terminal Classic, but the dating control is poor), albeit in a context of intense regional competition, making it difficult to postulate that the collapse was a pan-lowland phenomenon (an observation echoed by Ringle et al. [2004: 506]). The fortunes of individual sites differ significantly, though, and the leaders of Chichen Itza were clearly the most successful at expanding their regional influence and power during this drier period, especially during the tenth century.[3]

What role did climatic change play in the success of Chichen Itza? William Folan and Joel Gunn (Folan et al. 1983a; Gunn and Adams 1981) made the first strong argument for the role of climatic change in the Terminal Classic northeastern lowlands, but few archaeologists followed their lead. Today, most archaeologists working in the northern lowlands stress the role of Chichen Itza and emphasize the political and economic differences between Chichen Itza and its neighbors: its mul tepal political organization and ideology (Dunning and Kowalski 1994; Ringle et al. 1998); its more efficient market economy, based on mercantile instead of staple production (Kepecs et al. 1994); and its privileged links to circum-Caribbean canoe trade routes (Andrews and Gallareta 1986; Kepecs et al. 1994).

These sociopolitical characteristics would have provided the leaders of Chichen Itza considerable advantages in times of drought-induced staple shortage, as Nicholas Dunning (1992) points out. Mercantile production of salt and presumably cotton would have provided high-value goods that could have been traded for staples from areas more conducive to agriculture and/or unaffected by drier climatic conditions, and their access to the canoe networks would have allowed them to import those bulky staple goods more efficiently than sites that relied on overland trade routes (Drennan 1984). A market-oriented economic system would have permitted the more efficient distribution of staple and nonstaple goods (Kepecs et al. 1994), and a decentralized political organization could have responded more effectively to local shortfalls (Dunning 1992). Finally, an ideology based more on secular rulership and mul tepal governance would not be as susceptible as divine kingship to being undermined by natural disasters (Dunning and Kowalski 1994).

The correspondence of Chichen Itza's expansion with drier conditions in the proxy data from Lakes Chichancanab and Punta Laguna suggests that drier conditions were a factor in that polity's success. The region's other sites, however, do not experience the rapid decline and abandonment that many centers to the south witnessed, despite similarly high population densities, political authority based on divine kingship, and the lower average precipitation in this region compared to the southern and central lowlands. This regional variability demonstrates that the role of climatic change in the Terminal Classic Period must be understood in light of the sociopolitical and cultural aspects of Maya civilization.

The Northwestern Lowlands

The Paleoclimatic Record. The northwestern corner of the Yucatan Peninsula is the driest region of the entire peninsula and is highly suited to paleoclimatic study because of the steep rainfall gradient (fig. 2). Hodell and colleagues (Hodell et al. 2005b) studied a 5.1 m sediment core from a sinkhole lake named Aguada X'caamal. Unlike the Chichancanab and Punta Laguna

3 Even under the shorter chronology for Chichen Itza that some scholars advocate (e.g., Ringle et al. 1998; Schele and Mathews 1998; cf. Suhler et al. 2004), the site clearly prospered in the tenth century.

records, the oxygen isotopic signal at X'caamal shows relatively little change during the Terminal Classic Period (fig. 4). The ninth- and tenth-century drought, which is so prominent in the north-central and northeastern lowlands, either had little effect on E/P in northwestern Yucatan or was not recorded in the sediments of Aguada X'caamal. Nyberg and colleagues (2002) found evidence of a warming in the northeastern Caribbean off Puerto Rico around AD 700–950 and suggested that the pattern may have resulted from exceptionally severe or frequent El Niño events from ca. AD 750–950 (Ely et al. 1993; Quinn 1992). The X'caamal record shows an important climatic change in the fifteenth century AD, however, when $\delta^{18}O$ values increase by ~3‰. This pronounced increase in E/P coincided with the start of the Little Ice Age (Hodell et al. 2005b) and correlates to a prominent reduction in rainfall inferred from Ti in the Cariaco Basin (Haug et al. 2003).

The Archaeological Record. The area best known in the northwestern lowlands is the Puuc region (fig. 15). This area has deep, well-drained, fertile soils that are among the most productive in the lowlands (Dunning 1992). Although rainfall is relatively scarce, precipitation levels do permit maize agriculture, and *chultuns* stored rainwater for domestic use (McAnany 1990; Thompson 1897).

The Puuc region is characterized generally as witnessing a rapid boombust cycle in the ninth century, but there are significant earlier occupations at sites like Chac II (Smyth et al. 1998), Oxkintok (Rivera 1991), and Xkipche (Prem 1991). Oxkintok, for example, was a major urban center with significant monumental architecture in the Early-Classic Ichpá (AD 300–550) and Noheb I (AD 550–650) phases (Rivera 1991). After a period of decreased activity, Oxkintok saw renewed growth during the Ukmul phase (AD 740–850), and the ceramic assemblages and stone monuments provide evidence of interaction with Maya cities in the Usumacinta and the Peten. Stela 9, the last monument with a calendrical inscription, dates to AD 859, but many sculptures without texts probably date to AD 850–950 based on their stylistic characteristics (Pablo 1991; García 1991). The city as a whole, though, witnessed much less architectural investment after ca. AD 850, a trend that culminated in the site's abandonment by ca. AD 1000. Miguel Rivera Dorado (1991: 47) links this decline to the rise of Uxmal as a regional center of power.

Despite antecedents like Oxkintok, the Puuc region underwent a period of remarkable expansion of population densities in the late eighth and ninth centuries. This growth may have included some immigration from southern polities, although scholars working in the region find little evidence that this population growth was not largely autochthonous (Carmean et al. 2004). Regardless, by AD 800, the Puuc region was densely occupied, and closely spaced political centers were characterized by large palaces, pyramids, and broad plazas. As was the case in most of the central and southern lowlands in the eighth and ninth centuries, the population expansion in the Puuc brought increasing interpolity conflict: walls were built at many sites, including Uxmal, Chacchob, and Cuca (Kurjack and Andrews 1976; Webster 1979); murals at

Mulchik show a battle scene; and stone sculptures depict rulers in military costume. The Puuc polities were also similar to their southern counterparts in that their rulers' authority derived in part from an ideology of divine kingship (Carmean et al. 2004; Dunning 1992; Dunning and Kowalski 1994; Kowalski 1987, 1989, 1994; Kowalski and Dunning 1999). Rulers of powerful sites commissioned carved stone stelae depicting them wearing costume elements like K'awiil headdresses traditionally associated with divine rulership and inscribed with hieroglyphic texts naming them as *k'uhul ajawob'*.

Sayil is one of the Puuc sites that we know best, thanks to extensive mapping and excavation directed by Jeremy Sabloff (Sabloff and Tourtellot 1992). Survey around Sayil indicates that the city extended over 4.5 km² and had between 7,000 and 10,000 inhabitants, a figure that reached perhaps 16,000 if the site's hinterland is included (Dunning 1992). The core of the city is a series of architectural complexes, including elite residential compounds, temple-pyramids, and a ball court. Raised sacbes connect these major groups, which are surrounded by a mosaic of residential groups occupied by people of various statuses (Carmean 1991, 1998). Based on two radiocarbon dates, architectural mosaic styles, and obsidian hydration dates, Tourtellot and Sabloff (1994: 87–88) argue that the site was occupied between AD 800 and 950, and that the site center and most of the hinterland were abandoned prior to AD 1000, thus following the boom-bust pattern found at many Puuc sites.

The largest of the Puuc cities was Uxmal, which grew in power and influence to eclipse neighboring cities and become the capital of a regional state in the eastern Puuc by ca. AD 850 (Dunning and Kowalski 1994). The ruler Chan Chak K'ak'nal Ahaw ("Lord Chac") commissioned many of the largest structures at the heart of Uxmal, drawing on a large hinterland populace who built the most powerful statements of divine kingship known to the Puuc region (Kowalski 1987). The public monuments of Chan Chak K'ak'nal Ahaw's reign also show more militaristic themes than do earlier monuments (Dunning and Kowalski 1994: 81), and they can be dated by hieroglyphic texts to between AD 895 and 907 (Kowalski 1994). Some of the artistic canons and symbols employed on Chan Chak K'ak'nal Ahaw's art and architecture indicate influences from highland or Mexicanized Putun Maya peoples, as was also the case at Chichen Itza. Despite these "foreign" elements, the buildings and architectural sculpture of buildings of Chan Chak K'ak'nal Ahaw's reign, like the House of the Governor and the Nunnery Quadrangle, show clear connections to traditional Classic Maya architectural design and ideology (Kowalski 1987, 1989, 1994). There is little evidence of monumental buildings later than the reign of this powerful ruler, suggesting that the site ceased to be a major political force sometime between AD 925 and 975 (Kowalski 1994).

Explanations of Uxmal's fall from prominence tend to focus on either political or ecological factors. Political models generally emphasize the Terminal Classic growth of Chichen Itza, the increasing depictions of military themes at Puuc sites and Chichen Itza, and the appearance of the same personal names at both Chichen Itza and Uxmal (Kowalski 1985).

Furthermore, excavators at Uxmal and other sites in the region have found some Sotuta ceramics, usually associated with Chichen Itza (cf. Stanton and Gallareta 2001). Many scholars argue that this evidence betokens a regional conflict between Uxmal and Chichen Itza, in which Chichen Itza eventually won, resulting in the rapid decline of Puuc sites in the early- to mid-tenth century and the continued growth of Chichen Itza and its influence over the Puuc region (Carmean et al. 2004; Dunning and Kowalski 1994).

Nicholas Dunning suggested that the abandonment of many Puuc cities in the tenth century was due in part to environmental changes caused by a shift in climatic conditions.[4] He viewed the ninth-century demographic explosion, political centralization, and artistic and architectural florescence in the Puuc region as an "adaptive response to a period of increasing rainfall and regional population growth" (Dunning 1992: 153) that could not be sustained when rainfall began to decrease in the tenth century. Dunning did not see climatic change occurring in a vacuum, however, but contextualized it within the environmental conditions and political structures of the northwestern lowlands. He argued that the precarious ecological balance between burgeoning populations and the carrying capacity of the Puuc region was especially risky. Furthermore, given the ideology of divine kingship and the place of the ruler as the intermediary between his subjects and the spiritual realm, crop failure was likely perceived as a failure of individual rulers and, ultimately, of divine kingship itself. In contrast, he proposed, the mul tepal system of council government suggested for Chichen Itza entailed an ideology that was less susceptible to being undermined by agricultural downturns, thus helping explain the growth of that polity during the ninth through eleventh centuries.

North and east of the Puuc region, Dzibilchaltun went through a similar trajectory of Late Classic growth and Terminal Classic florescence, reaching a population maximum of perhaps 25,000, followed by a subsequent rapid population decline and cessation of monumental building activities (Andrews and Andrews 1980; Sharer 1994). During its heyday, the polity's rulers commissioned twenty-five carved monuments, including one with a period-ending date corresponding to AD 849, and the rulers show themselves using emblems of Classic divine kingship, such as K'awiil scepters.

Discussion. Assigning climatic change a role in the ninth- and tenth-century history of the northwestern lowlands is complicated by the lack of direct evidence for any drought during this period in the Aguada X'caamal proxy data. Even if one wished to discount the X'caamal record as not recording

4 Since Dunning originally formulated this model, paleoclimatic studies have advanced considerably, and the available data do not support the scenario he originally proposed. In a more recent treatment of the Puuc region during the Terminal Classic, Carmean and colleagues (2004: 441–442, 446) continue to suggest that drier conditions played an important role in the abandonment of most Puuc centers in the tenth century, although they note the apparent paradox that the highest population levels in the Puuc region correspond with evidence of drier conditions in the proxy records. The new data collected from Lake Chichancanab by Hodell and colleagues (2005a) suggest that the peak of political centralization at Uxmal corresponds with a period of *increased* rainfall, although much of the demographic growth that preceded Chan Chak K'ak'nal Ahaw's reign occurred under drier climatic conditions.

climatic change, the next best proxy records, the cores from Chichancanab (fig. 11), indicate early (ca. AD 770–870) and late (ca. AD 920–1100) drought phases with an intervening period (ca. AD 870–920) of relatively moister conditions. The archaeological data for the Puuc region during this same period demonstrate the complex relationships between climatic, environmental, and sociocultural changes: the dry ninth century witnessed significant population growth and political florescence in the Puuc region, a trend that continued with increased political centralization of the region under Uxmal as rainfall increased in the later ninth century. The region's sites underwent a decline subsequently in the tenth century after drier conditions resumed.[5]

The apparent paradox, though, between demographic growth and political centralization in the Puuc during drier conditions is even more striking at sites farther north. Dzibilchaltun is in the most arid part of the peninsula, and it receives significantly less precipitation than the Puuc region. One would expect that any lessening of rainfall would be felt most strongly here, given that even slight fluctuations in rainfall could limit maize agriculture. Dzibilchaltun, however, experiences roughly the same boom-bust trajectory as the Puuc region, including a ninth-century florescence during dry conditions.[6]

DISCUSSION AND CONCLUSIONS

The ninth and tenth centuries are a critical period for understanding the history of Maya civilization, but the complexities of this period are masked when it is framed as a unitary, event-like "collapse" (see Cowgill 1988). A growing body of archaeological data demonstrates that the Maya collapse can only be understood as a set of interrelated historical processes, not as a single event. Furthermore, contrary to popular conception, these processes did not bring about the end of Maya civilization. They did, however, correspond with a shift in the center of gravity of Maya civilization away from the central and southern lowlands, as new centers of political power and cultural innovation rose in the northern lowlands and, later, the highlands. New political structures and ideologies and changes in the economic foundations of these new centers mark the Postclassic Period of Maya civilization, distinct from but in no ways inferior to the Classic Period (Sabloff 1990).

What role did climatic change play in these remarkable transformations? Several observations are pertinent. First, the accumulating evidence demonstrates that there was a multigenerational period of significantly elevated evaporation-to-precipitation ratios (i.e., drier conditions) across much of

5 Dunning (1992: 153) suggests that there was a tenth-century decrease in rainfall due not to global climatic change, but instead to largely anthropogenic processes, "the product of population growth across the peninsula, increased deforestation, and reduced evapotranspiration." Although the proxy data from Lake Chichancanab support Dunning's hypothesis of drying in the tenth century, they do not support his suggestion that the ninth century was a period of increasing rainfall that facilitated the Puuc florescence.

6 It is possible that Dzibilchaltun's economy was based not on maize agriculture but on control of salt-producing areas of the nearby coast, which would be enhanced by drier climatic conditions (Andrews 1983). Furthermore, the water table in this region is closer to the surface than is the case in the Puuc zone, and thus it is accessible through *cenotes*.

Year AD	Northwest (X'caamal)	North-central (Chichancanab)	Northeast (Punta Laguna)	South-central (Peten-Itza)	South-central (Salpeten)	Maya Cultural Period
1900-2000						
1800-1900						
1700-1800						Historic
1600-1700						
1500-1600						
1400-1500						
1300-1400						Late Postclassic
1200-1300						
1100-1200						Early Postclassic
1000-1100						
900-1000						Terminal Classic
800-900						
700-800						
600-700						Late Classic
500-600						
400-500						Early Classic
300-400						
200-300						
100-200						Late Preclassic
1-100						

FIG. 16 Paleoclimatic summary for the Yucatan Peninsula during the past two millennia. Increasing gray scale represents increasing $\delta^{18}O$ values in cores shown in figure 4.

the Yucatan Peninsula during the Terminal Classic Period (AD 800–1000). Oxygen isotope ratios in cores from four of the five lakes studied show an increase in $\delta^{18}O$ values in the Terminal Classic Period beginning around AD 800 (figs. 4 and 16). These results are supported by the data from the Cariaco Basin that indicate an extended dry period from ca. AD 750 to 950. Moreover, the sediments from cores from Lake Chichancanab and the Cariaco Basin strongly suggest that this period of generally reduced precipitation was punctuated by distinct multiyear drought events with a 40- to 50-year periodicity (ca. AD 760, 810, 860, and 910; Haug et al. 2003; Hodell et al. 2005a).

Equally important, though, is the geographic variability in the paleoclimatic signal (fig. 16). The strongest evidence for reduced rainfall levels comes from the north-central and northeastern region of the Yucatan Peninsula, where cores suggest two phases of drier conditions during the Terminal Classic Period (ca. AD 770–870 and AD 920–1100) with an intervening period (ca. AD 870–920) of relatively moister conditions. In contrast, sediment cores from Aguada X'caamal in the northwestern part of the Yucatan Peninsula record no change during the same period. The $\delta^{18}O$ values in cores from both Lakes Peten Itza and Salpeten increased during the Terminal Classic Period, indicating either increased E/P (i.e., reduced rainfall) and/or reduced runoff as a consequence of reforestation of the lakes' drainage basins.

The general chronological correspondence between climatic shifts evident in the various proxy records and the diverse sociopolitical transformations experienced by different Maya polities in the ninth and tenth centuries that we have described above is striking, and it demands explanation. But it can also lead to simplistic models of cause-and-effect without modeling the complex relationships between climatic change, environmental transformation, and sociocultural processes. For example, Peter deMenocal (2001:

672, emphasis added) states that "available paleoclimate and archaeological data show that societal collapse and prolonged drought were coincident *within respective dating uncertainties.* Coincidence alone cannot demonstrate causality. . . . However, joint interpretation of the paleoclimatic and archaeological evidence now underscores the important role of persistent, long-term drought in the collapse" of Maya civilization. The simple fact that climatic changes and cultural changes are effectively contemporaneous in our observational record serves as a very weak empirical foundation for scientifically evaluating the causal relationship between the two, given the limitations described above in precisely dating either record, which are then amplified when trying to correlate the two records.

If we are to confidently move from temporal correlation to causality, we must model how climatic changes would have affected human populations and the environment in which they lived. Indeed, until that occurs, conclusions about the relative importance—or unimportance—of climatic change will be almost entirely determined by a scholar's a priori assumptions about the role of climate (see Erickson 1999; Webster 2002: 247). Two facts complicate this kind of modeling. First, obtaining a quantitative estimate of past rainfall reduction from the measurable changes in proxy records is problematic, requiring complex modeling and many assumptions about the relationships between proxy measures and past climatic conditions. Second, available proxy records lack the temporal resolution needed to assess intra-annual rainfall differences. This is critical, because there is no simple linear correlation between crop yields and annual precipitation. Annual rainfall levels can drop significantly before passing a threshold at which they adversely affect yields. The seasonal distribution of rainfall is the key factor, because the yields of many crops, including maize, are most affected by shortages in moisture availability during certain critical phases of crop development, such as germination and pollination. The same was likely true of water needed for domestic purposes like cooking: reduced rainfall during the rainy season would probably not prevent most households from storing enough water to last through the dry season, whereas reduced dry season rainfall could lead to critical shortages. Although the discovery of proxy records with seasonal resolution seems unlikely at present, we will likely be able to more securely retrodict seasonal patterns of past rainfall as models of regional climatic patterning improve.

Despite those limitations, we believe that the data currently available do permit some broad conclusions about the role of climate in the Terminal Classic Period. First, neither the paleoclimatic nor the archaeological data presently available support assertions that climatic change or drought was a primary cause for the Classic Maya collapse (cf. Gill 2000; Gill et al. 2007). The paleoclimatic data from the Maya lowlands show markedly distinct regional signals (fig. 16) that belie models of a monolithic, long-lasting, peninsula-wide drought. Although the proxy data suggest an overall reduced precipitation profile punctuated by some drought periods, it is not at all clear that the normal precipitation in the drier ninth century would have resulted in reduced agricultural yields during the rainy season or dry-season water shortages in the central and southern lowlands.

Although some scholars like Richardson Gill (2000) and Peter deMenocal (2001) argue that the scope of the Terminal Classic transformations were such that only significant environmental stress like that caused by climatic change could have triggered them, the variability observable in the archaeological record suggests otherwise. In the southern and central lowlands, the ninth century is *generally* characterized by the abandonment of many large cities, a shift away from the ideology of divine kingship, and significant demographic decline, but the details and timing of these transformations vary considerably between regions. The Puuc region, in contrast, experienced a demographic explosion in the ninth century before undergoing a broadly parallel decline in the later tenth century. Chichen Itza, in further contrast, apparently grew to dominate the northern lowlands in the tenth century and only later experienced a decline. To force these three cycles of political florescence and subsequent decline and decentralization, each separated by a century or more, into a unitary drought-driven model obscures more than it explains about the transformation of Maya civilization in the Terminal Classic Period.

At the regional level, significant differences from site to site also argue against any monocausal model. In many parts of the central and southern lowlands, for example, the processes that were integral to the collapse—intense interpolity competition, population growth, and increased warfare, for example—were well underway by the mid-eighth century. Similarly, at Xunantunich, on the other side of the central lowlands, significant demographic decline occurred prior to AD 780, although the site remained a political center until at least AD 850 (Le Count et al. 2002). In contrast, in other polities—Tikal and Calakmul, for example—political decentralization is most pronounced in the first half of the ninth century, although others like Caracol and Seibal experienced a political renaissance during that same period. One could attempt to correlate the various severe droughts suggested by the Cariaco data with the different sequences within the central and southern lowlands, but any correspondence would still beg the question of why the people in different sites responded in such distinct ways.

Many riverside and lakeside sites were completely abandoned during the ninth and tenth centuries, despite their prime locations in the putatively drier environment. This is true not only of sites that were the capitals of polities founded on the ideology of divine kingship, like Quirigua and Piedras Negras, but also of many small villages and rural populations like those of the Mopan River valley in Belize (Ashmore et al. 2004; cf. Lucero 2002). Furthermore, the population density in most of the central and southern lowlands remained low for centuries, even after a return to moister climatic conditions (Culbert 1988). These patterns force us to again recognize the strong roles that sociopolitical factors played in the collapse, and they highlight the utility of thinking of the collapse in terms of local responses to sociopolitical, climatic, and environmental conditions (see Palka 1997).

To conclude, there is no evidence that climatic change was any more of a "trigger" for the Terminal Classic transformation of Maya civilization than were population growth, interpolity conflict, soil erosion, and the other observable processes that began prior to or coeval with climatic changes in various regions. One could argue that climatic change is qualitatively different from these processes in that people cannot in most cases regulate its causes, but they certainly respond to adverse climatic changes to mitigate their effects. Indeed all climatic changes, even those we might consider negative, encourage some kinds of behaviors and practices and discourage others. At Chichen Itza, for example, drier conditions apparently favored economic strategies that emphasized long-distance trade and production of salt and other nonagricultural products, as well as a shift to a more decentralized political structure.

The rich and rapidly growing body of available archaeological, environmental, and climatological data leads us to offer three criteria that any model of the Maya collapse should meet. First, it must conceptualize the collapse not as a catastrophe or an event, but as a complex set of processes that were inherently social and cultural, and that transformed the demography, economy, and political organization of lowland Maya civilization over the course of several centuries. Second, it must include explicit discussion of the ways in which climatic change affects the natural and cultural contexts that shape people's decisions, past and present. Interdisciplinary research programs following the "conjunctive approach" (Marcus 1995; Fash and Sharer 1991) provide the diverse kinds of data we need to evaluate the interrelationships between climatic change, the environment, and Maya civilization. Finally, these models must pay close attention to local climatic, environmental, and cultural conditions. The results of decades of archaeological, paleoenvironmental, and paleoclimatic research leave no room to doubt that the factors that led to the collapse of Maya polities and the abandonment of Maya sites during the Terminal Classic Period, as well as the processes of collapse themselves, varied significantly in time and space across the Maya lowlands.

ACKNOWLEDGMENTS

We would like to thank Jeffrey Quilter and Daniel Sandweiss for inviting us to participate in the 2002 Pre-Columbian Studies symposium at Dumbarton Oaks. This chapter benefited from feedback from Minette Church, David Lentz, Joyce Marcus, Robert Sharer, and the volume's anonymous reviewers. The authors retain responsibility for any errors of fact and the interpretations presented herein.

REFERENCES CITED

Adams, Richard E. W.
1973 The Collapse of Maya Civilization:
A Review of Previous Theories. In
The Classic Maya Collapse (T. Patrick
Culbert, ed.): 21–34. University of New
Mexico Press, Albuquerque.

1997 *Ancient Civilizations of the New
World*. Westview, Boulder, CO.

Anderson, Patricia K.
1998 Yula, Yucatan, Mexico: Terminal
Classic Maya Ceramic Chronology
for the Chichen Itza Area. *Ancient
Mesoamerica* 9 (1): 151–166.

Andrews, Anthony P.
1983 *Maya Salt Production and Trade.*
University of Arizona Press, Tucson.

1990 The Fall of Chichen Itza: A
Preliminary Hypothesis. *Latin American
Antiquity* 1: 258–267.

**Andrews, Anthony P., and
Tomás Gallareta Negrón**
1986 The Isla Cerritos Archaeological
Project. *Mexicon* 8 (3): 44–48.

**Andrews, Anthony P., and
Fernando Robles Castellanos**
1985 Chichen Itza and Coba:
An Itza-Maya Standoff in Early
Postclassic Yucatan. In *The Lowland
Maya Postclassic* (Arlen F. Chase
and Prudence M. Rice, eds.): 62–72.
University of Texas Press, Austin.

**Andrews, E. Wyllys, IV, and
E. Wyllys Andrews V**
1980 *Excavations at Dzibilchaltun,
Yucatan, Mexico*. Publication 48. Middle
American Research Institute, Tulane
University, New Orleans.

**Andrews, E. Wyllys, V,
and Jeremy A. Sabloff**
1986 Classic to Postclassic: A Summary
Discussion. In *Late Lowland Maya
Civilization: Classic to Postclassic*
(Jeremy A. Sabloff and E. Wyllys
Andrews V, eds.): 433–456. University
of New Mexico Press, Albuquerque.

**Anselmetti, F. S., David A. Hodell,
David Ariztequi, Mark Brenner,
and Michael F. Rosenmeier**
2007 Quantification of Soil Erosional
Rates Related to Ancient Maya
Deforestation. *Geology* 35: 915–948.

**Ashmore, Wendy, Jason
Yaeger, and Cynthia Robin**
2004 Commoner Sense: Late and
Terminal Classic Social Strategies
in the Xunantunich Area. In *The
Terminal Classic in the Maya Lowlands:
Collapse, Transition, and Transformation*
(Arthur A. Demarest, Prudence M.
Rice, and Don S. Rice, eds.): 302–323.
University Press of Colorado, Boulder.

Ball, Joseph W.
1977 An Hypothetical Outline of the
Coastal Maya Prehistory: 300 B.C.–A.D.
1200. In *Social Process in Maya Prehistory:
Studies in Honour of Sir J. Eric S.
Thompson* (Norman Hammond, ed.):
167–196. Academic Press, London.

**Bey, George J., III, Craig A.
Hanson, and William M. Ringle**
1997 Classic to Postclassic at Ek Balam
Yucatán: Architectural and Ceramic
Evidence for Defining the Transition.
Latin American Antiquity 8: 237–254.

**Bey, George J., III, Tara M.
Bond, William M. Ringle, Craig A.
Hanson, Charles W. Houk, and
Carlos Peraza Lope**
1998 The Ceramic Chronology of Ek
Balam. *Ancient Mesoamerica* 9 (1):
101–120.

Braswell, Geoffrey E.
2003 Obsidian Spheres of Exchange
of Postclassic Mesoamerica. In *The
Postclassic Mesoamerican World*
(Michael E. Smith and Frances
Berdan, eds.): 131–158. University
of Utah Press, Salt Lake City.

**Braswell, Geoffrey E., Joel D. Gunn,
María del Rosario Domínguez
Carrasco, William J. Folan,
Laraine A. Fletcher, Abel Morales
López, and Michael D. Glascock**
2004 Defining the Terminal Classic
at Calakmul, Campeche. In *The
Terminal Classic in the Maya Lowlands:
Collapse, Transition, and Transformation*
(Arthur A. Demarest, Prudence M.
Rice, and Don S. Rice, eds.): 162–194.
University Press of Colorado, Boulder.

Brenner, Mark, Michael F. Rosenmeier, David A. Hodell, and Jason H. Curtis
2002 Paleolimnology of the Maya Lowlands: Long-Term Perspectives on Interactions among Climate, Environment, and Humans. *Ancient Mesoamerica* 13: 141–157.

Brenner, Mark, David A. Hodell, Jason H. Curtis, Michael F. Rosenmeier, Flavio S. Anselmetti, and Daniel Ariztegui
2003 Paleolimnological Approaches for Inferring Past Climate Change in the Maya Region: Recent Advances and Methodological Limitations. In *The Lowland Maya Area: Three Millennia at the Human-Wildland Interface* (Arturo Gómez-Pompa, Michael F. Allen, Scott L. Fedick, and Juan J. Jiménez-Osornio, eds.): 45–75. Haworth Press, Binghamton, NY.

Carmean, Kelli C.
1991 Architectural Labor Investment and Social Stratification at Sayil, Yucatan, Mexico. *Latin American Antiquity* 2: 151–165.

1998 Leadership at Sayil: A Study of Political and Religious Decentralization. *Ancient Mesoamerica* 9 (2): 259–270.

Carmean, Kelli, Nicholas Dunning, and Jeff Karl Kowalski
2004 High Times in the Hill Country: A Perspective from the Terminal Classic Puuc Region. In *The Terminal Classic in the Maya Lowlands: Collapse, Transition, and Transformation* (Arthur A. Demarest, Prudence M. Rice, and Don S. Rice, eds.): 424–449. University Press of Colorado, Boulder.

Chase, Arlen F.
1985 Troubled Times: The Archaeology and Iconography of the Terminal Classic Southern Lowland Maya. In *Fifth Palenque Round Table*, 1983 (Virginia M. Fields, ed.): 103–114. Pre-Columbian Art Research Institute, San Francisco.

1994 A Contextual Approach to the Ceramics of Caracol, Belize. In *Studies in the Archaeology of Caracol, Belize* (Diane Z. Chase and Arlen F. Chase, eds.): 157–182. Monograph 7. Pre-Columbian Art Research Institute, San Francisco.

Chase, Arlen F., and Diane Z. Chase
2004 Terminal Classic Status-Linked Ceramics and the Maya "Collapse": De Facto Refuse at Caracol, Belize. In *The Terminal Classic in the Maya Lowlands: Collapse, Transition, and Transformation* (Arthur A. Demarest, Prudence M. Rice, and Don S. Rice, eds.): 342–366. University Press of Colorado, Boulder.

Chase, Arlen F., Nikolai Grube, and Diane Z. Chase
1991 *Three Terminal Classic Monuments from Caracol, Belize*. Research Reports on Ancient Maya Writing No. 36. Center for Maya Research, Washington, D.C.

Chase, Arlen F., and Prudence M. Rice (eds.)
1985 *The Lowland Maya Postclassic*. University of Texas Press, Austin.

Cobos Palma, Rafael
2004 Chichén Itzá: Settlement and Hegemony during the Terminal Classic Period. In *The Terminal Classic in the Maya Lowlands: Collapse, Transition, and Transformation* (Arthur A. Demarest, Prudence M. Rice, and Don S. Rice, eds.): 517–544. University Press of Colorado, Boulder.

Cobos, Rafael, and Terance L. Winemiller
2001 The Late and Terminal Classic-Period Causeway Systems of Chichen Itza, Yucatan, Mexico. *Ancient Mesoamerica* 12 (2): 283–291.

Cooke, C. Wythe
1931 Why the Mayan Cities of the Peten District, Guatemala, Were Abandoned. *Journal of the Washington Academy of Sciences* 21 (13): 283–287.

Cowgill, George L.
1988 Onward and Upward with Collapse. In *The Collapse of Ancient States and Civilizations* (Norman Yoffee and George L. Cowgill, eds.): 244–276. University of Arizona Press, Tucson.

Craine, Eugene R., and Reginald C. Reindorp
1979 *The Codex Perez and the Book of Chilam Balam of Mani*. University of Oklahoma Press, Norman.

Culbert, T. Patrick
1988 The Collapse of Classic Maya
Civilization. In *The Collapse of Ancient
States and Civilizations* (Norman Yoffee
and George L. Cowgill, eds.): 69–101.
University of Arizona Press, Tucson.

Culbert, T. Patrick (ed.)
1973 *The Classic Maya Collapse.*
University of New Mexico Press,
Albuquerque.

Culbert, T. Patrick, Laura J.
Kosakowsky, Robert E. Fry,
and William A. Haviland
1990 The Population of Tikal, Guatemala.
In *Precolumbian Population History in the
Maya Lowlands* (T. Patrick Culbert and
Don S. Rice, eds.): 103–121. University of
New Mexico Press, Albuquerque.

Curtis, Jason H., Mark Brenner,
and David A. Hodell
1999 Climate Change in the Lake
Valencia Basin, Venezuela, ~12,600 yr
BP to Present. *The Holocene* 9: 609–619.

Curtis, Jason H., Mark Brenner, David
A. Hodell, Richard A. Balser, Gerald
A. Islebe, and Henry Hooghiemstra
1998 A Multi-Proxy Study of Holocene
Environmental Change in the Maya
Lowlands of Peten, Guatemala. *Journal of
Paleolimnology* 19: 139–159.

Curtis, Jason H., David A.
Hodell, and Mark Brenner
1996 Climate Variability on the Yucatan
Peninsula (Mexico) during the Past
3500 Years, and Implications for Maya
Cultural Evolution. *Quaternary Research*
46: 37–47.

Dahlin, Bruce H.
1983 Climate and Prehistory on the
Yucatan Peninsula. *Climatic Change* 5:
245–263.

Deevey, Edward S., Mark
Brenner, Michael S. Flannery,
and G. Habib Yezdani
1980 Lakes Yaxha and Sacnab, Peten,
Guatemala: Limnology and Hydrology.
Archiv für Hydrobiologie, Supplement 57:
419–460.

Deevey, Edward S., Don S. Rice,
Prudence M. Rice, Hague H. Vaughan,
Mark Brenner, and Michael S. Flannery
1979 Mayan Urbanism: Impact on a
Tropical Karst Environment. *Science* 206:
298–306.

Deevey, Edward S., and Minze Stuiver
1964 Distribution of Natural Isotopes
of Carbon in Linsley Pond and Other
New England Lakes. *Limnology and
Oceanography* 9: 1–11.

Demarest, Arthur A.
1997 The Vanderbilt Petexbatun
Regional Archaeological Project,
1989–1994: Overview, History, and Major
Results of a Multidisciplinary Study
of the Classic Maya Collapse. *Ancient
Mesoamerica* 8 (2): 209–227.

2004 After the Maelstrom: Collapse
of the Classic Maya Kingdoms and the
Terminal Classic in Western Petén.
In *The Terminal Classic in the Maya
Lowlands: Collapse, Transition, and
Transformation* (Arthur A. Demarest,
Prudence M. Rice, and Don S. Rice,
eds.): 102–124. University Press of
Colorado, Boulder.

Demarest, Arthur A., Matt
O'Mansky, Claudia Wolley, Dirk
Van Tuerenhout, Takeshi Inomata,
Joel Palka, and Héctor Escobedo
1997 Classic Maya Defensive Systems
and Warfare in the Petexbatun
Region: Archaeological Evidence and
Interpretations. *Ancient Mesoamerica* 8
(2): 229–253.

Demarest, Arthur A., Prudence M.
Rice, and Don S. Rice
2004 The Terminal Classic in the
Maya Lowlands: Hermeneutics,
Transitions, and Transformations.
In *The Terminal Classic in the Maya
Lowlands: Collapse, Transition, and
Transformation* (Arthur A. Demarest,
Prudence M. Rice, and Don S. Rice,
eds.): 545–572. University Press of
Colorado, Boulder.

Demarest, Arthur A., Prudence M.
Rice, and Don S. Rice (eds.)
2004 *The Terminal Classic in the Maya
Lowlands: Collapse, Transition, and
Transformation.* University Press of
Colorado, Boulder.

deMenocal, Peter
2001 Cultural Responses to Climate
Change during the Late Holocene.
Science 292: 667–673.

Diamond, Jared M.
2005 *Collapse: How Societies Choose to
Fail or Succeed.* Viking, New York.

Drennan, Richard D.
1984 Long-Distance Movement of
Goods in Prehispanic Mesoamerica: Its
Importance in the Complex Societies of
the Formative and the Classic. *American
Antiquity* 49: 27–43.

Dunning, Nicholas P.
1992 *Lords of the Hills: Ancient Maya
Settlement in the Puuc Region, Yucatán,
Mexico.* Prehistory Press, Madison, WI.

**Dunning, Nicholas P.,
and Timothy Beach**
2000 Stability and Instability in Pre-
Hispanic Maya Landscapes. In *Imperfect
Balance: Landscape Transformations in the
Pre-Columbian Americas* (David Lentz,
ed.): 179–202. Columbia University
Press, New York.

**Dunning, Nicholas P., Timothy
Beach, and David Rue**
1997 The Paleoecology and Ancient
Settlement of the Petexbatun Region,
Guatemala. *Ancient Mesoamerica* 8 (2):
255–266.

**Dunning, Nicholas P., and
Jeff Karl Kowalski**
1994 Lords of the Hills: Classic Maya
Settlement Patterns and Political
Iconography in the Puuc Region, Mexico.
Ancient Mesoamerica 5 (1): 63–95.

**Ely, Lisa L., Yahouda Enzel, Victor R.
Baker, and Daniel R. Cavan**
1993 A 5000-Year Record of Extreme
Floods and Climate Change in the
Southwestern United States. *Science* 262:
410–412.

**Emery, Kitty F., Lori. E. Wright,
and Henry Schwarcz**
2000 Isotopic Analysis of Ancient Deer
Bone: Biotic Stability in Collapse Period
Maya Land-Use. *Journal of Archaeological
Science* 27: 537–550.

Enfield, David B., and Eric J. Alfaro
1999 The Dependence of Caribbean
Rainfall on the Interaction of the
Tropical Atlantic and Pacific Oceans.
Journal of Climate 12: 2093–2103.

Erasmus, Charles J.
1968 Thoughts on Upward Collapse: An
Essay on Explanation in Anthropology.
Southwestern Journal of Anthropology 24:
170–194.

Erickson, Clark L.
1999 Neo-environmental Determinism
and Agrarian "Collapse" in Andean
Prehistory. *Antiquity* 73: 634–642.

Fagan, Brian M.
1999 *Floods, Famines and Emperors: El
Niño and the Fate of Civilizations.* Basic
Books, New York.

**Fash, William L., E. Wyllys
Andrews V, and T. Kam Manahan**
2004 Political Decentralization,
Dynastic Collapse, and the Early
Postclassic in the Urban Center
of Copán, Honduras. In *The
Terminal Classic in the Maya Lowlands:
Collapse, Transition, and Transformation*
(Arthur A. Demarest, Prudence M.
Rice, and Don S. Rice, eds.): 260–287.
University Press of Colorado, Boulder.

Fash, William L., and Robert J. Sharer
1991 Sociopolitical Developments
and Methodological Issues at Copán,
Honduras: A Conjunctive Approach.
Latin American Antiquity 2: 166–187.

Folan, William J.
1983a Archaeological Investigations of
Cobá: A Summary. In *Cobá, a Classic
Maya Metropolis* (William J. Folan,
Ellen R. Kintz, and Laraine A. Fletcher,
eds.): 1–10. Academic Press, New York.

1983b Ruins of Cobá. In *Cobá, a Classic
Maya Metropolis* (William J. Folan,
Ellen R. Kintz, and Laraine A. Fletcher,
eds.): 65–87. Academic Press, New York.

1983c Summary and Conclusions.
In *Cobá, a Classic Maya Metropolis*
(William J. Folan, Ellen R. Kintz, and
Laraine A. Fletcher, eds.): 211–217.
Academic Press, New York.

**Folan, William J., Betty Faust,
Wolfgang Lutz, and Joel D. Gunn**
2000 Social and Environmental
Factors in the Classic Maya Collapse.
In *Population, Development, and
Environment on the Yucatan Peninsula:
From Ancient Maya to 2030* (Wolfgang
Lutz, Leonel Prieto, and Warren
Sanderson, eds.): 2–32. International
Institute for Applied Systems Analysis,
Laxenburg, Austria.

Folan, William J., Joel D. Gunn,
Jack D. Eaton, and Robert W. Patch
1983a Paleoclimatological Patterning in
Southern Mesoamerica. *Journal of Field
Archaeology* 10: 454–468.

Folan, William J., and B. H. Hyde
1985 Climatic Forecasting and Recording
among the Ancient and Historic
Maya: An Ethnohistoric Approach to
Epistemological and Paleoclimatological
Patterning. In *Contributions to the
Archaeology and Ethnohistory of Greater
Mesoamerica* (William J. Folan, ed.):
15–48. Southern Illinois University
Press, Carbondale.

Folan, William J., Ellen R. Kintz,
and Laraine A. Fletcher (eds.)
1983b *Cobá, a Classic Maya Metropolis.*
Academic Press, New York.

Folan, William J., Joyce Marcus,
Sophia Pincemin, María del Rosario
Domínguez Carrasco, Laraine A.
Fletcher, and Abel Morales López
1995 Calakmul: New Data from an
Ancient Maya Capital in Campeche,
Mexico. *Latin American Antiquity* 6:
310–334.

Ford, Anabel
1986 Population Growth and Social
Complexity: An Examination of
Settlement and Environment in the
Central Maya Lowlands. *Anthropological
Research Papers* 35. Arizona State
University, Tempe.

Freidel, David A.
1986 Terminal Classic Lowland Maya:
Successes, Failures, and Aftermaths.
In *Late Lowland Maya Civilization:
Classic to Postclassic* (Jeremy A.
Sabloff and E. Wyllys Andrews V,
eds.): 409–430. University of New
Mexico Press, Albuquerque.

Freidel, David, and Justine Shaw
2000 The Lowland Maya Civilization:
Historical Consciousness and
Environment. In *The Way the Wind
Blows: Climate, History, and Human
Action* (Roderick J. McIntosh,
Joseph A. Tainter, and Susan Keech
McIntosh, eds.): 271–299. Columbia
University Press, New York.

García Campillo, José Miguel
1991 Edificios y dignatarios: La historia
escrita de Oxkintok. In *Oxkintok, una
ciudad maya de Yucatán: Excavaciones
de la misión arqueológica de España en
México, 1986–1991*: 55–76. Ministerio de
Asuntos Exteriores, Dirección General
de Relaciones Culturales y Científicas,
y Ministerio de Cultura, Dirección
General de Bellas Artes y Archivos,
Instituto de Conservación y Restauración
de Bienes Culturales, Madrid.

Giannini, Alessandra G., Yochanan
Kushnir, and Mark A. Cane
2000 Interannual Variability of
Caribbean Rainfall, ENSO, and the
Atlantic Ocean. *Journal of Climate* 13:
297–311.

2001a Seasonality in the Impact of
ENSO and the North Atlantic High
on Caribbean Rainfall. *Physics and
Chemistry of the Earth*, Part B: *Hydrology,
Oceans and Atmosphere* 26 (2): 143–147.

2001b Interdecadal Changes in the ENSO
Teleconnection to the Caribbean Region
and the North Atlantic Oscillation.
Journal of Climate 14: 2867–2879.

Giannini, Alessandra G., Richard
Seager, John C. H. Chiang, Mark A.
Cane, and Yochanan Kushnir
2001c The ENSO Teleconnection to the
Tropical Atlantic Ocean: Contributions
of the Remote and Local SSTs to
Rainfall Variability in the Tropical
Americas. *Journal of Climate* 24:
4530–4544.

Gill, Richardson B.
2000 *The Great Maya Droughts: Water,
Life, and Death.* University of New
Mexico Press, Albuquerque.

Gill, Richardson B., Paul A. Mayewski,
Johan Nyberg, Gerald C. Haug, and
Larry C. Peterson
2007 Drought and the Maya Collapse.
Ancient Mesoamerica 18: 283–302.

Gray, Calvin R.
1987 *History of Tropical Cyclones
in Jamaica, 1886–1986.* National
Meteorological Service, Jamaica.

1993 Regional Meteorology and
Hurricanes. In *Climatic Change in
the Intra-Americas Sea* (George A.
Maul, ed.): 87–99. United Nations
Environmental Program, Edward
Arnold, London.

Gunn, Joel D., and Richard E. W. Adams
1981 Climatic Change, Culture, and Civilization in North America. *World Archaeology* 13: 87–100.

Gunn, Joel D., William J. Folan, and Hubert R. Robichaux
1995 A Landscape Analysis of the Candelaria Watershed in Mexico: Insights into Paleoclimates Affecting Upland Horticulture in the Southern Yucatan Peninsula Semi-Karst. *Geoarchaeology* 10: 3–42.

Gunn, Joel D., Ray T. Matheny, and William J. Folan
2002 Climate-Change Studies in the Maya Area: A Diachronic Analysis. *Ancient Mesoamerica* 13: 79–84.

Harrison, Peter D.
1999 *Lords of Tikal: Rulers of an Ancient Maya City.* Thames and Hudson, London.

Hastenrath, Stefan
1966 On General Circulation and Energy Budget in the Area of the Central American Seas. *Journal of Atmospheric Sciences* 23: 694–711.

1967 Rainfall Distribution and Regime in Central America. *Archiv für Meteorologie, Geophysik und Bioklimatologie*, Ser. B 15: 201–241.

1976 Variations in Low-Latitude Circulation and Extreme Climatic Events in the Tropical Americas. *Journal of Atmospheric Sciences* 22: 202–215.

1984 Interannual Variability and the Annual Cycle: Mechanisms of Circulation and Climate in the Tropical Atlantic Sector. *Monthly Weather Review* 112: 1097–1107.

1991 *Climate Dynamics of the Tropics.* Kluwer Academic Publishers, Dordrecht, The Netherlands.

Haug, Gerald H., Konrad A. Hughen, Daniel M. Sigman, Larry C. Peterson, and Ursula Rohl
2000 Southward Migration of the Intertropical Convergence Zone through the Holocene. *Science* 293: 1304–1308.

Haug, Gerald H., Detlef Günther, Larry C. Peterson, Daniel M. Sigman, Konrad A. Hughen, and Beat Aeschlimann
2003 Climate and the Collapse of Maya Civilization. *Science* 299: 1731–1735.

Hodell, David A., Jason H. Curtis, Glenn A. Jones, Antonia Higuera-Gundy, Mark Brenner, Michael W. Binford, and Kathleen T. Dorsey
1991 Reconstruction of Caribbean Climate Change over the Past 10,500 Years. *Nature* 352: 790–793.

Hodell, David A., Jason H. Curtis, and Mark Brenner
1995 Possible Role of Climate in the Collapse of Classic Maya Civilization. *Nature* 375: 391–394.

Hodell, David A., Mark Brenner, Jason H. Curtis, and Thomas P. Guilderson
2001 Solar Forcing of Drought Frequency in the Maya Lowlands. *Science* 292: 1367–1369.

Hodell, David A., Mark Brenner, George Kamenov, and Rhonda Quinn
2003 Spatial Variation of Strontium Isotopes ($^{87}Sr/^{86}Sr$) in the Maya Region: A Tool for Tracking Ancient Human Migration. *Journal of Archaeological Science* 31: 585–601.

Hodell, David A., Mark Brenner, and Jason H. Curtis
2005a Terminal Classic Drought in the Northern Maya Lowlands Inferred from Multiple Sediment Cores in Lake Chichancanab (Mexico). *Quaternary Science Reviews* 24: 1413–1427.

Hodell, David A., Mark Brenner, Jason H. Curtis, Roger Medina-González, Enrique Ildefonso-Chan Can, Alma Albornaz-Pat, and Thomas P. Guilderson
2005b Climate Change on the Yucatan Peninsula during the Little Ice Age. *Quaternary Research* 63: 109–121.

2007 Climate and Cultural History of the Northeastern Yucatan Peninsula, Quitana Roo, Mexico. *Climate Change* 83: 215–240.

Houston, Stephen D.
1989 Archaeology and Maya Writing. *Journal of World Prehistory* 3: 1–32.

1993 *Hieroglyphs and History at Dos Pilas: Dynastic Politics of the Classic Maya.* University of Texas Press, Austin.

Islebe, Gerald A., Henry Hooghiemstra, Mark Brenner, Jason H. Curtis, and David A. Hodell
1996 A Holocene Vegetation History from Lowland Guatemala. *The Holocene* 6: 265–271.

Jones, Christopher
1991 Cycles of Growth at Tikal. In *Classic Maya Political History: Hieroglyphic and Archaeological Evidence* (T. Patrick Culbert, ed.): 102–127. Cambridge University Press, Cambridge.

Kepecs, Susan
1998 Diachronic Ceramic Evidence and Its Social Implications in the Chikinchel Region, Northeast Yucatan, Mexico. *Ancient Mesoamerica* 9 (1): 121–135.

Kepecs, Susan, Gary Feinman, and Sylviane Boucher
1994 Chichen Itza and its Hinterland: A World-Systems Perspective. *Ancient Mesoamerica* 5 (2): 141–158.

Kowalski, Jeff Karl
1985 Lords of the Northern Maya: Dynastic History in the Inscriptions of Uxmal and Chichen Itza. *Expedition* 27 (3): 50–60.

1987 *The House of the Governor: A Maya Palace at Uxmal, Yucatan, Mexico.* The Civilization of the American Indian 176. University of Oklahoma Press, Norman.

1989 Who Am I among the Itza? Links between Northern Yucatan and the Western Maya Lowlands and Highlands. In *Mesoamerica after the Decline of Teotihuacan, A.D. 700–900* (Richard A. Diehl and Janet C. Berlo, eds.): 173–185. Dumbarton Oaks Research Library and Collection, Washington, D.C.

1994 The Puuc as Seen from Uxmal. In *Hidden among the Hills: Maya Archaeology of the Northwest Yucatan Peninsula* (Hanns J. Prem, ed.): 99–120. Acta Mesoamericana 7. Verlag von Flemming, Möckmühl, Germany.

Kowalski, Jeff Karl, and Nicholas P. Dunning
1999 The Architecture of Uxmal: The Symbolics of Statemaking at a Puuc Maya Regional Capital. In *Mesoamerican Architecture as a Cultural Symbol* (Jeff Karl Kowalski, ed.): 274–297. Oxford University Press, New York.

Kurjack, Edward B., and E. Wyllys Andrews V
1976 Early Boundary Maintenance in Northwest Yucatan, Mexico. *American Antiquity* 41: 318–325.

Kurjack, Edward B., and Merle G. Robertson
1994 Politics and Art at Chichén Itzá. In *Seventh Palenque Round Table, 1989* (Merle G. Robertson and Virginia M. Fields, eds.): 19–23. Pre-Columbian Art Research Institute, San Francisco.

LeCount, Lisa J., Jason Yaeger, Richard M. Leventhal, and Wendy Ashmore
2002 Dating the Rise and Fall of Kunantunich, Belize. *Ancient Mesoamerica* 13: 41–63.

Leyden, Barbara W., Mark Brenner, and Bruce H. Dahlin
1998 Cultural and Climatic History of Coba, a Lowland Maya City in Quintana Roo, Mexico. *Quaternary Research* 49: 111–122.

Lowe, John W. G.
1985 *The Dynamics of Apocalypse: A Systems Simulation of the Classic Maya Collapse.* University of New Mexico Press, Albuquerque.

Lucero, Lisa J.
2002 The Collapse of the Classic Maya: A Case for the Role of Water Control. *American Anthropologist* 104: 814–826.

McAnany, Patricia A.
1990 Water Storage in the Puuc Region of the Northern Maya Lowlands: A Key to Population Estimates and Architectural Variability. In *Precolumbian Population History in the Maya Lowlands* (T. Patrick Culbert and Don S. Rice, eds.): 263–284. University of New Mexico Press, Albuquerque.

MacKie, Euan W.
1961 New Light on the End of Classic Maya Culture at Benque Viejo, British Honduras. *American Antiquity* 27: 216–224.

Magaña, Victor, Jorge A.
Amador, and Socorro Medina
1999 The Midsummer Drought
over Mexico and Central America.
Journal of Climate 12: 1577–1588.

Marcus, Joyce
1976 *Emblem and State in the Classic
Maya Lowlands: An Epigraphic Approach
to Territorial Organization.* Dumbarton
Oaks Research Library and Collection,
Washington, D.C.

1992 *Mesoamerican Writing Systems:
Progaganda, Myth, and History in
Four Ancient Civilizations.* Princeton
University Press, Princeton.

1993 Ancient Maya Political
Organization. In *Lowland Maya
Civilization in the Eighth Century
A.D.* (Jeremy A. Sabloff and John S.
Henderson, eds.): 111–183. Dumbarton
Oaks Research Library and Collection,
Washington, D.C.

1995 Where Is Lowland Maya
Archaeology Headed? *Journal of
Archaeological Research* 3: 3–53.

1998 The Peaks and Valleys of Ancient
States: An Extension of the Dynamic
Model. In *Archaic States* (Gary M.
Feinman and Joyce Marcus, eds.): 59–94.
School of American Research, Santa Fe.

Martin, Simon, and Nikolai Grube
2000 *Chronicle of the Maya Kings
and Queens: Deciphering the
Dynasties of the Ancient Maya.*
Thames and Hudson, London.

Mesoamerica Climate Outlook Forum
1998 Mesoamerica Climate Outlook
Forum. Conference held 18–19 May,
Panama City, Panama.

Messenger, Lewis C., Jr.
1990 Ancient Winds of Change:
Climatic Settings and Prehistoric Social
Complexity in Mesoamerica. *Ancient
Mesoamerica* 1: 21–40.

2002 Los Mayas y El Niño: Paleoclimatic Correlations, Environmental
Dynamics, and Cultural Implications for the Ancient Maya. *Ancient
Mesoamerica* 13 (1): 159–170.

Morley, Sylvanus G.
1937–38 *The Inscriptions of Petén.*
Carnegie Institution of Washington
Publication 477. Carnegie Institution of
Washington, Washington, D.C.

1946 *The Ancient Maya,* 1st ed. Stanford
University Press, Stanford.

Nyberg, Johan, Björn A. Malmgren,
Antoon Kuijpers, and Amos Winter
2002 A Centennial-Scale Variability
of Tropical North Atlantic Surface
Hydrography during the Late Holocene.
*Palaeogeography, Palaeoclimatology,
Palaeoecology* 183: 25–41.

Pablo Aguilera, María del Mar de
1991 El arte de la piedra: Evolución
y expresión. In *Oxkintok, una ciudad
maya de Yucatán: Excavaciones de la
misión arqueológica de España en México,
1986–1991:* 79–104. Ministerio de
Asuntos Exteriores, Dirección General
de Relaciones Culturales y Científicas,
y Ministerio de Cultura, Dirección
General de Bellas Artes y Archivos,
Instituto de Conservación y Restauración
de Bienes Culturales, Madrid.

Palka, Joel W.
1997 Reconstructing Classic Maya
Socioeconomic Differentiation and the
Collapse at Dos Pilas, Peten, Guatemala.
Ancient Mesoamerica 8 (2): 293–306.

Pendergast, David M.
1991 The Southern Maya Lowlands
Contact Experience: The View
from Lamanai, Belize. In *The
Spanish Borderlands in Pan-
American Perspective* (David H.
Thomas, ed.): 337–354. Columbian
Consequences 3. Smithsonian
Institution, Washington, D.C.

Prem, Hanns J.
1991 The Xkipché Archaeological Project.
Mexicon 13: 62–63.

Price, T. Douglas, Linda Manzanilla,
and William D. Middleton
2000 Immigration and the Ancient
City of Teotihuacan in Mexico: A
Study Using Strontium Isotope Ratios
in Human Bone and Teeth. *Journal of
Archaeological Science* 37: 903–913.

Puleston, Dennis E.
1979 An Epistemological Pathology and the Collapse; or, Why the Maya Kept the Short Count. In *Maya Archaeology and Ethnohistory* (Norman Hammond and Gordon R. Willey, eds.): 63–74. University of Texas Press, Austin.

1983 *The Settlement Survey of Tikal* (William A. Haviland, ed.). Tikal Report 13, University Museum Monograph 48. University Museum, University of Pennsylvania, Philadelphia.

Quinn, William H.
1992 A Study of Southern Oscillation–Related Climatic Activity for AD 622–1900 Incorporating Nile River Flood Data. In *El Niño: Historical and Paleoclimate Aspects of the Southern Oscillation* (Henry F. Diaz and Vera Markgraf, eds.): 119–149. Cambridge University Press, Cambridge.

Ramsey, C. Bronk
1995 Radiocarbon Calibration and Analysis of Stratigraphy: The OxCal Program. *Radiocarbon* 37: 425–430.

1998 Probability and Dating. *Radiocarbon* 40: 461–474.

Rice, Don S.
1993 Eighth-Century Physical Geography, Environment, and Natural Resources in the Maya Lowlands. In *Lowland Maya Civilization in the Eighth Century A.D.* (Jeremy A. Sabloff and John S. Henderson, eds.): 11–63. Dumbarton Oaks Research Library and Collection, Washington, D.C.

1996 Paleolimnological Analysis in the Central Peten, Guatemala. In *The Managed Mosaic: Ancient Maya Agriculture and Resource Use* (Scott L. Fedick, ed.): 193–206. University of Utah Press, Salt Lake City.

Rice, Prudence M.
2004 *Maya Political Science: Time, Astronomy, and the Cosmos.* University of Texas Press, Austin.

Rice, Prudence M., and Donald Forsyth
2004 Terminal Classic–Period Lowland Ceramics. In *The Terminal Classic in the Maya Lowlands: Collapse, Transition, and Transformation* (Arthur A. Demarest, Prudence M. Rice, and Don S. Rice, eds.): 28–59. University Press of Colorado, Boulder.

Rice, Prudence M., and Don S. Rice
2004 Late Classic to Postclassic Transformations in the Petén Lakes Region, Guatemala. In *The Terminal Classic in the Maya Lowlands: Collapse, Transition, and Transformation* (Arthur A. Demarest, Prudence M. Rice, and Don S. Rice, eds.): 125–139. University Press of Colorado, Boulder.

Ricketson, Oliver G., Jr., and Edith Bayles Ricketson
1937 *Uaxactun, Guatemala, Group E, 1926–1937.* Carnegie Institution of Washington Publication 477. Carnegie Institution of Washington, Washington, D.C.

Ringle, William M., Tomás Gallareta Negrón, and George J. Bey III
1998 The Return of Quetzalcoatl: Evidence for the Spread of a World Religion during the Epiclassic Period. *Ancient Mesoamerica* 9 (2): 183–232.

Ringle, William M., George J. Bey III, Tara Bond Freeman, Craig A. Hansen, Charles W. Houck, and J. Gregory Smith
2004 The Decline of the East: The Classic to Postclassic Transition at Ek Balam, Yucatán. In *The Terminal Classic in the Maya Lowlands: Collapse, Transition, and Transformation* (Arthur A. Demarest, Prudence M. Rice, and Don S. Rice, eds.): 485–516. University Press of Colorado, Boulder.

Rivera Dorado, Miguel
1991 Ruinas, arqueólogos y problemas. In *Oxkintok, una ciudad maya de Yucatán: Excavaciones de la misión arqueológica de España en México, 1986–1991*: 9–53. Ministerio de Asuntos Exteriores, Dirección General de Relaciones Culturales y Científicas, y Ministerio de Cultura, Dirección General de Bellas Artes y Archivos, Instituto de Conservación y Restauración de Bienes Culturales, Madrid.

Robles Castellanos, Fernando
1990 *La secuencia cerámica de la región de Cobá, Quintana Roo.* Instituto Nacional de Antropología e Historia, México, D. F.

Robles Castellanos, Fernando, and Anthony P. Andrews
1986 A Review and Synthesis of Recent Postclassic Archaeology in Northern Yucatan. In *Late Lowland Maya Civilization: Classic to Postclassic* (Jeremy A. Sabloff and E. Wyllys Andrews V, eds.): 53–98. University of New Mexico Press, Albuquerque.

Rosenmeier, Michael F., David A. Hodell, Mark Brenner, Jason H. Curtis, and Jonathan B. Martin
2002a A 3500 Year Record of Climate Change and Human Disturbance from the Southern Maya Lowlands of Peten, Guatemala. *Quaternary Research* 57: 183–190.

Rosenmeier, Michael F., David A. Hodell, Mark Brenner, Jason H. Curtis, Jonathan B. Martin, Flavio Anselmetti, Daniel Ariztegui, and Thomas P. Guilderson
2002b Influence of Vegetation Change on Watershed Hydrology: Implications for Paleoclimatic Interpretation of Lacustrine $\delta^{18}O$ Records. *Journal of Paleolimnology* 27: 117–131.

Sabloff, Jeremy A.
1973a Major Themes in the Past Hypotheses of the Maya Collapse. In *The Classic Maya Collapse* (T. Patrick Culbert, ed.): 35–40. University of New Mexico Press, Albuquerque.

1973b Continuity and Disruption during Terminal Late Classic Times at Seibal: Ceramic and Other Evidence. In *The Classic Maya Collapse* (T. Patrick Culbert, ed.): 107–131. University of New Mexico Press, Albuquerque.

1990 *New Archaeology and the Ancient Maya*. W. H. Freeman, New York.

1992 Interpreting the Collapse of Classic Maya Civilization: A Case Study of Changing Archaeological Perspectives. In *Metaarchaeology: Reflections by Archaeologists and Philosophers* (Lester E. Embree, ed.): 92–119. Kluwer Academic Publishers, Boston.

Sabloff, Jeremy A., and E. Wyllys Andrews V (eds.)
1986 *Late Lowland Maya Civilization: Classic to Postclassic*. University of New Mexico Press, Albuquerque.

Sabloff, Jeremy A., and William L. Rathje
1975 The Rise of a Maya Merchant Class. *Scientific American* 23 (4): 72–82.

Sabloff, Jeremy A., and Gair Tourtellot III
1992 Beyond Temples and Palaces: Recent Settlement Pattern Research at the Ancient City of Sayil. In *New Theories on the Ancient Maya* (Elin C. Danien and Robert J. Sharer, eds.): 155–160. University Museum Monograph 77, University Symposium Series 3. University Museum, University of Pennsylvania, Philadelphia.

Sanchez, W. A., and John E. Kutzbach
1974 Climate of the American Tropics and Subtropics in the 1960s and Possible Comparison with Climatic Variations of the Last Millennium. *Quaternary Research* 4: 128–135.

Sanders, William T.
1973 The Cultural Ecology of the Lowland Maya: A Re-evaluation. In *The Classic Maya Collapse* (T. Patrick Culbert, ed.): 325–365. University of New Mexico Press, Albuquerque.

Scarborough, Vernon L.
1996 Reservoirs and Watersheds in the Central Maya Lowlands. In *The Managed Mosaic: Ancient Maya Agriculture and Resource Use* (Scott L. Fedick, ed.): 304–314. University of Utah Press, Salt Lake City.

Schele, Linda, and David Freidel
1990 *A Forest of Kings: The Untold Story of the Ancient Maya*. William Morrow, New York.

Schele, Linda, and Peter Mathews
1998 *The Code of Kings: The Language of Seven Sacred Maya Temples and Tombs*. Scribner, New York.

Sharer, Robert J.
1977 The Collapse Revisited: Internal and External Perspectives. In *Social Processes in Maya Prehistory, Essays in Honour of Sir J. Eric S. Thompson* (Norman Hammond, ed.): 532–552. Academic Press, New York.

1988 Quirigua as a Classic Maya Center. In *The Southeast Classic Maya Zone* (Elizabeth H. Boone and Gordon R. Willey, eds.): 31–65. Dumbarton Oaks Research Library and Collection, Washington, D.C.

1994 *The Ancient Maya*, 5th ed. Stanford University Press, Stanford.

Sharer, Robert J., and
Charles W. Golden
2004 Kinship and Polity: Conceptualizing the Maya Body Politic. In *Continuities and Changes in Maya Archaeology: Perspectives at the Millennium* (Charles W. Golden and Greg Borgstede, eds.): 23–50. Routledge, New York.

Shaw, Justine M.
2003 Climate Change and Deforestation: Implications for the Maya Collapse. *Ancient Mesoamerica* 14 (1): 157–167.

Smith, James G.
2001 Preliminary Report of the Chichen Itza–Ek Balam Transect Project. *Mexicon* 23 (2): 30–35.

Smith, Robert E.
1965 *Ceramic Sequence at Uaxactun, Guatemala*. Publication 20. Middle American Research Institute, Tulane University, New Orleans.

1971 *The Pottery of Mayapan*. Papers of the Peabody Museum of Archaeology and Ethnology 66. Harvard University, Cambridge, MA.

Smyth, Michael P., José Ligorred Perramon, David Ortegón Zapata, and Pat Farrell
1998 An Early Classic Center in the Puuc Region: New Data from Chac II, Yucatan, Mexico. *Ancient Mesoamerica* 9 (2): 233–257.

Stanton, Travis W., and
Tomás Gallareta Negrón
2001 Warfare, Ceramic Economy, and the Itza: A Reconsideration of the Itza Polity in Ancient Yucatan. *Ancient Mesoamerica* 12 (2): 229–245.

Stuart, David
1992 Hieroglyphs and Archaeology at Copán. *Ancient Mesoamerica* 3 (1): 169–184.

1993 Historical Inscriptions and the Maya Collapse. In *Lowland Maya Civilization in the Eighth Century A.D.* (Jeremy A. Sabloff and John S. Henderson, eds.): 321–354. Dumbarton Oaks Research Library and Collection, Washington, D.C.

Stuiver, Minze, Paula J. Reimer, Edouard Bard, J. Warren Beck, George S. Burr, Konrad A. Hughen, Bernd Kromer, F. Gerry McCormac, Johannes van der Plicht, and Marco Spurk
1998 INTCAL98 Radiocarbon Age Calibration 24,000–0 cal BP. *Radiocarbon* 40: 1041–1083.

Suhler, Charles, Traci Ardren, and David Johnstone
1998 The Chronology of Yaxuna: Evidence from Excavations and Ceramics. *Ancient Mesoamerica* 9 (1): 167–182.

Suhler, Charles, Traci Ardren, David Freidel, and David Johnstone
2004 The Rise and Fall of Terminal Classic Yaxuna, Yucatán, Mexico. In *The Terminal Classic in the Maya Lowlands: Collapse, Transition, and Transformation* (Arthur A. Demarest, Prudence M. Rice, and Don S. Rice, eds.): 450–484. University Press of Colorado, Boulder.

Taschek, Jennifer T., and
Joseph W. Ball
1999 Las Ruinas de Arenal: Preliminary Report on a Subregional Major Center in the Western Belize Valley (1991–1992 Excavations). *Ancient Mesoamerica* 10 (2): 215–235.

Thompson, Edward H.
1897 *The Chultunes of Labná, Yucatan: Report of Explorations by the Museum, 1888–89 and 1890–91*. Memoirs of the Peabody Museum of American Archaeology and Ethnology 1 (3). Harvard University, Cambridge, MA.

Thompson, J. Eric S.
1966 *The Rise and Fall of Maya Civilization*, 2nd ed. University of Oklahoma Press, Norman.

Tourtellot, Gair, III
1988 *Excavations at Seibal, Department of Petén, Guatemala: Peripheral Survey and Excavation, Settlement and Community Patterns*. Memoirs of the Peabody Museum of American Archaeology and Ethnology 16. Harvard University, Cambridge, MA.

Tourtellot, Gair, and Jason J. González
2004 The Last Hurrah: Continuity and Transformation at Seibal. In *The Terminal Classic in the Maya Lowlands: Collapse, Transition, and Transformation* (Arthur A. Demarest, Prudence M. Rice, and Don S. Rice, eds.): 60–82. University Press of Colorado, Boulder.

Tourtellot, Gair, and Jeremy A. Sabloff
1994 Community Structure at Sayil: A Case Study of Puuc Settlement. In *Hidden among the Hills: Maya Archaeology of the Northwest Yucatan Peninsula* (Hanns J. Prem, ed.): 71–92. Acta Mesoamericana 7. Verlag von Flemming, Möckmühl, Germany.

Turner, Billie Lee, II
1990 The Rise and Fall of Population and Agriculture in the Central Maya Lowlands: 300 BC to Present. In *Hunger in History: Food Shortage, Poverty, and Deprivation* (Lucile F. Newman, ed.): 178–211. Basil Blackwell, Oxford.

United Nations Environment Programme
2002 *Global Environment Outlook 3: Past, Present and Future Perspectives.* Earthscan, London.

Valdés, Juan Antonio, and Federico Fahsen
2004 Disaster in Sight: The Terminal Classic at Tikal and Uaxactun. In *The Terminal Classic in the Maya Lowlands: Collapse, Transition, and Transformation* (Arthur A. Demarest, Prudence M. Rice, and Don S. Rice, eds.): 140–161. University Press of Colorado, Boulder.

Webster, David L.
1979 *Cuca, Chacchob, Dzonot Aké: Three Walled Northern Maya Centers.* Occasional Papers 11. Department of Anthropology, Pennsylvania State University, University Park.

2002 *The Fall of the Ancient Maya: Solving the Mystery of the Maya Collapse.* Thames and Hudson, London.

Webster, David L., and AnnCorinne Freter
1990 Settlement History and the Classic Collapse at Copán: A Redefined Chronological Perspective. *Latin American Antiquity* 1: 66–85.

Webster, James W., George A. Brook, L. Bruce Railsback, Cheng Hai, R. Lawrence Edwards, Clark Alexander, and Phillip P. Roeder
2007 Stalagmite Evidence from Belize Indicating Significant Droughts at the Time of the Preclassic Abandonment, the Maya Hiatus, and the Classic Maya Collapse. *Palaeogeography, Palaeoclimatology, Palaeoecology* 280: 1–17.

Weiss, Harvey, and Raymond S. Bradley
2001 What Drives Societal Collapse? *Science* 291: 609–610.

Weiss-Krejci, Estella, and Thomas Sabbas
2002 The Potential Role of Small Depressions as Water Storage Features in the Central Maya Lowlands. *Latin American Antiquity* 13 (3): 343–357.

Wilk, Richard R.
1985 The Ancient Maya and the Political Present. *Journal of Anthropological Research* 41: 307–326.

Willey, Gordon R., and Dmitri B. Shimkin
1973 The Maya Collapse: A Summary View. In *The Classic Maya Collapse* (T. Patrick Culbert, ed.): 457–502. University of New Mexico Press, Albuquerque.

Wren, Linnea H., and Peter Schmidt
1991 Elite Interaction during the Terminal Classic Period: New Evidence from Chichen Itza. In *Classic Maya Political History: Hieroglyphic and Archaeological Evidence* (T. Patrick Culbert, ed.): 199–225. Cambridge University Press, Cambridge.

Wright, Lori E.
1997 Biological Perspectives on the Collapse of the Pasión Maya. *Ancient Mesoamerica* 8 (2): 267–273.

Yaeger, Jason
2003 Internal Complexity, Household Strategies of Affiliation, and the Changing Organization of Small Communities in the Upper Belize River Valley. In *Perspectives on Ancient Maya Rural Complexity* (Gyles Iannone and Samuel V. Connell, eds.): 43–58. Monograph 49. Cotsen Institute of Archaeology, University of California, Los Angeles.

AND THE WATERS TOOK THEM:
CATASTROPHIC FLOODING AND CIVILIZATION
ON THE MEXICAN GULF COAST

S. Jeffrey K. Wilkerson

In October of 1999, Mexico's Gulf Coast experienced a disaster of truly staggering proportions. The calamity began with Tropical Depression #11, a cyclone that was expected to develop into a hurricane. Many such fall tempests had formed over the years in the Gulf of Mexico and this one did not initially have overt features that were considered cause for unusual alarm. What occurred in the next few days, however, went well beyond the experience of even the very oldest inhabitants of the affected regions of Veracruz and the adjoining states of Puebla and Tabasco (figs. 1 and 2). Those in the direct path of the storm, along the 150 km long segment of coast drained by the Tecolutla, Nautla, Cazones, and Tuxpan rivers, perceived its destructive impact as being of biblical proportions (fig. 3). But the legacy transcended the maelstrom of natural destruction, entailing also the ample demographic and social alteration that followed. To the bewildered survivors the prolonged aftermath with its profoundly disruptive social stress was simply beyond imagination.

Although the region is a tropical hurricane coast with frequent impacts, the scope, flooding, and devastation of this particular storm greatly exceeded any in living memory, surpassing any similar recorded event in the region since the sixteenth century. Hundreds were killed, entire communities destroyed, thousands left homeless, and the natural habitats of a very large geographical area were greatly degraded or profoundly altered. This tropical depression caused more deaths than all the hurricanes of the 1999 season combined.[1] In the days and weeks that followed, particularly near the central axis of the catastrophe along the Tecolutla River, many stunned survivors openly interpreted the disaster as apocalyptic. There was also astonishment and incredulity at the still-developing repercussions. Paralyzing shock was both individual and collective. Three years later, in spite of considerable official effort, the region remained greatly changed, partially depopulated, and with no real expectation of a full revival in the immediate future.[2]

---

1 From review of monthly and seasonal summaries of the National Weather Service, June–November, 1999.

2 This inauspicious panorama remains unchanged in the fall of 2007 as this study goes to press; it is highly likely to persist for years to come. The unsettling aftermath of such disasters is enduring.

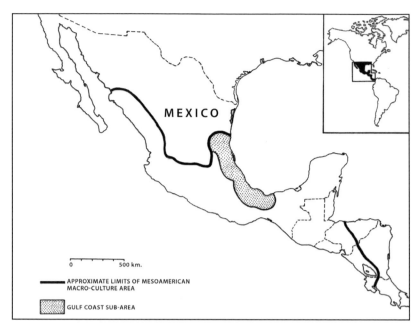

FIG. 1 Location of the Gulf Coast culture area in relation to ancient Mesoamerica and modern Mexico

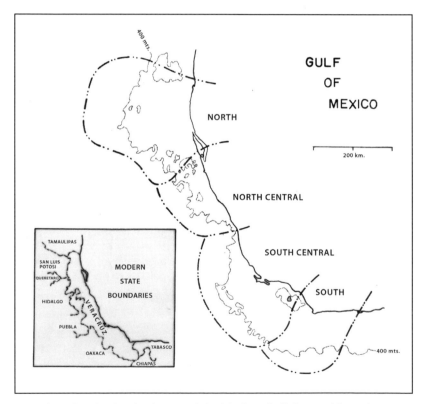

FIG. 2 Natural and cultural subareas of the Mexican Gulf Coast with modern state boundaries shown in insert

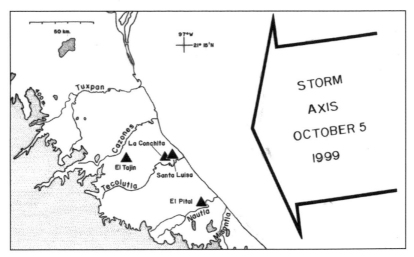

FIG. 3 The North-Central Gulf area showing archaeological sites and rivers mentioned in the text. The catastrophic storm of 1999 exceeded the size of the entire north central subarea and extended far into adjacent districts.

What occurred on the western shores of the lower Gulf of Mexico in that Niña year offers many lessons for interpreting the past, as well as for planning for the future. The parameters of both natural and cultural processes—in spite of the illusion of insulation derived from contemporary technologies and the enveloping arrogance of modernity—are essentially the same today as in antiquity. The disastrous 1999 cyclone and its shattering consequences were not attributable to an isolated climatic peak. The implications for both natural environmental fluctuations and the course of civilization are extensive indeed. It is instructive for the time-space consideration of the Gulf of Mexico, as well as civilization in eastern Mesoamerica, to follow the sequence of events, prior, during, and after this particular storm.

First, however, a brief overview of the topography and culture history of the Gulf Coast is important. Four natural districts along the western margins of the Gulf correspond to cultural subareas of great antiquity (Wilkerson 1972, 1974, 1987, 1994a, n.d.a). Each of these scallop-shaped zones is largely ringed by a series of mountain ranges. In most instances the coastal plain is narrow, marshy, and transected by rivers that descend from the high Sierra Madre Oriental to the west. Here the altitudes mostly range from 2,400 to over 5,700 m. From the Gulf lowlands to the central plateau of Mexico, and to even higher elevations, there is an astonishingly wide range of exploitable resources from distinct microenvironments.

Each of these districts has different annual rainfall averages and soil types. Excessive rainfall in any of the units has the potential of considerable impact throughout the entire natural-cultural area due to the great disparity of elevation and the limited number of rivers that drain each region. The runoff from such steep terrain, especially when deforestation is extensive, is extremely rapid. The rivers of the north-central area have the most runoff per hectare of catchment of any along the entire Gulf Coast (National Oceanic and Atmospheric Administration [NOAA] 1985; Wilkerson 2005).

All four regions have long cultural chronologies that led to successive apogee points over time. The flow of florescence was largely from south to north (Wilkerson 1981, 1987). The South Gulf area was the initial focus of the Olmecs at the very end of the Early Formative Period (ca. 2400–1000 BC), and, with the lower reaches of the South-Central area, was a region of Olmec polities in the Middle Formative Period (ca. 1000–300 BC). In the Late Formative Period (ca. 300 BC–AD 100), the South-Central area appears to have been increasingly dynamic, and by the Protoclassic Period (ca. AD 100–300) and most probably from the preceding Late Formative, there were large urban centers in the North-Central area. Both central areas may have had important seaborne migrations in these periods (Wilkerson 1994b, 1997b).

The North-Central area was also a major focus of complex urbanism during the ensuing Classic (ca. AD 300–900) and local Epi-Classic periods (AD 900–1100). Destructive migrations ended the florescence here. The North Gulf area, following some truncated localized apogees in the Protoclassic and Early Classic (ca. AD 300–600) periods, reached its zenith in the Late Postclassic Period (ca. AD 900–1520). Prior to the Aztec conquests of the fifteenth century, this region impacted neighoring peoples to the south and west. At the very end of that period, large urban centers were also re-forming in both the North-Central and South-Central areas. There were sudden changes and hiatuses in all these areas over time. Some of these discontinuities are very likely related to disruptive weather phenomena.[3]

THE 1999 STORM

September, normally the rainiest time in northern Veracruz, was unusually dry in 1999. Uncharacteristically, the first really heavy rains came only at the end of the month, starting on the afternoon of the 28th and continuing through the morning of October 1. This was the result of a front that had moved southward from the central United States into the Gulf. Nearly 400 mm of rain fell, more than most years see for the entire month. It left the ground totally saturated and the rivers of the region near flood stage. Many low terrace zones and tributaries actually did flood. There followed a little over two days of gradually clearing weather. By the evening of October 3, however, the skies were once more cloudy and the air humid from the gathering tropical depression over the Gulf.

The destructive storm ultimately derived from an extensive low-pressure system in the western Caribbean on September 30 that, in turn, had come from a tropical wave (front) that had left the coast of Africa eight days before (Beven 1999). The now-diffuse system slowly traversed the Yucatan Peninsula to the southern portion of the Gulf of Mexico, known as the Bay of Campeche. There it became slightly more organized and formed a tropical depression on October 4. Its trajectory, however, was very unpredictable and confined to the southern Gulf (fig. 4).

3 Summaries of these areas with a focus on their successive florescence can be found in Wilkerson 1981, 1987, 1994, and 2001a, b, c.

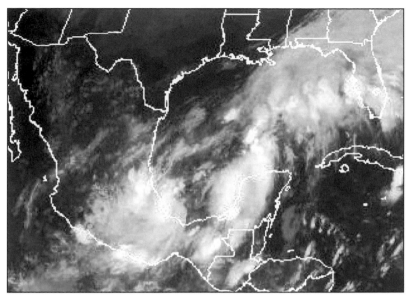

FIG. 4 Tropical Depression #11 going ashore on October 5, 1999. In this view, taken from an infrared satellite image (courtesy of NOAA), note how the storm essentially covers the entire Veracruz coast and fills the southern Gulf of Mexico (Bay of Campeche).

Heavy rains began on the upper Veracruz coast that same day, and we registered another 300 mm at our research center on the Tecolutla River by the following morning. By the afternoon of October 5, we were receiving an average of 41 mm per hour. With the loss of instrumentation and records, it is difficult to be precise concerning the total rainfall since the last days of September, but it was almost certainly between 1000 and 1300 mm. Some other parts of the lower Tecolutla River valley, however, appear to have received more than 1500 mm.[4] Such quantities are equivalent to average *annual* precipitation.

The Tecolutla River as well as the flanking Tuxpan, Cazones, and Nautla rivers were now carrying extraordinary quantities of water, and all began to flood. The Tecolutla River returned to its ancient course from the beginning of the Holocene, flowing northward rather than eastward to the sea. It would open five new distributary channels, including a huge one at least 800+ m wide and larger than the original pre-flood river mouth. The field research center and botanical garden of the Institute for Cultural Ecology of the Tropics were directly in the path of the new flow of water.

INUNDATION OF THE TECOLUTLA VALLEY

The Tecolutla, a broad, generally tranquil river, is normally the ninth largest waterway by volume in Mexico. What happened next in its delta can be illustrated by what occurred at the institute facility. Over a period of nearly thirty years the center had been built with a specific eye to occasional flooding. All new buildings had heavily reinforced foundations and heights

4 Data derived from interviews at ranches in the area that compiled rainfall data.

calculated to withstand an inundation fifty percent greater than that demonstrated by the huge floods of 1888 and 1955.[5] Both of these floods occurred in conjunction with late hurricanes in strong Niña years.

Situated on the highest point of the delta, the research station had for decades provided rain and climatic data to the regional radio stations and government authorities. By the afternoon of Tuesday, October 5, rainfall was simply torrential. Electricity ceased early in the afternoon, followed soon thereafter by the telephone system and the radio stations. There was little wind, just rain falling in long continuous sheets punctuated by occasional lapses of silence in which it was possible to see through the mist to the dark gray sky of low clouds that appeared not to be moving. The river, brown with mountain silt from the previous days of rain, had increased considerably in speed but remained largely at the edge of its banks until late in the day.

By mid-afternoon it was obvious that there would be some flooding, but the magnitude was totally unclear. In fact, the irregular breaks in the rainfall suggested to many that the storm might even stop. Nevertheless, there was never any clearing of the sky and the entire day was spent in virtual penumbra. Precise television or radio coverage was lacking, in part because there was little accurate data available, no centralized reporting system, and sparse experience at projecting the consequences. Most unfortunately, while the radio stations were still on the air, they had given no specific notice about the extraordinary degree of flooding immediately upstream.

At the Remolino canyon, some 33 km upriver from the research center, the turbid river was rising to unprecedented levels. The annual average highest volume for nonflood years is 2,739 m³ per second. The greatest previous flow ever measured, during a flood in 1981, was 8,300 m³ of water per second. On this day it surpassed 20,000 m³ per second before instrumentation was lost and the station abandoned.[6] The rushing water rose at least 20 m above the mean river level, covering the accesses to a large suspension bridge and threatening the span itself.

The inconceivable sight terrified people living on the high bluffs over the canyon. Many attempted to flee on foot across the tops of the steep hills that were now collapsing with gigantic mudslides. Inevitably, such an incredible volume of water would have a cataclysmic effect downstream. However, there existed no studies of crest projection, no on-site communication system, and no established procedures for emergencies at the monitoring station.[7]

5 These were the two greatest floods in living memory in north-central Veracruz prior to the 1999 event.

6 These figures are from the Comisión Federal de Electricidad station at Remolino, Veracruz, where data have been kept since 1961 (Comisión Federal de Electricidad 1961–99). Following a sudden peak at 7,491 m³ per second during the 1974 storm, the application of three frequency models (Nash, Lebediev, Gumbel) to the drainage conditions of the Tecolutla River suggested that 6,800–8,500 m³ per second flows would occur only once every 20 years, and that 14,000–17,900 m flows would happen every 10,000 years. All models were widely wrong on both counts.

7 Nine years after the event, this situation had not changed.

At this point in time, well back from the coast, the river and its tributaries were gorged beyond capacity by the extreme runoff from the largely deforested slopes of the Sierra Madre Oriental, the piedmont, and the hilly coastal plain. Decades of indiscriminate clearing of even near-vertical inclines had made rapid runoff inevitable. A massive amount of water was advancing rapidly downstream in the growing darkness, racing through the many riverbank communities and carrying all before it. The magnitude of the deluge bearing down on the already partially submerged river delta was colossal.

Personal Perception

At the research center, the author and a skeleton crew of five, including two students, was rapidly moving through the center's standard flood plan. Scientific materials and specimens were placed in more secure locations, objects that might float away were tied down, emergency supplies were packed, and vehicles were moved to higher ground. These efforts took on further urgency when the water, just moments before nightfall, began to enter the surrounding botanical garden at a rate far faster than in previous floods. After a few minutes of reprieve, it began to rain hard once again. A quick reconnaissance of the raised, causeway-like highway showed that the only road out was already largely underwater between high ground and us.[8] We were cut off.

Soon the water was flowing swiftly out from the river, carrying flotsam of all kinds. The staff, as well as three dogs, were now concentrated at the new laboratory building, the highest structure at the center. As the moving water increased in speed, it began a shrieking sound that was to grow notably louder as the floodwaters continued to augment in velocity. This high-pitched howl was in part due to the reappearance of the wind, which seemed to be mostly at water level. But the water itself was the principal source of the noise, for it contained great quantities of sand, earth, and wreckage that were being projected against the center's buildings. There were constant grinding sounds as well as the gathering roar of rushing water from upstream.

Large segments of brick and cement sidewalks from around the building and from walkways in the botanical garden were torn from the ground, lifted, and then hurled against the outer walls of the laboratory. Soon we heard water hitting the top of the roof at intervals as if it were especially hard sheets of rain. Only later did we realize that the 35 m portion of the botanical garden separating the lab from the river was now totally gone and the loud reverberation on the roof was from waves breaking over it. This meant that the tops of the crests were at least 10 m above the normal river level. When water reached waist height inside the structure, it was time to depart. As we did so, a glance at a large reinforced picture window gave the impression that we were in a descending submarine.

---

8 At this juncture, some people did try to move along the highway by the research station, some taking refuge by the botanical garden wall. All disappeared in the torrent. Their bodies were washed to sea and some were found a week or so later on the beach at the base of the Gulf of Mexico, some three hundred miles to the south.

Everyone moved to a fiberglass boat that we had tied to the door on the lee side of the building. It was like attempting to leave a fast-sinking ship. Life vests were handed out, strapped on, and checked. As we assembled in the boat with supplies, cell phone, and camera equipment, it was possible to illuminate with our flashlights the debris that rushed by the two sides of the building. It included trees, large parts of houses, plastics of all types, and the bodies of both drowned people and farm animals. We then got into the craft and set out in total darkness with the intention of reaching the broad roof of the oldest building at the center, which was protected in part from the torrent by the thick vegetation of the surrounding botanical garden. We never made it.

Working our way slowly through the botanical garden, from submerged treetop to treetop, we found that the water was not rising steadily. Rather it spurted upward in increments, sometimes as high as 10 or even 20 cm at a time. At moments it would drop back a few centimeters only to rise again quickly. There were also ferocious, stream-like currents that ran within the strong movement of the flood. Often these ran around a building or clump of tees and violently converged in a great churning movement. At one of these points we were caught in extreme turbulence and capsized.

We lost all our supplies and equipment except for a hand light. The group was violently washed to a tree line at the edge of the botanical garden, except for one high school student. With the light we found that she had been pushed in a different direction up against another tree some 15 m away. There she remained clutching a dog in mute fear. Attempts by all three men in the party to swim the frothing current to reach the girl, now in growing shock, were fruitless. At this point the dog, the station's golden retriever mascot, oriented to our direction and began to swim with the girl holding on to her collar. She actually managed at an extraordinary expenditure of energy to cross the torrent and reach us with the stunned student. It was truly an incredible display of selfless valor and strength. Then, totally exhausted, she was swept away in the direction of the sea.[9]

Shortly afterward there was another sudden rise in the water and we were forcefully picked up and washed toward the raised highway. The surge slammed us into the stones of the disintegrating road. There we climbed on top of one of our own vehicles, a pickup truck that had been swept along. But the respite was short, for the water again rose with a jump, and we were propelled into the deep water on the other side of the causeway. The vehicle soon sank beneath us, leaving us being pushed along in the dark with the flow.

9 Hers is a tale of true courage under duress, interspecies bonding in adversity, and flood-borne disease. Olwin was ultimately swept over five km to the sea and then out into the Gulf of Mexico. She swam to shore, traveled up the beach transected by new river mouths, and eventually crossed a huge inundated swamp. Much battered, after six days she reached the ruins of the research center. Eight months after this exceptional feat of endurance she died of *coccidioidomycosis*, which affects both humans and animals. The causal fungal spores had been carried in silt from the highlands via the floodwaters. The high school senior she saved is now married with children. Olwin's niece, Arwin (ZZ), today guards the research station. The flood context, as well as the valorous actions and untimely fate of Olwin, are described in Wilkerson (2000).

We linked arms and eventually I was able to grab a limb of a largely submerged tree. Climbing up we found a dog and snake already there. We held on to the limbs and, as the water rose, climbed higher into the canopy until the branches started to break. As the night wore on, the water reached us and the tree shook violently with the current. The wind and rain picked up. Bodies, broken wood from houses, and uprooted trees bumped into our perch. At daylight the water dropped back slightly but rose again by mid-morning. This was the runoff water from the high mountains beginning to reach the coast.

The rain slackened and we tried to visually locate the nearest town some 4 km away but were puzzled because we could not see even the highest six-story building. Rain and water vapor were not the only problem. Later, however, we found out that waves as high as 11.5 m were obscuring our view upriver. The wind picked up again, chilling us. We could see portions of the research center that had not been carried off the first night being destroyed by the new onslaught. During the second night, it appeared that the rising water would reach us again, but just before daylight it began to recede and by mid-day we could cautiously climb down.

We found mud, glass, and debris everywhere. We also then realized that the current had ripped most of our clothes and shoes from us. The highway was mostly gone where the current had been strongest. Mud banks covered the rest. Huge tree trunks, carried great distances downriver until their roots caught on obstructions, littered the remains of the roadway. Huge piles of debris, some up to 10 m high, were piled against fence lines and the remnants of citrus groves.

 It was slow going picking our way over and around obstacles. When we finally reached the town, it was strangely quiet. People stood in a stupor on the mud bars talking to themselves. Others, speechless, stared blankly at the damage. A few spoke in whispers. But the paralyzing astonishment at the fate of the town and region was generalized. Food was at a premium. Limping down the street, our life vests still on, we apparently attracted some attention. A longtime friend asked us in and prepared us food from his second-floor larder. Although we had not had a meal in over forty-eight hours, we simply could not eat. But all of us enjoyed a few swallows of coffee. In fact, it would be several days before we could stomach more than a token amount of nourishment.

Survival in a catastrophe is often a matter of impulsive decisions taken in an instant. Some call it luck, others say following the plan made it happen. Still others evoke supernatural protection or guidance. In such episodes some perform below their demonstrated capacity in unstressed times, while others perform far above what they could have imagined for themselves. The realization that immediate actions have life or death consequences can traumatize, shock, petrify, or expedite decision-making. While I can recount what my crew and I did to survive, it cannot be precisely defined why we survived while hundreds of people around us died.

AFTERMATH

The following months of looting, economic stagnation, inflation, and aid efforts, as well as out-migration, were a succinct experience in the modern, but also timeless, difficulties of dealing with a massive natural disaster in the tropics.

In the three days of primary flooding, the field research center and botanical garden of the Institute for Cultural Ecology of the Tropics were utterly devastated. Some structures disappeared entirely. Reinforced masonry buildings were razed, including the deep stone foundations. Large fragments of edifices, in some cases weighing many tons, were carried up to a half a kilometer. Lighter debris was encountered on the beach some 10 km distant.

Most of the vegetation was gone, and immense silt and sand deposits up to 2 m deep covered much of the delta and river terraces. Immense piles of wreckage and flotsam, including parts of wooden houses, were stacked against groups of trees that had survived. Carcasses of thousands of drowned cattle, horses, hogs, domestic fowl, and pets, as well as bodies of people, were strewn about the terrain, generally where they had been snagged by trees or caught in the receding water. All wells for drinking water in the region were fouled.

The weather remained threatening for nearly ten days, with the river only barely constrained at the top of its banks. Great expanses of land were still covered in water or viscous quick sands. The entire region was without communication, road access, potable water, provisions, gasoline, or cooking gas. Shops were closed or eventually opened with exorbitant prices. Food was at a premium and often stolen. We were forced to bury most of our salvaged supplies.

People were collectively dazed and, in a few cases, individually crazed. Quite a few walked about in the streets staring at the wreckage in speechless shock. Some were barely clothed. A few individuals had heart attacks or strokes. Others stood and, to any that would listen, gave their angry interpretations of why this occurred as a result of negligence with the highland dams on the river. Later, once roads were partially cleared, people drove at breakneck speeds without regard for normal traffic procedures, and major and minor accidents were daily occurrences.

Within twenty-four hours looting started. Shots were fired. The army placed a curfew. The focus of pillaging soon moved out of the towns to the surrounding communities, ranches, and even the research center. Gangs, increasingly organized and in some cases including women, fanned out into the disaster zone. The plundering occurred initially during the day. Then it became chiefly a nocturnal pursuit that left people cowering in the surviving ranch houses when the bands tried repeatedly to knock down the barred doors. Armed guards became necessary. The looters were rarely the flood survivors or refugees, but rather people of all ages who lived on higher ground and had suffered very little damage from the storm. Looting was to continue for over a year, and an aura of lawlessness has continued where there are less frequent police patrols.[10]

10 The nature of this persisting violence, relaxation of social norms, and bizarre behavior is examined in a number of field summaries (Wilkerson 1999–2003). Guards at the research station were necessary for five years.

Out-migration began at once. People at first just walked out carrying a few possessions on their back. As soon as bridges were reconstructed, roadways cleared, and the highways reopened, small trucks were loaded with salvaged household items and more families left. Most of those observed to leave initially were from the ranches, hamlets, and villages surrounding the towns. However, many also left from the larger communities. In the towns of the disaster zone, and those on the fringes of the north-central area, many people left in short order to search for jobs in the United States.[11] Most of these people have never returned and movement out of the region continues. Today some bigger towns have a similar, if somewhat reduced, population than they had before the flood, but this is due to the arrival of the survivors from entire rural communities. Long lines form at the telegraph offices following paydays in the United States as emigrants send back earnings for family members still in the disaster zone.

Farms and ranches were gravely affected by the climatic event. Farm animals that survived the flooding often died afterward from infections and disease. Looters and cattle rustlers wantonly slaughtered hundreds of animals in the first weeks after the flood. Three years after the event, many ranches were still without cattle or crops. Quite a few were unable even to replace the miles of barbwire fences required and, lacking bank credits, had no possibility of replacing animals or infrastructure. Numerous ranches were for sale with no takers.

Following the flood, surviving animals (usually young adults) adapted unusual behavior patterns. Snakes, carried great distances by the waters, were disoriented and aggressive. Coyotes, felines, otters, coatimundi, anteaters, and others changed their habits. Many roamed by day rather than by night. In some cases they approached damaged houses in groups searching for food. To these wild animals were added numerous domesticated dogs and cats that had been carried by the flood or were abandoned by their fleeing owners afterward. Except for vultures, birds were totally absent at first.

The swollen river disgorged for months a gigantic plume of sediment-laden fresh water into the Gulf. The sea essentially became the depository of everything, ancient or modern, natural or man-made that was carried off by the storm. This powerful expulsion extended for many kilometers beyond the normal discharge zone, raising and altering the seabed of a wide sector of the continental shelf. With massive modifications of the sea bottom, salinity, and visibility, the marine ecosystems of the entire region were extensively disrupted. Commercial and sport fishing dropped to almost zero yields. Large migrating fish, such as tarpon and some sharks, appeared to bypass the impacted zone. Shellfish populations in the estuaries, inundated with silt-laden fresh water, also radically plummeted. Three years after the event, the marine, estuarine, and riverine environments, with their respective flora and fauna, still had not recovered even a fraction of their preflood diversity or abundance.

11 Border authorities (anonymous personal communications) in south Texas, the closest part of the United States, indicated that they began to note an increase in illegal crossings by people from Veracruz as soon as 72 hours after the climatic event ended.

Most of the coastal plain terrain is composed of lateritic clays, which are subject to compression despite being heavy soils. The earth was compacted considerably further by the immense quantity of water that moved over it for days with more weight than a steamroller. In many cases the land reached the consistency of a cement sidewalk, in others it was covered by vast amounts of waterborne sand and clay. Three years after the event, there had been no successful crops harvested in major fields since the flood. The soil hardness is such that root structures of many domestic plants are altered and growth stunted.

All preexisting crops were destroyed, including plantations of citrus and other fruiting trees. Much of the river delta looked like a lunarscape. The region remained without vegetative growth for nearly five months after the flood. The waters had also brought in spoors, seeds, weeds, and exotic plants. Only the most hardy of grasses took hold. When some attempts at planting were later attempted, it was found that moisture, mold, and insect plagues gravely affected crops. Corn, beans, squash, and bananas rotted from water perched in silt deposited above the original and now compressed soils, or simply never matured before dying. Three years later, some ranches had contracted bulldozers to remove the recent flood deposits, only to find the highly compressed soil beneath.

Health was a major issue following the disaster. As soon as the sun appeared about two weeks later, the skies filled with literally thousands of vultures. Most of these carrion birds were on their fall migration southward. Nevertheless, they found thousands of carcasses on which to feed and remained in the region for many weeks. So numerous were they that helicopters and small planes could not land in the river delta area. As the remains decomposed, virtual swarms of aggressive flies appeared and lingered for two months or so. The quantity of insects exceeded any such plague in regional memory.

With all wells contaminated, the initial water crisis was momentarily resolved by trucking in bottled water after the roads were opened. Wells had to be cleaned by hand and treated with chlorine tablets. Epidemics were averted only by virtually blanketing the area with medical teams for over two months. Still, there were outbreaks of many unusual afflictions, including many skin ailments from the water and silt. Additional illnesses were caused by the new spoors and pollen carried into the region by the flood.

After a month much of the surface moisture evaporated and the flood-borne silt turned to fine powder. The choking clouds of dust were so thick and pervasive that people used facemasks or wrapped their heads in cloths. The fog of clay particles lasted for over two months, gradually tapering off over the following three months. Only after vegetative cover began to grow and the roads were repaired did the dust diminish.

Local costs in the region soared immediately—inflation was instantaneous and enduring. Not only did it persist beyond the initial emergency, it continued in the disaster region for three years at a rate well beyond the national average. This factor, coupled with out-migration and the destruction of the rural infrastructure, contributed greatly to a persisting economic stagnation throughout the disaster zone. Another factor was local people's lack of interest in regional employment as long as federal and state aid programs were functioning. This actually led to the necessity of bringing in workers from as far as 250 km away for many reconstruction and refugee housing projects. The same situation persists today, largely because most prospective workers have gone to the United States or major Mexican cities. Their relatives are subsisting on funds sent to them and, more often than not, feel no necessity to seek local employment.

ARCHAEOLOGICAL CONNOTATIONS

Salvage operations, lasting over a year and utilizing some archaeological controls, were initiated at the ruins of the field station to remove up to 2 m of silt and recover scientific materials. The hand removal of over 230 dump-truck loads of earth resulted in the recovery of unexpected objects and also the detection of modern artifact distributions that directly paralleled examples the author had previously encountered in excavations of ancient riverine sites. In subsequent months these particular circumstances were checked out at other locations in the deltas and found to recur consistently.

The research center is situated on an edge of the vast Santa Luisa site. The salvage excavations showed that as the river had eaten away some 30 to 40 m of the terrace down to a depth of 6 to 10 m, it had lifted all artifacts, ancient and modern, large and small, that had been in that zone up onto the surviving terrace and carried them as far as several kilometers back from the river. That is, they were propelled along and dropped out of the current when it slowed down or when they hit trees and other obstacles. Thus, objects from the modern laboratory, glass from late nineteenth-century containers, Classic-Period ceramics, and Early Formative–Period stone grinders were often repositioned together.

Such mixed deposits are common in archaeological contexts along the Gulf Coast. They can sometimes be used to determine the riverward or upstream presence of older site components. In a number of cases, once the objects had fallen out of the strong driving-surface, or immediately subsurface, currents, they moved along the bottom in narrow, streamlike flows. Exactly such distributions are readily visible in archaeological contexts.

Soil deposition is extremely important. Because the force of the water was very strong in the 1999 inundation, peaked triangular banks of sand and silt were formed. These features sometimes contained large-grained sand, fine-grained sand, and clay or silt. Their appearance and configuration was such that had they appeared in an archaeological trench the tendency would be to associate the various components with distinct events. Yet here the deposition diverged because, during the flood, the subsurface currents pulsated and changed the direction and distance from which they brought the sand, clay, or silt. With this evidence it is possible to reinterpret archaeological profiles along the Gulf Coast and much more easily pinpoint the occurrences of megafloods.

Soil hardness was also highly variable. Preexisting levels that were compacted by the immense weight of water were much harder than before the flood, at times extremely so. Also, the deepest new deposits were laid down where the current tended to slack for various reasons, not where it had been strongest. Objects in these distinct layers fared differently. Perishable objects, including paper and photographs, were best preserved where they were immediately covered by silt. Paper without kaolin coatings survived the best. Wood items for the most part floated off.

There are still other lessons to be learned from this flood. Similar deposits have been found in many archaeological probes of the Gulf Coast, indicating major cycles that transcend a single climatic event or even a related series of inundations.

IMPLICATIONS FOR INTERPRETING CULTURE HISTORY

The ongoing study of the most recent cataclysmic event in historical perspective—from the standpoint of direct experience, salvage excavations, and intensive study of the region over four decades—reinforces the author's earlier interpretation of a long, cyclic sequence of megafloods with direct influence upon the course of cultural evolution in eastern Mesoamerica (Wilkerson 1994a, 1994b, 1999, 2001b). It also suggests that such a flood, however calamitous in the sense of loss of human life and destruction of physical infrastructure, is less socially and culturally threatening than the aftermath.

Survivors are subject to profound social disruption, potentially prolonged undernourishment, unfamiliar diseases, erratic conduct, and the impossibility of immediately resuming normal intra- or interregional activities. Formerly productive agricultural fields are often rendered infertile by numerous factors, including landslides in hill zones, extensive deposition of sands in alluvial areas, extreme soil hardening, waterborne introduction of exotic plants and spoors, as well as insect plagues and mold. Out-migration becomes an almost immediate reaction, further debilitating the recovery potential of the region.

Unquestionably, such a sudden calamity would result in truly appalling casualties. The official figures for the 1999 storm described above vary from 400 to 1100.[12] More than four hundred people, most of whom almost certainly would have drowned, were saved by helicopters on the second day of the flood. In ancient times the river deltas and alluvial plains were the primary zones of high population. An event of this nature, greatly exceeding human memory, could be expected to cause substantial, perhaps horrendous, fatalities.

A storm of this magnitude would also have swift and profound effects upon the foundations of any ancient society. As most extensive and intensive agriculture was situated on the alluvial terraces of the rivers, there would inevitably be a direct and immediate impact upon the subsistence base. The massive silting of the raised fields and canals would render inoperable the

12 Federal and state government officials publicly provided the former figure for the entire coast and local authorities, anonymously, provided the latter just for the Tecolutla River catchment alone. Official estimates are widely thought to reflect election-year concerns and commonly accepted popular estimates in the disaster zone are many times these totals. Precision seems impossible, except to say fatalities were unquestionably very high and greatly exceeded those from any previous cyclone on this coast in modern times.

primary intensive food-production zones. Also impacted would be the cash crops like cotton grown in similar fields and cacao plantations. Both food and commerce, two mainstays of society, would suffer immediately.

Stored grains and foodstuffs that somehow escaped the water would soon be subject to mold and rot, as well as plagues of both rodents and insects. As these storms tend to occur in the early fall, it is normally too late to replant for the year's second growing season (*temporal* planting: late June–early December). Additionally, the land would be highly moist and the ground unusually hard.[13] It would be unlikely that any substantial crop could be planted until the winter for harvest in late spring (*tonamil* planting), some eight or nine months after the event. Even then the yield would likely be noticeably below normal.

Fruit trees would lose their fruit. Fishing as well as the collection of shell-fish would be largely unproductive. Available animal protein would also be diminished, as both dogs, the major source of domestic animal protein, and hunted wild animals, such as deer and *tepeizcuintle*, would also have been greatly diminished. Famine for the surviving populations would be extremely probable under such adverse conditions.

Out-migration could be expected to be a prompt response to severe food distress. As this proceeds, the region loses its labor pool and the ability to repair and respond rapidly to the disaster. Social control would tend to be stressed to the point of collapse. Additionally, the demise of the polity or polities in the affected area opens the possibility for the expansion of neighboring, less affected groups. Discontinuities in the archaeological record can be expected under such circumstances and, in fact, are detectible.

REGIONAL DIMENSIONS

Review of the 1999 storm season suggests that lesser flooding, both preceding and following the event, heavily damaged key adjoining coastal areas (fig. 5). This situation in antiquity, in conjunction with the regional megaflood, would have destabilized societies along a considerable expanse of coastal Mexico. This pulsating effect of storm cycles is due in part to the topography of the Gulf area, where a single storm can exceed in size any one natural-cultural area, and to the Niña-year tendency to retain storm systems within the Gulf of Mexico. Under such circumstances, the magnitude of the flooding in the impact zone or adjoining regions has much to do with the meander or stationary time of the storm prior to dissipation over land or rapid movement away from the southern Gulf.[14]

For instance, in 1999 the center of the storm's impact was the north-central area but considerable rains stretched northward into the lower northern Gulf

13 Unlike the European use of plows that open soils, the Mesoamerican preference for dibble sticks compacts soil around the precise spot to be planted. Such implements are not well suited to loosening highly compressed soils.

14 Hurricane Mitch in Central America the year before the 1999 storm in the Gulf of Mexico was made all the more devastating by the semistationary nature of the cyclone near the coast and later over the coastal plain. Strength descriptions can be found in Tassara and Grando (1999).

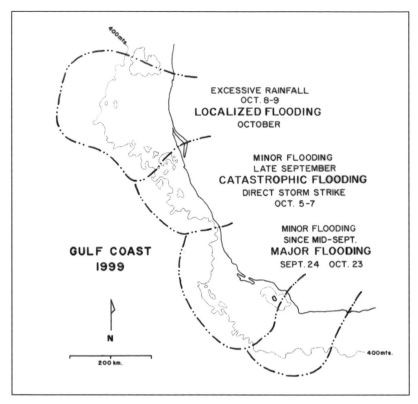

FIG. 5 Distribution of successive flooding along the Gulf
Coast in the months of September and October 1999

area. Additionally, there were heavy rains from first Tropical Depression #11, and then a storm two weeks afterward in both the south-central and southern Gulf areas. There was ample flooding, although not anywhere as violent as in the north-central area where Tropical Depression #11 impacted. Again there was very pervasive crop loss and widespread displacement of people. To deal with this additional danger, substantial equipment and manpower utilized in the aid and reestablishment of services in the northern reaches of the Gulf were transferred to the south. It was readily apparent that the resources of even a modern nation-state were considerably taxed to deal with such a situation over a broad region.

In ancient times such drastic circumstances would have weakened many polities along an enormous portion of the Gulf Coast, in addition to the severe, and probably crushing, effect upon the societies in the core impact zone. It is fully possible that in the acute cases numerous peoples, with distinct cultural traditions, were set in motion and that populations once more coalesced where resources were more accessible or were available for the taking.[15] No Mesoamerican area was too isolated or sufficiently secure to be immune to marauding or migrating groups.

15 This would certainly be a significant contributory process to the formation of multiethnic polities that are likely to have been common on the Gulf Coast since early times (Wilkerson 1994a, 1994b, 1997b).

FREQUENCY AND INTENSITY DYNAMICS

Another significant regional aspect of these megafloods is the grouping of severe events. The archaeological data from the north-central Gulf Coast suggests that there were often groups of major storms in the years immediately preceding or following a megaflood. This is also implied by mid-sixteenth-century historical data, from the early 1550s through the mid-1580s, and the current weather cycle in the same area. The years immediately prior to the 1999 event constitute a good example.

In early August 1990, a non-Niño year, Hurricane Diana, with a category 2 intensity, moved rapidly across the Gulf and hit the upper portion of the North-Central area. There and in the adjoining North Gulf area it caused much wind damage and regional inundation. Rivers such as the Tecolutla also rose in the south but flooding in that area was localized.

In late October 1995, a Niña year, Hurricane Roxanne entered the Gulf of Mexico and meandered for a week, unable to move northward out of the greater Bay of Campeche. It ultimately lessened to a tropical depression and was pushed back southward for a direct impact on the southern reaches of the North-Central area (Avila 1995).[16] This resulted in both the Nautla and Misantla rivers flooding to a degree not seen in hundreds of years. Nevertheless, the magnitude of the 1999 event in the same catchments was at least twice the 1995 magnitude. What prevented still greater landslides and inundations was the fact that the storm had retrogressed rapidly, not permitting a prolonged accumulation of precipitation prior to impact as occurred in 1999. Nor was there a prior high rainfall peak associated with normal seasonal precipitation as occurred in 1999.

The immediate post-1999 years have similar examples. In 2005, a Niño year of devastating impacts all around the Gulf of Mexico, there were various cyclones on the upper Veracruz coast. Tropical storms Bret, Gert, and José in June, July, and August of 2005 were all rapid in movement but left localized and regional flooding, including the Tecolutla catchment. Hurricane Stan (category 1), in early October, caused some flooding in the North-Central area including the Tecolutla delta and catchment. Nevertheless, its major coastal impact was in the Southern Gulf and South-Central areas. Essentially all coastal areas of Veracruz were affected that year and there was ample crop and infrastructure damage. However, the impacts were spread out in time and space, and the level of devastion never reached the catastrophic.

The Niña year 2007 had two significant hurricanes that once more damaged the research station. Dean (category 2) in late August and Lorenzo (category 1) in late September both hit the Tecolutla River delta head on. The first, a rapidly moving hurricane, caused considerable high wind damage. The second, after lingering over the Gulf for days and then suddenly strengthening and increasing in velocity, provoked primarily water damage. In both instances, wide-scale major flooding in the North-Central area was averted

16 The Gulf of Mexico movement pattern and impact zone of this cyclone is comparable to the 1999 storm up until the final two days of retrocession, which moved it quickly into the coastal plain and mountains.

Year	Month	Preceding Niño	Niño year	Niña year
1915	Sept.	1914 (M+)	no	?
1933	Sept.	1932 (W/M)	no	yes
1944	Sept.	1943 (M+)	no	?
1955	Sept.	1953 (M+)	no	yes
1974	Sept.	1972–73 (S)	no	yes
1981	Aug.	1976 (M)	no	no
1986	Nov.	1983 (VS)	yes	no
1995	Oct.	1994	no	yes
1999	Oct.	1997–98 (VS)	no	yes

Total flood events	9	100 %
Total associated with Niña years:	5	55.6 %
Total associated with post-Niño years	2	22.2 %
Total associated with Niño years	1	11.1 %
Total not associated with Niño or Niña	1	11.1 %

TABLE 1 Twentieth-century flood years: Tecolutla and Nautla rivers, Veracruz, Mexico. Year and Niño strength references from Quinn et al (1987), Santos (1999), Caviedes (2001). [W = weak, M = medium, S = strong, VS = very strong]

by the final rapid ground velocity, which prevented prolonged precipitation and massive runoff as in 1999.

Nevertheless, with the second hurricane there was severe localized flooding as well as very substantial inundations in adjacent regions, particularly the North Gulf area. Had the conditions of this cyclone been just a little different, the potential for catastrophic damage would have been present. The current extreme storm cycle appears still active and potentially very dangerous. There is no guarantee that there is only one mega-flood per cycle peak.

The Tecolutla and Nautla rivers, the two closest rivers in the area and the epicenter of the 1999 event, have had ten major floods since 1888. Most of these inundations were in a Niña year, or in the year following a Niño (ENSO) event. These seem to fit two patterns of time intervals. From 1888 until 1974 the recurrence interval for six floods ranged from twenty-seven to eleven years (average 16.16 years). From 1974 to 1999 the recurrence interval ranged only from seven to four years (average 6.25 years). There is definitely a quickening of the flood occurrences on the Gulf Coast in the late twentieth century (table 1).

The greatest flood of the first set was 1955.[17] To date, the greatest flood of the second set is 1999, which was four to five times greater in magnitude

17 Oral history in the region suggests that the early September 1888 flood may also have been of considerable magnitude.

than that of 1955. The 1995 inundation, primarily in the Nautla and Misantla drainages, was also greater than that of 1955. There appears to be a marked increase in both intensity (up to 5x) as well as frequency (2.59x) of flooding since the mid-1970s for the upper Veracruz coast.[18]

The rough correlation with Niña years and post-Niño years must be taken with much caution. As has been observed in South America (Santos 1999), strong Niño events are not necessarily followed by strong Niña occurrences. In fact, a Niña can occur without a Niño. Flood magnitude does relate to these processes, but is unlikely to be fully regulated by them. There are additional important parameters, some influenced by human activities.

Among the factors that directly affect the intensity of flooding is land use. Chief among them is deforestation of the hilly coastal and piedmont districts, as well as the high sierra slopes of the North-Central area catchments. Deforestation considerably increases vulnerability, as has been shown in the studies of flooding and casualties from Hurricane Mitch in Central America (Winograd 2002). While this practice is acute in modern times, it also occurred in antiquity and has been strongly suggested by archaeological evidence in the Tecolutla drainage (Wilkerson 1997c, 2001b, n.d.b). Insufficient vegetation to retain or delay the runoff leads to massive erosion, widespread mudslides, greater water velocity, higher downstream crests, and the carving of new distributary channels in the deltas.

As happens today, the greater the population in antiquity, as well as the greater the degree of social centralization for large-scale agricultural projects, the more frequently land was cleared. But what would the impact of such catastrophic events be upon ancient societies?

SOCIAL IMPACT

In remote antiquity, say in the base camps, hamlets, or early villages of the Archaic Period, as well as the still-larger villages at the beginning of the Early Formative Period, the climatic devastation would overwhelm the settlements and shatter the social units. Nevertheless, their technological level, utilizing primarily local resources, was not so complex that it would not have been largely duplicated by less affected peoples farther up or down the coast. The demographic void resulting from the disappearance of settlements in the core impact zone would eventually be filled by the immigration of other peoples with similar social organization and technological skills.

Chiefdoms at the end of the Early Formative and during the Middle Formative Period, sometimes drawing on resources from beyond the immediate catchment, might fare slightly better. They would certainly be subject to much pressure from aggressive neighbors even if not directly disintegrated by the disaster itself. Because many of these polities were situated in low areas by rivers, they were subject to extreme stress in major flood years. In the event they were devastated, they would often have been eventually replaced by a similar social unit. Nevertheless, there are many discontinuities at the end of both these periods.

18 It is also largely true for the entire Mexican Gulf Coast, but this paper focuses on the north-central area.

In the Late Formative and Early Classic periods, complex chiefdoms, kingdoms, and early states had larger geographic districts, far more diverse resources, and much greater populations at their disposal. Nevertheless, their centralization may have made them even more vulnerable to catastrophic events. Most of their principal population centers and capitals were situated in the lower river valleys and deltas.

An excellent example of this is El Pital in the Nautla River delta of north-central Veracruz. Following a huge flood at the very end of the Middle Formative Period (ca. 400–300 BC), the city was laid out with a grid early in the Late Formative Period. The site grew rapidly to become the largest urban center on the Gulf Coast, concentrating much of the region's population at the city and around the adjacent canal systems covering nearly a 100 km² area. It may have suffered some flood damage about AD 100 but appears to have been impacted considerably by major flooding about AD 500.[19] After-ward, El Pital dropped dramatically in population and regional dominance. It was later conquered by peoples associated with rival El Tajín (Wilkerson 1993, 1994a, 1997a, 1999, 2001b).

Following this event El Tajín, which had only localized importance prior to the eclipse of El Pital, rapidly became the dominant city of the northern Gulf Coast during the Late Classic and Epiclassic periods (Wilkerson 1987, 1990, 1994a, 2001b). This site is situated on higher ground in a narrow valley in a chain of hills separating the Tecolutla River valley from that of the Cazones. It became a grand city filled with palaces, ball courts, and temples. Classic Veracruz art and architecture reached its apogee here, the greatest metropolitan center of Classic Veracruz civilization. It was clearly a capital city of a wide district that, at a minimum, dominated the majority of the North-Central area of the Gulf Coast. Its influence in architecture, art, and ball-game ritualism went far beyond these geographic bounds, and even well beyond Mesoamerica itself (Wilkerson 1984, 1985, 1990, 1991, 1997a, 1999).

El Tajín also was conquered following a major climatic event during which excessive precipitation resulted in the collapse of part of the valley where the site was built. Massive erosion of similar age is also found at agricultural locations around the hill site of La Conchita (Wilkerson n.d.b) and coin-cides with flood evidence at the riverbank site of Santa Luisa. Much like the 1999 event, this eleventh-century occurrence appears to have been regional, provoking massive mudslides and an inundation of significant proportions. The event in all probability heavily damaged the downstream canal systems, the food base, and commercial crops of the region. Without these the city could not maintain its prominence, labor pool, or subject populations, or adequately defend itself.

In this case the great storm may well have been accompanied by subse-quent droughts on the upper Gulf Coast that helped set in motion south-ward migrations of groups that ultimately sacked the city (Wilkerson n.d.c).

19 The dating here is approximate. The flood might have occurred as much as a century earlier. Indicators of weather damage, in the form of collapsed temple structures, are also apparent at Santa Luisa from about the same date (Wilkerson n.d.a).

The demise of El Tajín heralded the end of Classic Veracruz civilization and its form of intensive ball-game ritualism. The very centralized nature of the Tajín society made for rupture and dispersal once the hinterland was degraded and the capital itself damaged. Any chance of renovation or resurgence was ultimately curtailed by the incessant, and bellicose, long-range migrations set in motion by the sequential climatic catastrophes along the Gulf Coast.[20]

It is doubtful that any ancient society could easily survive a storm and aftermath similar to the 1999 event, much less even larger ones that did in fact occur.

CYCLICAL ASPECTS

Although incomplete in many districts, the archaeological record of the Mexican Gulf Coast is highly suggestive of multicentury cycles of megafloods with dynamic impact upon human populations. Information from sites such as El Tajín, El Pital, La Conchita, and Santa Luisa strongly suggest that these events occurred at approximately 400- to 700-year intervals stretching well into the past. They often appear to mark important junctures of cultural or city demise, migration, and periods of significant change. There is also some evidence that these broad regional events roughly correspond with droughts in still other coastal zones, such as the extreme upper Gulf Coast (the present-day states of Tamaulipas and Texas).[21] The cycles also appear to have trans-Gulf and intercontinental correlations.

Utilizing the chronological column of Santa Luisa (Wilkerson 1980, 1981, n.d.a) as a central reference in the North-Central area, the core impact zone of the 1999 event, we can derive a series of approximate times for ancient megafloods. In some cases there are relevant radiocarbon dates, in others the dates are based upon artifact comparisons or historical documentation. Of particular importance for cyclical considerations are mid-sixteenth century AD, AD 1100, AD 400–500, AD 1–100, and 400–300 BC.[22] There are other, still earlier dates that are also quite plausible, among them 1000 BC, 1800 BC, 2400 BC, 4000 BC, and approximately 5600 BC.[23]

20 The Postclassic Period after AD 1100 is not discussed here given the current lack of indications of major flooding between that date and the middle of the sixteenth century. Also, references to the early Colonial-Period evidence are much abbreviated, given space limitations and the primary emphasis on archaeological manifestations in this volume.

21 There are also very likely to be some important correlations southward and eastward with the Yucatan Peninsula, but this complex interrelationship is not addressed in the present paper. There is a growing body of evidence of severe droughts heavily impacting Maya civilization (Gill 2000; Haug et al. 2003; see Yaeger and Hodell, this volume). Some of the proposed dates do appear to overlap with times of culture change and/or climatic alteration on the Veracruz coast.

22 The dates, except for those in the twentieth and sixteenth centuries, are approximate. AD 1100 is likely to be accurate to within 50 years. The earlier dates are within approximately a century of variation, with the probability that if they are in error the correct date is earlier. The approximate date 300 BC may also correspond with severe volcanic ash fall in the South-Central and Southern Gulf areas (Wilkerson 1994b, 1997b).

23 These points in time are currently based upon uncorrected radiocarbon dates.

Approximate date	Episode strength relative to 1999*
1974–Present	Benchmark
AD 1550–1585	Slightly more intense
AD 1100 (AD 1050–1150)	About same
AD 500 (AD 400–500)	About same?
AD 100 (AD 1–100)	Stronger ~ 2x
300 BC (400–300 BC)	Much stronger ~ 3x

* Estimates based upon observable geologic strata and archaeological excavations of event deposits

TABLE 2 Megaflood cycles 300 BC–present along the Mexican Gulf Coast

While it is highly probable that specific data can be amassed for all these dates, the evidence for the first group of dates is much stronger at the moment. Based upon the thickness and extent of the flood deposits it would appear that, if we utilize the 1999 flood as a benchmark, the first set can be provisionally ranked with respect to magnitude (table 2).[24] Just as the 1999 event and the sixteenth-century sequence suggest, it is very likely that all of these exceptionally large inundations were part of a succession of floods over a few decades. These sequences in turn are likely to have corresponded with an intensification of Niña year and/or post-Niño year effects in the southern Gulf of Mexico.

The earliest megaflood of the first group, around 400–300 BC, may well have been a factor in the final decline or collapse of Olmec or Olmecoid chiefdoms along the central Gulf Coast. Such inundations in the North-Central area, in conjunction with probable volcanic ash fall in the South-Central and Southern Gulf areas, may have hastened the transition to the Late Formative Period and the migrations along the coast that led to the foundation of major multiethnic centers such as El Pital (Wilkerson 1994b, 1997b).

The AD 1–100 event again appears to come at a time of transition to the Protoclassic Period, and at sites such as El Pital heralds an even greater centralization of population, perhaps in response to regional devastation and the labor required to extend or repair the ample canal systems that surround the city. The AD 400–500 flood is the time of El Pital's demise and the rise of El Tajín, while the AD 1100 date marks the end of El Tajín and the collapse of Classic Veracruz civilization.

The mid-sixteenth-century sequence of floods and heightened humidity could not help but considerably exacerbate the recurrent epidemics of European diseases that devastated the regional populations of the Gulf Coast. In fact, by 1573 within 50 km of El Pital and Nautlan, its Late Postclassic

24 These magnitudes do reflect real distinctions but are not meant to be final estimates of the force and destructive power of these huge inundations.

descendent center, only eleven families remained (Wilkerson 1993). Population in the North-Central area continued to decline drastically and did not level out until approximately 1620 (Wilkerson 1979, n.d.a).

The 1999 megaflood, perhaps part of an episode of floods that began in 1974, is certain to have a lasting impact upon the entire region. This flood sequence is probably not over and future Niña, or post-Niño years, may bring still more forceful storms. The 1999 occurrence may or may not be the strongest event of the current episode. Its use as a benchmark in this study is strictly a convenience for reference only. Although truly catastrophic and, as I can affirm personally, almost beyond imagination during its impact, it could very easily have been still more devastating. The same cyclone, at tropical storm or hurricane force, and following a similarly slow, meandering track, would have wrought still greater damage. With such magnitudes nothing man-made, ancient or modern, would have survived it in the river deltas or low alluvial terraces.

TRANS-GULF AND INTERCONTINENTAL CORRELATIONS

Given the ambiguity of dating for some of the megaflood episodes, as well as the seeming variation in ENSO correlation of the Gulf Coast events, it is still premature to affirm precise intercontinental correlations or unquestionable teleconnections. Also, in the context of this initial review of Gulf Coast flooding as a major factor in civilization discontinuity, such correspondence is not the primary objective. Nevertheless, there are some points of overlap with interpretations elsewhere. These include the northern reaches of the Gulf of Mexico and especially South America, for which there is much data available.[25] However, precise year-to-year correlations are not likely to be derived. While Mexican Gulf Coast flooding appears to be primarily associated with post-Niño-year or Niña-year timing, its parameters are not yet totally defined. The correlations that do exist are not necessarily always with other floods but rather with distinct types of climatic extremes.

From the more than 2 km deep Orca Basin, well off the Mississippi delta at the edge of the continental shelf, comes strong proxy evidence from cores for seven major megafloods. Stretching over nearly five thousand years and apparently correlating hydrographically with Gulf Current modifications, a series of biotic and sedimentary indices reflects calamitous episodes of extraordinary inundations in central North America (Brown and Kennett 1999). Although flooding both along and back from the northern reaches of the Gulf Coast has a much higher association with Niño-year peaks than the southern zone where this study is focused,[26] there is an approximate correlation of centuries

25 Caviedes (2001) provides a useful summary of many archaeological aspects of mega-Niño impacts.

26 There are recent examples of this tendency. In 2005, a diminishing Niño year, seven hurricanes and storms, including major hurricanes Emily, Katrina, Rita, and Wilma, tore into the northern Gulf of Mexico, but only Stan (then at category 1 intensity) and three tropical storms were present in the southern reaches of the Gulf of Mexico. Along the western shoreline of the Gulf in 2007, a weak Niña year, two hurricanes, Dean (category 2) and Lorenzo (category 1), impacted the Mexican Gulf coast but one hurricane, Humberto (category 1), hit the uppermost Texas coastline.

of occurrence for the earlier portion of the Orca sequence and even some possible rough correspondences in later dates. Regardless, these data reveal considerable potential for the convergence of archaeological and geological evidence in discerning megafloods around the Gulf of Mexico.

South America has many indications of cyclic climatic modification. At least two of the four major drought peaks identified by Meggers (1994a, 1994b) for the Amazon Basin, 1500 and 400 BP, appear to correspond with heightened flood episodes in coastal Mexico. On the north coast of Peru, early sixth-century AD and AD 1100 mega-Niños (Caviedes 2001) also appear to match the eastern Mesoamerican data roughly. Sediment cores demonstrate direct climatic correlations between western South America and Mexico, particularly for the first, tenth, and eleventh centuries AD (Chepstow-Lusty et al. 1996), the first and last being significant flood times on the Gulf Coast. In general, the major flood episode dates for the Gulf of Mexico do equate with significant climatic events elsewhere.

Undoubtedly, there will be many correspondences of climatic extremes that can be worked out with future chronological refinement of the Mexican Gulf Coast data, as well as a better definition of the nature of each episode. Massive flooding is a symptom of larger natural processes. It is associated with ENSO events in the context of multicentury cycles, which in turn have profound effects on human populations and their habitats. The general lack of consideration of the significance of climate upon culture history to the contrary, neither ancient Mesoamerica nor modern Mexico is exempt from the progression of climate fluctuations that have impacted the rest of the Americas.

CATASTROPHIC FLOODING IN PERSPECTIVE

These events, or rather this sequence of events, certainly did adversely affect the human populations and their societies. But how detrimental were these occurrences in the course of culture history? The answer for eastern Mesoamerica is neither precise nor uniform from event to event. Certainly, the course of regional cultural evolution would have been quite distinct without them. There are indications that the more centralized the affected society, the greater the impact may have been on the overall course of regional cultural evolution.

The abundant resources from contiguous and diverse environments fostered early state development in Veracruz on the narrow coastal plain drained by sizeable rivers. It also made these polities especially vulnerable to cataclysmic episodes. Their sudden demise altered not only population distribution in the immediate disaster zone, but also the commercial, political, and military arrangements that had formalized the interrelationships among the often diverse cultures in and around the Gulf of Mexico. Most agricultural societies can absorb gradual change, but not the rapid, overwhelming devastation that destroys the very basis for social cohesion and cultural continuity.

The Veracruz coast was the obligatory transit zone for the multiple contacts between the central highlands and the southeastern lowlands of Mesoamerica. Its social disruption and disintegration had far-reaching consequences for the culture area as a whole. The triggering device for such episodes of severe to cataclysmic flooding in eastern Mexico appears to be primarily in the Niña- or post-Niño–year variation of the Southern Oscillation.[27] At such times, storms ranging from tropical depressions to hurricanes are confined in the Gulf of Mexico by North American continental fronts.[28]

The key aspect of these occurrences is not the strength of the storm, although wind and tide damage can cause considerable havoc in coastal areas, but rather the stalling or meandering patterns that permit massive rainfall on steep—and often deforested—slopes adjoining torrential rivers. Hence Tropical Depression #11 of 1999 was far more detrimental to the Gulf Coast than Hurricane Roxanne in 1995, or Hurricanes Dean and Lorenzo in 2007, although all hit essentially the same area.

Nevertheless, had Tropical Depression #11 had just slightly stronger winds, the storm tide in the Tecolutla estuary would have been still greater, the volume of water retained over the delta considerably more, and the duration of flooding in the lower drainage much longer. Under such conditions I would not have survived to make this analysis.

Also, the event around 300 BC, seemingly at least three times the magnitude of the 1999 storm, may have been a stalled hurricane of truly epic proportions. Its extreme devastation must have literally exceeded the imaginable and would have justifiably been considered the end of their world by the townspeople and villagers of the Tecolutla River delta. The remains of their communities cover a wide zone in the terrace strata around the research center where we were engulfed by the 1999 flood.[29]

These events in eastern Mexico tend to occur most often in Niña years, followed by post-Niño years and then strong Niño years. The megafloods are in turn grouped in episodes of acute flooding, seemingly several decades in length and with short recurrence intervals. Around this sequence of flood events is a greater cycle in which such severe episodes reoccur. It seems to last about 400 to 700 years, but since at least AD 1100 appears to be quickening its pace. When the level of severity reaches the cataclysmic, the course of human events on the Gulf Coast is altered, sometimes dramatically.

In this preliminary effort to analyze one of the many processes by which weather and climate influence the essential parameters of human civilization,

---

27 Not all flooding is caused by such phenomena. Hurricanes in any year can hit the Mexican Gulf Coast and cause localized, and even some regional flooding. However, based on modern and historical data, most severe and catastrophic events appear ENSO related. When the flooding is in a Niña year or, in the absence of a Niña, a post-Niño year, the Niño event tends to be a strong one (example: 1914, 1943, 1983).

28 Not infrequently, as in 1999, the remnants of the storms merged with fronts descending the Gulf from the central U.S. Usually such a merger pushes them farther back onto the Veracruz coast. The southern movement of the jet stream in such years may also play a role in this Gulf pattern.

29 There is also considerable stratigraphic evidence of this event southward in the Nautla catchment preceding the rise of the great city of El Pital. Additionally, there are preliminary indications of ample corresponding flood deposits in the Cazones and Tuxpan catchments to the north. The data in aggregate suggest a broad impact zone that covered at least the entire North-Central area of the Mexican Gulf Coast. Flooding could be expected to have occurred in adjoining areas also.

it is clear that the past has much to say to the future. The environment is not now static and never has been. We must take this factor into far greater consideration, be it in our archeological interpretations of antiquity, or in generating models for the future. There can be little doubt that civilization in the coastal regions of eastern Mesoamerica was much impacted by the cyclical recurrence of catastrophic flooding.

Are we immune to similar fates today? My personal experience and studies of the Mexican Gulf Coast would lead me to reply categorically, *absolutely not*. On the contrary, highly centralized modern societies, characterized by rapid communications, are especially vulnerable to calamitous events that alter the fabric of society in unanticipated and uncontrollable ways.[30] Cataclysmic flooding is just one of many such calamities but one that, while it is happening, makes for rapid reflection on the minuscule size of man in the natural world.

When I spent several days and nights in an immersed treetop during the enormous flood of 1999, two things occurred to me nearly as soon as we managed to swim in the darkness to a tree. The first was that survival literally depended upon what I held onto. Put another way, my entire universe was reduced to my arm's reach. The second thought was that we were not alone in this desperate action, because people since remote antiquity had certainly done exactly the same, probably at, or very near, this very same spot. There are indeed continuities to be found in human adversity as well as in culture history.

ACKNOWLEDGMENTS

The portion of this paper derived from post-1999 research has been made possible in part by grants from the National Geographic Society, an anonymous foundation, and the Institute for Cultural Ecology of the Tropics. I would also like to thank the Veracruz State Government as well as Mexico's Secretariat of Defense for facilitating portions of the study.

---

30 The man-made tragedies of September 2001 stunned more than the localized areas of devastation. Many types of catastrophic events can paralyze societies.

REFERENCES CITED

Avila, Lixion A.
1995 Hurricane Roxanne,
1–21 October 1995, Tropical Cyclone
Report (Preliminary Report), 29
November. National Hurricane
Center, NOAA, Miami.

Beven, Jack
1999 Tropical Depression Eleven,
4–6 October 1999, Tropical Cyclone
Report (Preliminary Report), 1
December. National Hurricane
Center, NOAA, Miami.

Brown, Paul, and James P. Kennett
1999 Marine Evidence for Episodic
Holocene Megafloods in North America
and the Northern Gulf of Mexico.
Paleoceanography 14: 498–510.

Caviedes, César N.
2001 *El Niño in History: Storming through
the Ages.* University Press of Florida,
Gainesville.

**Chepstow-Lusty, Alex J.,
Keith D. Bennett, V. Roy
Switsur, and Ann Kendall**
1996 4,000 Years of Human Impact
and Vegetation Change in the Central
Peruvian Andes—With Events
Paralleling the Maya Record? *Antiquity*
70: 824–33.

Comisión Federal de Electricidad
1961–99 Gasto medio diario en metros
cúbicos por segundo y resumen anual,
Estación Remolino, Río Tecolutla.
México, D.F.

Gill, Richardson B.
2000 *The Great Maya Droughts:
Water, Life and Death.* University of
New Mexico Press, Albuquerque.

**Haug, Gerald H., Detlef Günther,
Larry C. Peterson, Daniel M.
Sigman, Konrad A. Hughen,
and Beat Aeschlimann**
2003 Climate and the Collapse of Maya
Civilization. *Science* 299: 1731–1735.

Meggers, Betty J.
1994a Archaeological Evidence for
the Impact of Mega-Niño Events
on Amazonia during the Past Two
Millennia. *Climatic Change* 28: 321–338.

1994b Biogeographical Approaches
to Reconstructing the Prehistory of
Amazonia. *Biogeographica* 70 (3): 97–110.

National Weather Service
1999 Monthly Tropical Weather
Summary for the North Atlantic,
Caribbean Sea, and Gulf of
Mexico, December 1. National
Weather Service, Miami.

NOAA
1985 *Gulf of México, Coastal and
Oceanic Zones Strategic Assessment:
Data Atlas.* National Oceanic and
Atmospheric Administration,
National Oceanic Service, United
States Department of Commerce,
Washington, D.C., December.

Santos D., José Luis
1999 El Niño y La Ñiña: Una oscilación
climática. In *El fenómeno de El Niño
en Ecuador, 1997–1999, del desastre a la
prevención* (Enrico Gasparri, Carlos
Tassara, and Margarita Velasco, eds.):
13–19. CISP, Quito.

Tassara, Carlo, and Luigi Grando
1999 Análisis y prioridades operativas
para enfrentar los efectos del huracán
Mitch en materia de vivienda e
infraestructura. In *El fenómeno de
El Niño en Ecuador, 1997–1999, del
desastre a la prevención* (Enrico Gasparri,
Carlos Tassara, and Margarita Velasco,
eds.): 159–181. CISP, Quito.

Wilkerson, S. Jeffrey K.

1974 Sub-Culture Areas of Eastern Mesoamerica. In *Primera Mesa Redonda de Palenque: A Conference on the Art, Iconography, and Dynastic History of Palenque, Palenque, Chiapas, Mexico, December 14–22, 1973*, vol. 2 (Merle G. Robertson, ed.): 89–102. Robert Louis Stevenson School, Pre-Columbian Art Research, Pebble Beach, CA.

1979 Eastern Mesoamerica from Pre-Historic to Colonial Times: A Model of Cultural Continuance. *Actes du XLIIe Congrès International des Americanistes: Congrès du centenaire, Paris, 2–9 septembre 1976* (André Marcel d'Aris, Arnaud Castel, and Georgette Soustel, eds.). J. LaFaye, Paris.

1980 Man's Eighty Centuries in Veracruz. *National Geographic Magazine* 158 (2): 202–231.

1981 The Northern Olmec and Pre-Olmec Frontier on the Gulf Coast. In *The Olmec and Their Neighbors: Essays in Memory of Matthew W. Stirling* (Elizabeth P. Benson, ed.): 181–194. Dumbarton Oaks Research Library and Collection, Washington, D.C.

1984 In Search of the Mountain of Foam: Human Sacrifice in Eastern Mesoamerica. In *Ritual Human Sacrifice in Mesoamerica* (Elizabeth H. Boone, ed.): 101–132. Dumbarton Oaks Research Library and Collection, Washington, D.C.

1985 *Sacrifice at Dusty Court: Evolution and Diffusion of the Ritual Ballgame of Northeastern Mesoamerica.* Institute for Cultural Ecology of the Tropics, Tampa.

1987 Cultural Time and Space in Ancient Veracruz. In *Ceremonial Sculpture of Ancient Veracruz* (Marilyn M. Goldstein, ed): 6–17. Hillwood Art Gallery, Long Island University, New York.

1990 El Tajín, Great Center of the Northeast. In *Mexico: Splendors of Thirty Centuries*: 155–181. The Metropolitan Museum of Art, New York.

1991 And They Were Sacrificed: The Ritual Ballgame of Northeastern Mexico through Time and Space. In *The Mesoamerican Ballgame* (Vernon L. Scarborough and David R. Wilcox, eds.): 45–72. University of Arizona Press, Tucson.

1993 Escalante's Entrada, The Lost Aztec Garrison of the Mar del Norte in New Spain. *National Geographic Research and Exploration* 9 (1): 12–31.

1994a The Garden City of El Pital: The Genesis of Classic Civilization in Eastern Mesoamerica. *National Geographic Research and Exploration* 10: 56–71.

1994b Nahua Presence on the Mesoamerican Gulf Coast. In *Chipping Away on Earth: Studies in Prehispanic and Colonial Mexico in Honor of Arthur J. O. Anderson and Charles E. Dibble* (Eloise Quiñones-Keber, ed.): 177–186. Labyrinthos Press, Lancaster, CA.

1997a El Tajín und der Höhepunkt der Klassischen Veracruz-Kultur. In *Präkolumbische Kulturen am Golf von Mexiko* (Judith Rickenbach, ed.): 61–76. Museum Rietberg, Zurich.

1997b In Booten über das Meer: Die Herkunft der Nahuas. In *Präkolumbische Kulturen am Golf von Mexiko* (Judith Rickenbach, ed.): 55–59. Museum Rietberg, Zurich.

1997c Die intensive Felderbewirtschaftung an der mexikanischen Golfküste in vorspanischer Zeit: Eine Neubewertung. In *Präkolumbische Kulturen am Golf von Mexiko* (Judith Rickenbach, ed.): 77–82. Museum Rietberg, Zurich.

1999 Classic Veracruz Architecture, Cultural Symbolism in Time and Space. In *Mesoamerican Architecture as a Cultural Symbol* (Jeff Karl Kowalski, ed.): 110–139. Oxford University Press, New York.

1999–2003 *From the Field 1–15.* Institute for Cultural Ecology of the Tropics, Gutiérrez Zamora, Veracruz, Mexico.

2000 Olwin—A True Canine Heroine in Roiling Waters. *Golden Retriever News* 57 (4, July–August): 22–29.

2001a Gulf Lowlands. In *Archaeology of Ancient Mexico and Central America: An Encyclopedia* (Susan Toby Evans and David L. Webster, eds.): 323–24. Garland Publishing, New York.

2001b Gulf Lowlands: North Central Region. In *Archaeology of Ancient Mexico and Central America: An Encyclopedia* (Susan Toby Evans and David L. Webster, eds.): 324–32. Garland Publishing, New York.

2001c Gulf Lowlands: North Region. In *Archaeology of Ancient Mexico and Central America: An Encyclopedia* (Susan Toby Evans and David L. Webster, eds.): 329–34. Garland Publishing, New York.

2005 Rivers in the Sea: The Gulf of Mexico as a Cultural Corridor in Antiquity. In *Gulf Coast Archaeology, The Southwestern United States and Mexico* (Nancy Marie White, ed.): 56–67. University of Florida, Gainesville.

n.d.a Ethnogenesis of the Huastecs and Totonacs: Early Cultures of North-Central Veracruz at Santa Luísa, Mexico. Ph.D. dissertation, Department of Anthropology, Tulane University, New Orleans, 1972.

n.d.b Report on the 1976 Season of the Florida State Museum—National Geographic Society Cultural Ecology Project. Ms. on file, University of Florida, Gainesville, 1976.

n.d.c El Tajín, A Handbook for Visitors. Government of Veracruz, Xalapa.

Winograd, Manuel
2002 Natural disasters in Honduras. *Tiempo* 43 (March). Online at http://www.cru.uea.ac.uk/tiempo/portal/archive/issue43/t43a2.htm.

NOTES ON CONTRIBUTORS

Brian Billman is an associate professor in the Department of Anthropology at the University of North Carolina at Chapel Hill. His research interests include political economy and change, human violence, the origins of social stratification, household archaeology, settlement pattern analysis, and heritage preservation. He has worked in Peru and the western United States for the last twenty-six years. Major publications include the coedited volume *Settlement Pattern Studies in the Americas* (1999) and *Irrigation and the Origins of the Southern Moche State* (2002).

David Hodell is a professor in the Department of Earth Sciences at the University of Cambridge. He is a paleoclimatologist whose research utilizes speleothems and sediment cores collected from lake and ocean bottoms to reconstruct past changes in Earth's climate, oceans, and environment. He has a longstanding interest in how ancient civilizations affected their environment and, in turn, how environmental and climate change may have influenced cultural evolution. His research has focused mainly on the Maya region of Mesoamerica.

Gary Huckleberry is a geoarchaeological consultant and adjunct associate professor in the Department of Anthropology at Washington State University in Pullman. Much of his research involves the application of soils, stratigraphy, and geomorphology to understanding the dynamics of cultural and environmental change, particularly in arid environments. He has conducted over twenty years of research in western North America and coastal Peru and has published in *American Antiquity, Geology, Geoarchaeology, Quaternary Research*, and the *Journal of Archaeological Sciences*.

David Keefer is a senior research geologist with the Earthquake Hazards Team of the U.S. Geological Survey in Menlo Park, California. He has conducted research on landslides, earthquakes, El Niño, other geological hazards, and geoarchaeology for thirty-three years and is the author of more than 130 publications in these fields. He has conducted fieldwork throughout the United States, England, Italy, Taiwan, Japan, Mexico, Costa Rica, Argentina, and Peru. Since 1995, he has been working in Peru on problems related to El Niño history, earthquake and landslide occurrence, and the geoarchaeology of the Paleoindian site of Quebrada Tacahuay.

James B. Kiracofe is director of the Inter-American Institute for Advanced Studies in Cultural History and owner of Jericho Rock Works, Ltd. He has conducted field research in Mexico, Peru, and the Dominican Republic. His research interests include cultural transmission and transformation during the contact period and the impact of epidemic disease on that process. His articles include "Was the *Huey Cocoliztli* a Hemorrhagic Fever?" (coauthored with John S. Marr, 2000) in the journal *Medical History*, "Sobrevivencia de la arquitectura de los establecimientos españoles en los Estados Unidos" in *Arquitectura Colonial Iberoamericana* (edited by Graziano Gasparini, 1997), and "Architectural Fusion and Indigenous Ideology in Early Colonial Teposcolula" in *Anales del Instituto de Investigaciones Estéticas* (1995).

Kirk Allen Maasch is a professor in the Climate Change Institute and Department of Earth Sciences at the University of Maine. He has over twenty years of experience using climate models and statistical methods to investigate the causes of climate change across a wide range of time scales. He has worked toward developing a comprehensive theory for climate change. His most recent interests include interannual to decadal scale climate variability in the Holocene and the relationship between climate change and human activities. Along with David Anderson and Dan Sandweiss he co-edited *Climate Change and Cultural Dynamics: A Global Perspective on Holocene Transitions* (2007).

John S. Marr is a retired physician. Previously, he was Virginia's state epidemiologist, Medical Director for Exxon's Latin American affiliates, and Principal Epidemiologist for the New York City Health Department. He has published extensively on infectious diseases of public health importance, coauthored a text on human parasitic diseases, and has written three novels dealing with contemporary, catastrophic health disasters and bioterrorism. Recently he published articles on the ten plagues of Egypt (1996), the *Huey Cocolitztli* (2002), and the death of Alexander the Great (2003). He is presently researching the cause of massive Amerindian die-off along the Massachusetts coast from 1616 to 1619.

Paul Andrew Mayewski is Director of the Climate Change Institute at the University of Maine. He has led more than forty-five scientific expeditions to Antarctica, the Arctic, the Himalayas—Tibetan Plateau, and other regions. He serves on national and international scientific committees and as chief scientist for several major national and international research programs such as the Greenland Ice Sheet Project Two and the International Trans Antarctic Scientific Expedition. He has published more than 300 articles. His most recent book is *The Ice Chronicles: The Quest to Understand Global Climate Change* (with Frank White, 2002).

Michael Moseley is Distinguished Professor at the University of Florida, Gainesville, who has conducted archaeological fieldwork in Peru and adjacent Andean nations since 1966. His 1977 discovery of prehistoric canal systems destroyed by massive flooding led to an enduring fascination with ancient and modern El Niños and with the impacts of other natural hazards. A broad range of research interests is reflected in his books, including *The Incas and Their Ancestors* (2001); *The Northern Dynasties: Kingship and Statecraft in Chimor* (edited with Alana Cordy-Collins, 1991); *Chan Chan: Andean Desert City* (edited with Kent C. Day, 1982); and *The Maritime Foundations of Andean Civilization* (1975).

Jeffrey Quilter is Deputy Director for Curatorial Affairs and Curator for Intermediate Area Archaeology at the Peabody Museum, Harvard, and former Director of Pre-Columbian Studies at Dumbarton Oaks. He has conducted archaeological fieldwork in Peru and Costa Rica, including recent work at El Brujo, in the Chicama Valley, Peru. His publications include *Treasures of the Andes* (2005); *Cobble Circles and Standing Stones: Archaeology at the Rivas Site, Costa Rica* (2004); and *Life and Death at Paloma: Society and Mortuary Practices at a Preceramic Peruvian Village* (1989), as well as journal and book articles on the Moche and other topics in archaeology and art history.

James B. Richardson III is Curator Emeritus of Anthropology, Carnegie Museum of Natural History, and Professor of Anthropology, University of Pittsburgh. He has conducted archaeological research in Peru, Pennsylvania, and on Martha's Vineyard. His research interests include maritime adaptations; the rise of civilization during the Peruvian Preceramic; the impact of natural catastrophes, particularly El Niño, on Andean cultural development; and Pre-Columbian trans-Pacific contact with the Central Andes. His publications include *People of the Andes* (1994) and "Looking in the Right Places: Maritime Adaptations in Northeastern North America and the Central Andes" in *From the Arctic to Avalon*, BAR International Series 1507 (2006).

Paul ("Jim") Roscoe is Professor of Anthropology and Co-operating Professor of Climate Change at the University of Maine. He has conducted ethnographic fieldwork in New Guinea. His research interests include warfare, the evolution of political leadership and social organization, and the hunter-gatherers of New Guinea. Recent publications include "New Guinea Leadership as Ethnographic Analogy" in *The Journal of Archaeological Method and Theory* (2000), "The Hunters and Gatherers of New Guinea" in *Current Anthropology* (2002), "Fish, Game, and the Foundations of Complexity in Forager Society: The Evidence from New Guinea" in *Cross-Cultural Research* (2006), and "Intelligence, Coalitional Killing, and the Antecedents of War" in *American Anthropologist* (2007).

Daniel H. Sandweiss is Dean and Associate Provost for Graduate Studies and Professor of Anthropology and Quaternary and Climate Studies at the University of Maine. His research interests include climatic change (especially El Niño) and its relation to prehistoric cultural development, and ancient maritime adaptations. He has carried out research in Peru since 1978 and has also worked in Guatemala, Honduras, and Cuba. Sandweiss is the founder and first editor of the journal *Andean Past*. His publications include *The Archaeology of Chincha Fishermen: Specialization and Status in Inka Peru* (1992), *The Pyramids of Túcume* (1995, with Thor Heyerdahl and Alfredo Narváez), and journal and book articles on varied topics in Andean prehistory and paleoclimatology.

Payson Sheets is a professor in the Department of Anthropology at the University of Colorado in Boulder. His research interests include the unanticipated consequences of repeated commoner activities in Mesoamerica and the Intermediate area, ancient adaptations and cosmologies, and the impact of explosive volcanic eruptions on ancient societies, and how they coped with those sudden massive stresses. Major books include *Before the Volcano Erupted: The Ancient Ceren Village in Central America* (2002), *Archaeology, Volcanism, and Remote Sensing in the Arenal Region, Costa Rica* (edited with Brian McKee, 2nd ed., 2006), and *An Ancient Village Buried by Volcanic Ash in Central America* (2nd. ed., 2006).

S. Jeffrey K. Wilkerson is Director of the Institute for Cultural Ecology of the Tropics and based in Veracruz, Mexico. For over forty years he has undertaken multidisciplinary field work in the lowland tropics of Mexico and Central America. He has written extensively on the art, architecture, and archaeology of eastern Mesoamerica, focusing especially on the cultural ecology of the pivotal ancient cities of El Tajín, El Pital, and Santa Luisa. His ongoing research into catastrophic natural phenomena and the impact on Gulf Coast culture history is the subject of the Discovery/Science Channel documentary, *Flooded Civilization* (2006–7). Recent publications include: "Rivers in the Sea: The Gulf of Mexico as a Cultural Corridor in Antiquity" in *Gulf Coast Archaeology* (edited by N. H. White, 2005); *El Tajin: The Royal City*, in press; and *El Tajin: The Archaeological City*, in press.

Jason Yaeger is Associate Professor of Anthropology at the University of Wisconsin—Madison. His research examines how past social and political institutions were constituted through interpersonal interactions and everyday practices, with special interest in the roles of material culture and the built environment. He has directed projects studying Classic and colonial Maya society in Belize and the Inca settlement at Tiwanaku, Bolivia. Recent publications include "Untangling the Ties That Bind: The City, the Countryside, and the Nature of Maya Urbanism at Xunantunich, Belize" (in *The Social Construction of Ancient Cities*, edited by Monica L. Smith, 2003) and the co-edited volume (with Marcello A. Canuto) *The Archaeology of Communities: A New World Perspective* (2000).

INDEX